IoT and Analytics in Renewable Energy Systems (Volume 1)

Smart grid technologies include sensing and measurement technologies, advanced components aided with communications and control methods along with improved interfaces and decision support systems. Smart grid techniques support the extensive inclusion of clean renewable generation in power systems. Smart grid use also promotes energy saving in power systems. Cyber security objectives for the smart grid are availability, integrity and confidentiality.

Five salient features of this book are as follows:

- AI and IoT in improving resilience of smart energy infrastructure
- IoT, smart grids and renewable energy: an economic approach
- AI and ML toward sustainable solar energy
- Electrical vehicles and smart grid
- Intelligent condition monitoring for solar and wind energy systems

IoT and Analytics in Renewable Energy Systems (Volume 1)

Sustainable Smart Grids & Renewable Energy Systems

Edited by
O. V. Gnana Swathika, K. Karthikeyan, and
Sanjeevikumar Padmanaban

CRC Press
Taylor & Francis Group
Boca Raton London New York

CRC Press is an imprint of the
Taylor & Francis Group, an **Informa** business

MATLAB® is a trademark of The MathWorks, Inc. and is used with permission. The MathWorks does not warrant the accuracy of the text or exercises in this book. This book's use or discussion of MATLAB® software or related products does not constitute endorsement or sponsorship by The MathWorks of a particular pedagogical approach or particular use of the MATLAB® software.

Cover image © Shutterstock

First edition published 2024
by CRC Press
2385 Executive Center Drive, Suite 320, Boca Raton FL 33431

and by CRC Press
4 Park Square, Milton Park, Abingdon, Oxon, OX14 4RN

ISBN: 9781032362816 (hbk)
ISBN: 9781032362823 (pbk)
ISBN: 9781003331117 (ebk)

DOI: 10.1201/9781003331117

Typeset in Times
by codeMantra

Contents

Editors

O. V. Gnana Swathika (Member'11–Senior Member'20, IEEE) earned a BE in Electrical and Electronics Engineering from Madras University, Chennai, Tamil Nadu, India, in 2000; an MS in Electrical Engineering at Wayne State University, Detroit, MI, USA, in 2004; and a PhD in Electrical Engineering at VIT University, Chennai, Tamil Nadu, India, in 2017. She completed her postdoc at the University of Moratuwa, Sri Lanka, in 2019. Her current research interests include microgrid protection, power system optimization, embedded systems, and photovoltaic systems. She is currently serving as Associate Professor Senior, Centre for Smart Grid Technologies, Vellore Institute of Technology, Chennai, India.

K. Karthikeyan is an electrical and electronics engineering graduate with a master's in personnel management from the University of Madras. With two decades of rich experience in electrical design, he has immensely contributed toward the building services sector comprising airports, Information Technology Office Space (ITOS), tall statues, railway stations/depots, hospitals, educational institutional buildings, residential buildings, hotels, steel plants, and automobile plants in India and abroad (Sri Lanka, Dubai, and the UK). Currently, he is Chief Engineering Manager – Electrical Designs for Larsen & Toubro (L&T) Construction, an Indian multinational Engineering Procurement Construction (EPC) contracting company. Also, he has worked at Voltas, ABB, and Apex Knowledge Technology Private Limited. His primary role involved the preparation and review of complete electrical system designs up to 110 kV. Perform detailed engineering stage which includes various electrical design calculations, design basis reports, suitable for construction drawings, and Mechanical Electrical Plumbing (MEP) design coordination. He is the point of contact for both client and internal project team, leads and manages a team of design and divisional personnel, engages in day-to-day interaction with clients, peer review of project progress, manages project deadlines and project time estimation, and assists in staff appraisals, training, and recruiting.

Sanjeevikumar Padmanaban (Member'12–Senior Member'15, IEEE) received a PhD degree in Electrical Engineering from the University of Bologna, Bologna, Italy, in 2012. He was an Associate Professor at VIT University from 2012 to 2013. In 2013, he joined the National Institute of Technology, India, as a Faculty Member. In 2014, he was invited as a Visiting Researcher at the Department of Electrical Engineering, Qatar University, Doha, Qatar, funded by the Qatar National Research Foundation (Government of Qatar). He continued his research activities with the Dublin Institute of Technology, Dublin, Ireland, in 2014.

Further, he served as an Associate Professor in the Department of Electrical and Electronics Engineering, University of Johannesburg, Johannesburg, South Africa, from 2016 to 2018. From March 2018 to February 2021, he served as an Assistant Professor in the Department of Energy Technology, Aalborg University, Esbjerg, Denmark. He continued his activities from March 2021 as an Associate Professor with the CTIF Global Capsule (CGC) Laboratory, Department of Business Development and Technology, Aarhus University, Herning, Denmark. Presently, he is a Full

Professor in Electrical Power Engineering in the Department of Electrical Engineering, Information Technology, and Cybernetics, University of South-Eastern Norway, Norway.

S. Padmanaban has authored over 750+ scientific papers and received the Best Paper cum Most Excellence Research Paper Award from IET-SEISCON'13, IET-CEAT'16, IEEE-EECSI'19, and IEEE-CENCON'19 and five best paper awards from ETAEERE'16-sponsored Lecture Notes in Electrical Engineering, Springer book. He is a Fellow of the Institution of Engineers, India, the Institution of Electronics and Telecommunication Engineers, India, and the Institution of Engineering and Technology, UK. He received a lifetime achievement award from Marquis Who's Who – USA 2017 for contributing to power electronics and renewable energy research. He is listed among the world's top two scientists (from 2019) by Stanford University, USA. He is an Editor/Associate Editor/Editorial Board for refereed journals, in particular the *IEEE Systems Journal*, *IEEE Transaction on Industry Applications*, *IEEE Access*, *IET Power Electronics*, *IET Electronics Letters* and *Wiley-International Transactions on Electrical Energy Systems*; Subject Editorial Board Member—*Energy Sources*—*Energies Journal*, MDPI; and the Subject Editor for the *IET Renewable Power Generation*, *IET Generation, Transmission and Distribution*, and *FACETS Journal* (Canada).

Contributors

Ifeanyi Michael Smarte Anekwe
School of Chemical and Metallurgical
    Engineering
University of the Witwatersrand
Johannesburg, South Africa

Edward Kwaku Armah
School of Chemical and Biochemical Sciences
Department of Applied Chemistry
C. K. Tedam University of Technology and
    Applied Sciences
Navrongo, Ghana

V. Barath
Renewable Energy Conversion Laboratory
Department of Electrical and Electronics
    Engineering
Sri Sivasubramaniya Nadar College of
    Engineering
Chennai, India

Somudeep Bhattacharjee
Department of Electrical Engineering
Tripura University
Tripura, India

Rabbiraj Chinnappan
Vellore Institute of Technology
Chennai, India

Ruhul Amin Choudhury
Lovely Professional University
Punjab, India

Luke Gerard Christie
School of Social Sciences and Languages
Vellore Institute of Technology
Chennai, India

I. William Christopher
Department of EEE
Loyola-ICAM College of Engineering and
    Technology
Chennai, India

Milind Shrinivas Dangate
Chemistry Division, School of Advanced
    Sciences
Vellore Institute of Technology
Chennai, India

Prantika Das
School of Electrical Engineering
Vellore Institute of Technology
Chennai, India

Rupan Das
Tripura Sundari H.S. School
Tripura, India

Uttara Das
Department of Electrical Engineering
Tripura University
Tripura, India

S. Devi
Renewable Energy Conversion Laboratory
Department of Electrical and Electronics
    Engineering
Sri Sivasubramaniya Nadar College of
    Engineering
Chennai, India

Varun Gopalakrishnan
School of Mechanical Engineering
Vellore Institute of Technology
Chennai, India

N. Harish
Renewable Energy Conversion Laboratory
Department of Electrical and Electronics
    Engineering
Sri Sivasubramaniya Nadar College of
    Engineering
Chennai, India

S.S. Harshad
Renewable Energy Conversion Laboratory
Department of Electrical and Electronics
    Engineering
Sri Sivasubramaniya Nadar College of
    Engineering
Chennai, India

K.T.M.U. Hemapala
Department of Electrical Engineering
University of Moratuwa
Moratuwa, Sri Lanka

V. Berlin Hency
School of Electronics Engineering
Vellore Institute of Technology
Chennai, India

Yusuf Makarfi Isa
School of Chemical and Metallurgical
    Engineering
University of the Witwatersrand
Johannesburg, South Africa

Swapnil Jain
Vellore Institute of Technology
Chennai, India

K. Jamuna
School of Electrical Engineering
Vellore Institute of Technology
Chennai, India

V. Jamuna
Department of EEE
Jerusalem College of Engineering
Chennai, India

Harsh Jethwai
School of Electrical Engineering
Vellore Institute of Technology
Chennai, India

R. Lalitha Kala
School of Electrical Engineering
Vellore Institute of Technology
Chennai, India

G. Kanimozhi
Centre for Smart Grid Technologies
Vellore Institute of Technology
Chennai, India

Ramani Kannan
Department of Electrical and Electronics
    Engineering
Universiti Teknologi Petronas
Petronas, Malaysia

Aayush Karthikeyan
Schulich School of Engineering
University of Calgary
Calgary, Canada

Devasish Khajuria
School of Electrical Engineering
Vellore Institute of Technology
Chennai, India

J. Anwesh Kumar
Department of Electrical and Electronics
    Engineering
VNR Vignana Jyothi Institute of Engineering
    and Technology
Hyderabad, India

Kura Ranjeeth Kumar
Software Testing Engineer
EPAM Systems India Private Limited
Hyderabad, India

A. Adhvaidh Maharaajan
School of Electrical Engineering
Vellore Institute of Technology
Chennai, India

Apurv Malhotra
School of Electrical Engineering
Vellore Institute of Technology
Chennai, India

P. Mirasree
School of Electrical Engineering
Vellore Institute of Technology
Chennai, India

V. Muralidharan
School of Electrical Engineering
Vellore Institute of Technology
Chennai, India

B. Nagaraja Naik
Department of Electrical and Electronics
    Engineering
VNR Vignana Jyothi Institute of Engineering
    and Technology
Hyderabad, India

T.V. Narmadha
Department of EEE
St. Joseph's College of Engineering
Chennai, India

Morteza Azim Nasab
Department of Business Development and
    Technology
CTiF Global Capsule
Herning, Denmark

Mostafa Azimi Nasab
Department of Business Development and
    Technology
CTiF Global Capsule
Herning, Denmark

S. Nithin
School of Electrical Engineering
Vellore Institute of Technology
Chennai, India

R. Nivedhithaa
School of Electrical Engineering
Vellore Institute of Technology
Chennai, India

Sanjeevikumar Padmanaban
University of South-Eastern Norway
Notodden, Norway

Shreyaa Parvath
School of Electrical Engineering
Vellore Institute of Technology
Chennai, India

S.S. Harish Raaghav
School of Electrical Engineering
Vellore Institute of Technology
Chennai, India

R. Rajapriya
School of Mechanical Engineering
Chemistry Division, School of Advanced
    Sciences
Vellore Institute of Technology
Chennai, India

C. Sankar Ram
School of Electrical Engineering
Vellore Institute of Technology
Chennai, India

P. Ramesh
Department of EEE
University College of Engineering
Ramanathapuram, India

P. Sathvik Reddy
Department of Electrical and Electronics
    Engineering
VNR Vignana Jyothi Institute of Engineering
    and Technology
Hyderabad, India

Y. Rekha
Department of EEE
Jerusalem College of Engineering
Chennai, India

Anjan Kumar Sahoo
College of Engineering Bhubaneswar
Bhubaneswar, India

Tina Samavat
CTiF Global Capsule
Department of Business Development and
    Technology
Herning, Denmark

P.R. Sai Sasidhar
Department of Electrical and Electronics
    Engineering
VNR Vignana Jyothi Institute of Engineering
    and Technology
Hyderabad, India

R. Seyezhai
Renewable Energy Conversion Laboratory
Department of Electrical and Electronics
    Engineering
Sri Sivasubramaniya Nadar College of
    Engineering
Chennai, India

Nasrin I. Shaikh
Department of Chemistry
Nowrosjee Wadia College
Pune, India

Mohit Sharan
School of Electrical Engineering
Vellore Institute of Technology
Chennai, India

Aditya Basawaraj Shiggavi
School of Electrical Engineering
Vellore Institute of Technology
Chennai, India

Mandeep Singh
Lovely Professional University
Punjab, India

P. Srividya
RV College of Engineering
Bangalore, India

P. Shyam Sundar
School of Electrical Engineering
Vellore Institute of Technology
Chennai, India

O. V. Gnana Swathika
Centre for Smart Grid Technologies, School of
    Electrical Engineering
Vellore Institute of Technology
Chennai, India

Emmanuel Kweinor Tetteh
Green Engineering Research Group
Department of Chemical Engineering
Faculty of Engineering and the Built
    Environment
Durban University of Technology
Durban, South Africa

M. Thirumaran
School of Electrical Engineering
Vellore Institute of Technology
Chennai, India

R. Atul Thiyagarajan
School of Electrical Engineering
Vellore Institute of Technology
Chennai, India

Poonam Upadhyay
Department of Electrical and Electronics
    Engineering
VNR Vignana Jyothi Institute of Engineering
    and Technology
Hyderabad, India

R. Bharath Vishal
Renewable Energy Conversion Laboratory
Department of Electrical and Electronics
    Engineering
Sri Sivasubramaniya Nadar College of
    Engineering
Chennai, India

Mohammad Zand
Department of Business Development and
    Technology
CTiF Global Capsule
Herning, Denmark

1 Policies for a Sustainable Energy-Dependent India

Luke Gerard Christie and Rabbiraj Chinnappan
VIT University

CONTENTS

ABSTRACT

India has aimed since the late 1990s for growth and development and in its desire for economic success has created a void in ecological conservation and protection. With the world accelerating towards a deadline of being less carbon-dependent by 2050, India, on the other hand, has vouched for the same by 2070. The efforts need celebration, but the extension of almost two decades has been where the country has not focused on a holistic perspective when it comes to inclusive and sustainable development. In the current era, the challenges are immense and the raging effects of global warming and climate change having negative repercussions with the onslaught of newer disease outbreaks brought on by global warming will only imperil India's focus in the coming years ahead. We have witnessed the spread of Covid-19 and its disastrous consequences on the social life as well as economic consequences, and also the current war between Russia and Ukraine which had a major impact on economic, social and ecological security. Governments seem to have lost vision in the current atmosphere with their limited aim of growth as only being economics *forgetting* that in the 21st century, growth is less about economics but ecological sustainability. The fundamental questions all stakeholders from policy makers to industry and education need to ask is how are they sensitized towards creating an inter-generational equity and a fair distribution of resources equally for all without inhibiting anyone's progress to live in a socially resilient and fair world. This chapter aims to discuss global policies that will influence countries and the policies that governments must work towards being enabler of change for a cleaner, low-carbon environment with the necessary stop-gaps. The policies that Global South countries must work towards despite exacerbating poverty, lack of resources and agrarian distress with the influx of new-age emergent technologies to cushion the inevitable crisis we have pushed ourselves to where the outcomes could be disastrous with no returns.

KEYWORDS

Policies; Emergent Technologies; Ecological Protection; Global Warming; Poverty Alleviation

1.1 INTRODUCTION

India is at the cross-roads on the Russia-Ukraine crisis due to its heavy dependence on fossil fuels. Though the country may be experimenting with alternate sources of energy, that still remains in the making as formidability and familiarity do not remain anywhere in the near horizon. India imports a significant percentage of its crude oil from Russia and relies more on Russia for military hardware

DOI: 10.1201/9781003331117-1

and defence technologies but still remains the world's second largest importer of crude oil. India's oil import requirement was 85.5% in 2021–2022, whereas in 2020–2021, its requirement was 84.4% of its crude oil. This specifically speaks about the volume of oil importation and equally in the current era where opportunities can be created in greening the economy. It is here that governments working with global policy makers must work on fresh policies to strategically tackle global warming and climate change challenges by shifting to alternate sources of fuel and harvesting emergent technologies to create a sustainable and inclusive planet. In the past decade, there have been many changes to India's buying oil. A country like India that is pivoted to be the most populated country in the world by 2023 has some serious challenges from education, healthcare to employment and accessibility to resources. The primary reason is that majority of India's populace come from the bottom and middle of the economic pyramid. It is a definite challenge fighting for resources for a sustainable and inclusive ecosystem. India, with its growing population, is driven by its heavy dependence on energy. Businesses, healthcare facilities and the entire gamut of services in the formal and informal sector are engineered and driven by energy. In fact, a few trends in the recent past reveal India's demand for energy.

Iraq, Saudi Arabia and UAE are the leading sources of crude oil imports for India. Iraq emerged as the leading exporter of crude oil to India in recent years, replacing Saudi Arabia. Some other nations, which exported crude oil to India over the years, are as follows:

i. UAE has in the last 4 years emerged as the third major exporter of crude oil to India. It was among the top five countries earlier. Kuwait remains to be one of India's biggest suppliers as well.
ii. Latin-American country Venezuela was among the leading exporters of crude oil to India.
iii. Mexico has emerged as a leading exporter of crude oil among the Latin-American countries.
iv. Nigeria continues to be a leading exporter to India; the share of imports from Angola has reduced in recent years.
v. Iran was among the top three countries from which India imported crude oil until 2018–2019. India stopped importing crude from Iran in 2019 onwards due to alliance-building efforts with the USA in clamping down on Iran on uranium enrichment.
vi. The USA remains India's prominent supplier of crude oil. In fact, the countries Iraq, Saudi Arabia and UAE make up 63% of the total value of crude oil imports to India.

However, it is only in 2021–2022 do we see Brazil and Russia being the leading countries to supply crude oil to India. As the third largest consumer of oil in the world, India's consumption is that of 5% of the global consumption.

When we analyse these facts, we realize that India faces greater troubles in the energy sector and the solutions should be immediate as its is a developing country and requires greater policy insight to address a difficult scenario due to the growing population and traffic congestion on roads that indirectly affect the growth of its economy. Today, due to its population and its heavy reliance on fossil fuel-based electricity, India remains the third largest electricity producer in the world with 1,393 TWh generation in FY 2019–2020, and almost 99.99% of the population have access to power supply. Even on the dependence on electricity, India has surpassed China and Russia with electric grids having an installed capacity of 399.470 GW as of 31 March 2022. Renewable energy plants, which include large hydroelectric plants, constitute 38.2% of total installed capacity. The energy sector in India is strengthened by fossil fuels such as coal, which make up the largest percentage of India's energy needs. Though the government has come up with plans of de-commissioning coal-powered plants, coal remains India's bedrock for generation of energy. With the ratification of the Paris Climate Accord, India being an unwilling partner to transition its economy to an economy of low carbon by 2050 has made a promise of becoming a totally low-carbon country by 2070. It is to be observed that India aims to transform almost 50% of its economy by relying on non-fossils or alternate sources of energy by 2030 with the International Energy Agency briefly estimating that due to India's population, the increase of an additional 1,200 GW seems evident where this amount is more than what the European Union had utilized in 2005. The demand for cooling too is increasing tremendously. It has always remained a big challenge for India to shift to electricity, as the country in the early days relied on firewood, cow dung and agricultural waste. These traditional uses of fuels were found to be highly toxic and carcinogenic, not only causing illness but also affecting air quality. The particulate matter such as NOX, polycyclic aromatic hydrocarbons (PAHs), carbon monoxide, formaldehyde and other pollutants poses a severe threat to health of individuals and damage to the environment. The World Health Organization asserts that almost more than 400,000 people die or suffer health hazards due to inhaling indoor pollution caused by traditional sources of fuel, further leading to climate change and damage to forests and ecosystems. With newer policies being framed did the governments realize the dangers and shifted slowly to electricity or combustion technologies or clean burning fuel, creating a more sustainable alternative to traditional burning of fuel. Another primary cause for individual cases of disease is the inability to treat sewage, being a primary reason for pollution of surface and groundwater. The major sewage treatment plants in India that are owned by the government are not operational many times. It is these unfiltered wastes that remain stagnant and flood urban areas releasing contaminated particles into the groundwater. Had India have effective policies to address the water pollution and environmental issues with a full onslaught of electricity supply, a few of India's socio-economic and environmental problems may remain a myth but we find them becoming an unnecessary hazard to contend with. A large country like India must not only have electricity supply round the clock to guarantee its economy is running, but it is also equally important to ensure that water woes, sewage treatment plants and pollution monitoring sites have a full-scale supply of electricity.

India as an emerging economy with deepening infrastructure drive by policymakers, accelerating exports and cost of living going up makes it more challenging for the citizen with the hunger for more coal as shifting to alternate sources of fuel is not anywhere close in the horizon. With our current energy needs only growing with the current statistic of 4.7 times every year for the next two decades speaks of population explosion and the demand for energy for household appliances and a growing economy that seems to be faltering in comparison to the richer countries. Though policy makers are working for long-term securities in energy supply with environmental and economic needs, providing uninterrupted supply remains a Herculean task. An economy that is playing catch up with the rich countries and our competitors is scaling up employment opportunities, developing human skill for jobs, and manufacturing abilities with multi-lateral automobile makers vying to get a percentage of the India clientele and to elevate their profit margins. There have been challenges in attracting international investment in domestic hydrocarbon as NELP failed to incentivize large international energy corporations. In this particular area, India will have to work hard to acquire hydrocarbon reserves abroad. India endeavoured to invest in countries like France and the USA by acquiring nuclear reactors to fuel domestic power plants and ensuring access to critical technologies but failed due to logistic and economic issues in setting up foreign-built nuclear reactors. It has also been reported that a large percentage (307 million Indians) do not have access to electricity and an even larger percentage (505 million Indians) are still dependent on biomass. We need policies in infrastructure development and skill that battles macrocosmic external issues, e.g. the Indo-USA nuclearization deal, import of oil from the Gulf countries or the Middle East. In the current times due to skirmishes among India's energy partners, e.g. America and Iran, India has had to diminish oil imports or trading with Iran.

1.2 THE NEED FOR POLICIES ON ALTERNATE SOURCES OF ENERGY TO POWER INDIA'S ECONOMY

India faces strategic challenges to meet its energy needs and lacks policies to shift to alternate sources of energy, which is the need of the hour. China's One Belt, One Road initiative has given the country the due fillip in disturbing India's access to energy. It is primarily for this reason that we should shift to alternate sources of energy or non-renewable sources like solar, wind and cleaner fuels. In this way, pollution can be curbed, our cities and towns can be cleaner and greener, and we can easily accomplish the ratification of a truly low-carbon economy by 2070, though we were asked to commit to 2050. The policies that India needs to focus on also require constant innovation and foster dialogue with our alliance partners, as all countries have to work with each other to accomplish a net-zero emissions and to decelerate global warming and climate change. Policy makers have to analyse that India's energy security is fragile and under immense pressure due to its dependence on fossil fuels that are imported, international monopolies, opaque pricing policies and regulatory uncertainty. India's policy makers measures to amplify energy security by changing to clean energy, consistent policies require the country to sustain the momentum especially where one of its biggest policies is guaranteeing electricity to every city and remote part of the country by the end of 2022. Though work has been initiated, and almost half of India's villages today receive electricity and the others have more ground to cover, the outcome is yet to be studied as supplying electricity on a round-the-clock basis will not only prove costly but a colossal challenge. A few policies that are being worked out are as follows:

    i. Installation of biomass pelletizing units.
    ii. Distribution of 'efficient biomass chullahs'.
    iii. Distribution of solar irrigation pumps and being financed through NABARD and government financial support.
    iv. Alternate non-conventional energy sources need to be explored to make them accessible and technologically economical like tidal or geothermal energy.
    v. The NMEEE (National Mission for Enhanced Energy Efficiency) has a tremendous responsibility to organize a comprehensive cost analysis of the accessible efficient technologies and technological products in a diversified ecosystem, especially in healthcare, housing, agriculture and transportation.
    vi. Policy makers have to ensure at the state or national level that they should reach out to personnel and engineers to ensure that energy efficiency and standards and regulatory practices are nimble and remain well established with constant checks and balances. In this way, the informal sector has opportunities to grow and more employment opportunities can be created, reducing economic pressures.
    vii. The auto-fuel efficiency should be constantly upgraded for cleaner, greener and sustainable planet.
    viii. Solar Energy Corporation of India Limited aims to grow and develop storage solutions to bring down prices through demand aggregation.

The temporary solution will be for India to increase its dependence on coal and disincentivize investment on imports. Policy makers should have a clear understanding of regulatory clearances and have to ensure that they work towards improving labour productivity in order to increase coal manufacture and develop efficiency of distribution before phasing out coal and simultaneously drafting newer policies and finding solutions as India invests on renewable energy. As of now, the Hydrocarbon Exploration and Licensing aims to offer policy makers leeway in decision making with discretion, diminish administrative or bureaucratic struggles, reduce delays and encourage the idea of revenue sharing and freedom of marketing and motivating its citizenry to shift to alternate sources of energy by reducing a portion of tax for its payees. We have witnessed with the Russia-Ukraine war that policy makers are in no uncertain terms waking up to the reality of being energy-dependent discussing new tax structure in the sale of energy and import to enhance the competitive aspect of the economy. The challenges and struggles with demand getting hoarse for the world to be low carbon or no carbon by 2050 and 2070, the India Energy Security Scenarios is proposed to effectively address energy concerns from coal to renewable energy. With new digital tools being developed to study and analyse for future policies on diversifying energy demand and supply that will resolve energy conundrums with the demand growing due to population pressure and employment creation in the informal and formal sector. The Energy Security Scenario analyzes and explores India's energy scenario in supply sectors such that all includes all fuel generation ecosystems to suit demands accordingly. The new digital framework and tools make allowance for interactions from an individual level to a societal and industrial level to make prudent energy choices, and to discover a vast or multi-dimensional gradient of results for the country – from carbon emissions and imports reliance to land utilization. India has established energy relationships with Myanmar, Vietnam and Gulf countries. The Indo-US Nuclear Deal opens up energy opportunities in nuclear energy facilitated through advanced technology and nuclear fuel. With the demand more like a clarion call, India has sought to engage with Australia and China.

1.3   CONCLUSION

There is a growing clamour in Global South countries in energy and security due to existing economic and social conditions where we are witnessing significant energy transformations from complete electrification to a paradigmatic shift to renewable energy with severe disturbances due to the energy price rise in the globalization and increasing of prices of oil and natural gas. In the current era, policy makers in India have to realize that India needs to augment its research with skills construing innovatively to deal with constant alterations in the energy region. Though there remains much scope, we fail to see return on investment as carbon emissions are yet high and the only possible or viable solution will be when the aggressive push for renewables becomes a reality, which will help us ratifying the goal of 2070, i.e. a carbon-free world. Policy makers have to shift attention from short-term goals and aim for the future in a strategic, balanced manner, as the planet's life is at stake and more significantly poor countries will have to confront the worst of it and endure the hardest struggles from economics to social improbabilities. It is time for renewables and emergent technologies to transform our energy sector with innovation at the heart of all energy policies.

BIBLIOGRAPHY

Barreto, L., 2001: Technological learning in energy optimisation models and deployment of emerging technologies. PhD thesis, Swiss Federal Institute of Technology, Zurich, Switzerland.

Beg, N., 2002: *Ancillary Benefits and Costs of GHG Mitigation: Policy Conclusion*. Publication Service, OECD, Paris, France, p. 57.

Burtraw, D. and M. Toman, 2000: Ancillary benefits of greenhouse gas mitigation policies. Climate Change Issues Brief, 7, Resources for the Future, Washington, DC.

Chae, Y. and C. Hope, 2003: Integrated assessment of CO_2 and SO_2 policies in North East Asia. *Climate Policy*, 3, Supplement 1, pp. S76–83S.

CIEP, 2004: Study on energy supply security and geopolitics. Final report, Clingendael International Energy Programme, Institute for International Relations 'Clingendael', The Hague, The Netherlands.

Datta, K.E., T. Feiler, K.R. Rábago, N.J. Swisher, A. Lehmann, and K. Wicker, 2002: Small is profitable. Rocky Mountain Institute, Snowmass, Colorado. www.rmi.org, accessed 13/7/2022.

Gibbins, J., S. Haszeldine, S. Holloway, J. Pearce, J. Oakey, S. Shackley, and C. Turley, 2006: Scope for future CO_2 emission reductions from electricity generation through the deployment of carbon capture and storage technologies. In H.J. Schellnhuber (ed.), *Avoiding Dangerous Climate Change*. Cambridge University Press, Cambridge, 379 pp. http://www.defra.gov.uk/environment/climatechange/research/dangerous-cc/index.htm accessed 13/7/2022.

Haas, R., 2002: Building PV markets: the impact of financial incentives. *Journal Renewable Energy World*, 5(4), pp. 18–201.187–257.

IAEA, 1997: Joint convention on the safety of spent fuel management and on the safety of radioactive waste management. INFCIRC/546, 24 December 1997 (entered into force on 18 June 2001); http://www.iaea.org/Publications/Documents/Infcircs/1997/infcirc546.pdf accessed 13/7/2022.

Kamman, D.M., K. Kapadia, and M. Fripp, 2004: Putting renewables to work: how many jobs can the clean energy industry generate? Renewable and Appropriate Energy Laboratory (RAEL) report, University of California, Berkeley, 35 pp.

Karl, T.L. and I. Gary, 2004: The global record. Petropolitics special report, January 2004: Foreign Policy in Focus (FPIF), Silver City, New Mexico and Washington, DC, 8 pp.

Koster, A., H.D. Matzner, and D.R. Nichols, 2003: PBMR design for the future. Nuclear Engineering and Design, 222(2–3), pp. 234–247, June.

Larson, E.D., Z.X. Wu, P. Delaquil, W.Y. Chen, and P.F. Gao, 2003: Future implications of China-technology choices. *Energy Policy*, 31(12), pp. 1177–1275.

McDonald, K. (ed.), 2006: *Solar Generation*. Greenpeace International, Amsterdam.

Mishra, V., 2003: Indoor air pollution from biomass combustion and acute respiratory illness in preschool age children in Zimbabwe. *International Journal of Epidemiology*, 2003(32), pp. 857–867.

Newman, J., N. Beg, J. Corfee-Morlot, G. McGlynn, and J. Ellis, 2002: *Climate Change and Energy: Trends, Drivers, Outlook and Policy Options*. OECD Publication Service, Paris, France, 107 pp.

Philibert, C. and J. Podkanski, 2005: *International Energy Technology Collaboration and Climate Change Mitigation Case Study 4: Clean Coal Technologies*. International Energy Agency, Paris.

Sawin, J.L., 2003: Charting a new energy future. In L. Starke (ed.), *State of the World 2003*. W.W. Norton and Company, New York, pp. 85–157.

UN, 2000: General Assembly of the United Nations Millennium Declaration. 8th Plenary Meeting, September, New York.

UN, 2004: Integration of the economies in transition into the world economy. A report on the activities of the Economic Commission for Europe, Department Policy and Planning Office, http://www.un.org/esa/policy/reports/e_i_t/ece.pdf, accessed 13/7/2022.

UNDESA, 2002: *The Johannesburg Plan of Implementation*. United Nations Department of Economic and Social Affairs, New York.

UNDP, 2000: Energy and the challenge of sustainability. World Energy Assessment, United Nations Development Programme, UNDESA and World Energy Council, New York, http://www.undp.org/energy/activities/wea/drafts-frame.html, accessed 13/7/2022.

UNDP, 2004: Access to modern energy services can have a decisive impact on reducing poverty. http://www2.undp.org.yu/files/news/20041119_energy_poverty.pdf, accessed 13/7/2022.

UNEP, 2004: Energy technology fact sheet, Cogeneration. UNEP Division of Technology, Industry and Economics - Energy and Ozone Action Unit, http://www.cogen.org/Downloadables/Publications/Fact_Sheet_CHP.pdf, accessed 13/7/2022.

UNDP, 2004: World energy assessment, overview 2004: an update of 'Energy and the challenge of sustainability, world energy assessment'. Published in 2000: UNDESA/WEC, United Nations Development Program, New York.

UNEP, 2006: The UNEP Risoe CDM/JI pipeline. October, www.cd4cdm.org/Publications/CDMPipeline.xls, accessed 13/7/2022.

UNESCO, 2006: Workshop on GHG from freshwater reservoirs. 5–6 December, United Nations Educational, Scientific and Cultural Organization Headquarters, Paris, France, UNESCO IHP-VI website, www.unesco.org/water/ihp/, accessed 13/7/2022.

UNFCCC, 2004: Synthesis and assessment of the greenhouse gas inventories submitted in 2004 from Annex-I countries. Pp. 66, 17, 783, and 94.

UNFCCC, 2006: Definition of renewable biomass. Minutes of CDM EB23, Annex 18, http://cdm.unfccc.int/EB/Meetings/023/eb23_repan18.pdf, accessed 13/7/2022.

USCCTP, 2005: United States Climate Change Technology Program, http://www.climatetechnology.gov/library/2003/tech-options/, accessed 13/7/2022.

2 A Review on Internet of Things with Smart Grid Technology

P. Shyam Sundar, S. Nithin, V. Berlin Hency,
G. Kanimozhi, and O.V. Gnana Swathika
Vellore Institute of Technology

Aayush Karthikeyan
University of Calgary

K.T.M.U. Hemapala
University of Moratuwa

CONTENTS

ABSTRACT

Nowadays we get to see the rapid expansion of technologies with the influence of Internet of Things (IoT) and Smart Grid (SG). So, whatever IoT devices we are using with some sort of protocols and parameters, we should be able to manage and know the measures and estimation of power consumption, interoperability, price forecasting, etc. One of the major constraints and possible challenges of IoT-SG is the security and uneven distribution of processing data from transmission lines, distribution substations, smart meters, and other sources. In the present era, smart grid IoT technologies are beneficial energy management solutions. It's a two-way data interchange between networked gadgets and hardware that recognizes and responds to human wants. It's also less expensive than the present power system. In this paper, we focus on how smart grid is an efficient way to estimate and manage energy efficiency and power demand, which will give a clear picture to the end users in an elaborate way.

KEYWORDS

Internet of Things; Smart Grid; Interoperability; Data Analytics; Power Demand

2.1 INTRODUCTION: GENERAL

This paper reviews the solutions for challenges faced by the end users and consumers in power-based grids, namely, power quality and reliability. This paper focuses on advanced metering infrastructure (AMI) technologies and smart meters (SM) that enable the analyzing of the power grid by displaying a two-way communication scheme. Reference [1] gives the overview about the SMs and sweep the attainability of SM for power quality and reliability monitoring. It kickstarts with the wireless technology of SMs and such advanced communication techniques that enable AMI [1]. The main point of this article is to look at the smart grid's communication necessities and to launch all IoT protocols and their standards. First, it covers the fundamentals of communication technologies, such as the Manufacturing Message Specification (MMS) protocol in local area networks (LANs) and Management Plane Protection (MPP) in wide-area network (WAN) will use this technique to expose flaws of the practice, such as IoT protocol performance employing multi-agent systems (MAS) and micro-grid. With the help of internet services, they analyze the data and display them on the dashboard, which is very cost-effective and caters to high-end communication requirements. Most importantly, in the case of security, they consider IoT protocol Quality of service (QoS) [2]. This paper conveys that in order to make the IoT-based smart grid–enabled devices more efficient and reliable, high-quality uninterrupted electricity is needed. A cyberattack on IoT devices has the effect of bringing the energy grid to a halt. As we all know, the main difficulty with this approach is security, as the gadgets are all connected to the internet, leaving the smart grid vulnerable to catastrophic assaults. Due to the enormous number of nodes and domains in an IoT-SG, it has a broad attack surface for denial-of-service attacks. And, this leads to huge economic and financial losses [3]. In this paper, IoT is presented as the key application technology for the smart grid. Because SG is a data communications network connected to the power grid to acquire and analyze data from transmission lines, distribution substations, and customers, they've demonstrated the link between IoT and SG. Some IoT designs in Singapore are examined, as well as the requirements for IoT adoption, IoT applications, and services, as well as challenges and future developments [4]. This paper reviews the requirements and the role of IoT and SG for a smart city. Since there is a huge surge in IoT demand for communication between devices, energy and internet efficiency have become more important. So, by these kinds of vital requirements, many developed countries came up with their IoT projects. This summarizes the architecture for the smart city for the correct usage of infrastructure and to overcome the challenges faced by IoT. This paper also discusses about the economic issues faced while equipping smart city with basic communication [5]. This paper states the applications of IoT in smart grid with its challenges and solutions. The vast development of energy interfaced with IoT leads to smart electrical power grid which in turn increases the efficiency in power plants. This also talked about the IoT-SG challenges regarding security and data issues, but has improved a new level with interoperability and such technical cases. Reference [6] focuses on the avenues where IoT-SG are deployed along with the key points on challenges faced in this process. This paper states the applications of IoT in smart grid with its challenges and solutions. The vast development of energy grid and with IoT leads to the electrical power grid which caused in a way that its technology increases the efficiency in power plants. This also talked about the IoT-SG challenges regarding security and data issues, but has improved a new level with interoperability and such technical cases. So, this patent focused on the investigation of IoT-SG and discussed the key points and challenges that lie ahead [7]. The IoT-based power regulating system for smart grid applications is the subject of this article. As we all know, IoT devices are used to analyze and regulate data as well as to provide effective power supply to the system. This system is able to capture the power used and analyze the electrical parameters. This paper also uses the "Things-speak" software to obtain the real-time data used by the consumer and the source company, which is providing the power to reduce their billing costs. And, the consumers can be aware about the updated amount of power usage [8]. The aim of the paper is to minimize the costs of power usage along with smart energy management controlling the demand and supply side management. This trend follows reducing costs without implementing limits on consumption by instead choosing the reduction of power in the peak hours with the consumer's preferences and to avoid a *blackout*. The above try makes the balance in the demand response side. The estimation of cost algorithm will be based on usage and time with sensory information from the end users [9]. This idea of the paper performs with the one of the fundamental demanding situations of IoT, i.e., security. In this paper, the threats and capacity answers of the IoT primarily based on totally clever grid are analyzed. As statistics and conversation technologies (ICT) evolved and were carried out in conventional electricity systems, the development of clever grid cyber-physical-systems (CPS) will increase too. They focus on cyberattack and make us understand the vulnerabilities and solutions to the problem and give us the next task of research with cyber-SG [10]. This paper reviews that the smart grid in the power system wants a proper utilization of energy usage and has a vital role. This is monitored and analyzed by the SMs. But these SMs have some problems to sort out. And one of them is no point-to-point interaction between devices. So these energy SMs control and survey with the help of WIFI module and calculate the energy usage. These collected data are then uploaded to the cloud, which will be utilized by the consumers. This finally conveys that this is a great step for IoT automation toward digital India [11]. This paper works with power quality management in wireless sensor network (WSN) in smart grid technology. To get flow with efficiency and reliability, sufficient power supply should be defined. This patent reviews the grid sharing model using WSN-based communication and its implementation. This makes the system an environment based on power quality. Appropriate structures and controls were established and analyzed for managing the overall performance of a tracking machine within the smart grid [12]. This blockchain primarily based totally get entry to protocol makes the unique add-directly to the IoT-SG technology. Through this proposal, the information is securely introduced to the provider companies from their respective SMs. The node-to-node (P2P) communique collects information from the SMs of customers and transfers it to the blocks after verification. This literally works as the broker between the power stations and users and thus makes it private and confidential. It is mainly used for recording information in a secure way [13]. This plays with the smart grid infrastructure and its architecture in a field area network (FAN). The utilization, tracking, and control of diverse renewable energy sources are the maximum outstanding functions of smart grid infrastructure for agriculture applications. This paper describes the implementation of an IoT primarily based on totally wireless strength control gadget and the tracking of climate parameters of the usage of a smart grid–based infrastructure. The concepts depicted in this paper give an answer for all three clever grid hierarchical networks in subject location networks to attach a big variety of devices [14]. In this article, a secure machine learning engine is proposed for the Internet of Things (IoT)-enabled grid. There is a lot of power distribution involved with the smart grid, so it is important to secure it properly. Advanced energy management and interface controlling agents are proposed to optimize energy utilization in difficult situations, like when the power system is strained. A flexible Demand Side Management (DSM) is proposed to adapt quickly in these situations. A specific resilience model is proposed to control intrusions in smart grids and used with Machine Learning (ML) classifiers from unwanted entities [15]. This paper discusses about the collaboration of IoT with the smart grid environment. Its contribution of IoT-SG has gained immense potential due to its various innovation features. This mainly focuses on research ideas with IoT with SG and encourages scientists to make more and more innovations by understanding the role of IoT with SG, which together demonstrates enormous growth potential that smart grids open up regarding awareness for new interdisciplinary research [16]. In this paper, we provide a comprehensive overview of IoT-aided smart systems, covering the current architecture, applications, and prototypes of these systems. Since the traditional grids are unidirectional, they can't adjust with some new features. SG offers bi-directional flow between providers and consumers. So, IoT enables SG, which requires connectivity and tracking along with upcoming research directions [17]. Reference [18] initially talks about the basic features of cloud computing and explains how cloud computing can be extended to the scenario where it is possible to realize fog computing adaptation for real-time analysis. Here it is possible to examine the data facilities at the premises of the users and to pre-collect the cloud data which aids in enabling the SG consumers with faster speed of communications. Cloud computing fails in the terms of quality of experience like low latency, etc. They also illustrate various algorithm models such as architecture, cost for real-time situations, etc. [18]. In this paper, they propose a new type of computing strategy to give solutions to location-aware, latency optimization. To regulate this, a fog-based architecture and programming is enabled. So, existing technologies and techniques hardly utilize the fog computing nodes in the smart grid. This coordinator periodically measures the information that is metered from the resources. This fog computing is also able to manage all the given tasks at a time. The obtained results are to be different from traditional grids [19]. This article discusses a privacy-preserving framework for energy management on the smart grid. The data collected from the users are very confidential and to be protected, which is revealed to the cloud servers for Q-learning strategy. During the process of Q-learning, the data which split at the beginning stage are kept in random shares to avoid privacy issues. Along with this, edge computing is also being processed to implement an additive secret-sharing protocol [20].

2.2 IoT-ENABLED SMART GRID WITH ENERGY EFFICIENCY IN VARIOUS ASPECTS

2.2.1 Radio Networking

They develop and integrate both energy harvesting (EH) and cognitive radio (CR) approaches in this research to overcome SG difficulties including severe channel circumstances and low battery power. IoT connects with the smart grid environment to get benefits such as interoperability, inter-connectivity, etc. EH and CR constitute a new phase of networking with IoT-SG. CR, which is a wireless communication, enables the IoT with detecting which channels are in use and which are not and prevent unwanted connection between devices [21].

2.2.2 Cyberattacks

In this paper, they announce the framework for SG, i.e., dependable time series analytics. It has the ability to provide dependable data transforming from the cyber physical system to aim for refining raw data and further rebuilding on it based on the matrix factorization method. This makes it possible to exchange raw time series data between the CPS and the objective system. It primarily involves a huge number of smart items that are connected via networks [22]. In this paper, they had an investigation regarding their collected infrastructure, and they proposed a general scalable mitigation approach called minimally invasive attack. This algorithm is able to withstand the timely detection. As we know that one of the key challenges of smart grid cyber security is their characteristic that leads to technical power failure. They also displayed some simulation to show the reliability for this approach [23]. The gist of this paper mainly conveys the authentication between smart grid users and service providers against various security attacks. These attacks are simulated with the help of quantum key that are generated for the verification of the proposed smart grid system and explains why security is a major concern in SGs. IoT has been expanded with large number of branches with the ability of transmission and communication, which has entered various levels of internet. With the above justification, smart grid faces various security-related issues and is withstood by the mutual authentication of users and providers [24]. This paper gives awareness on the vulnerabilities to cyberattacks, and they test smart gird systems in a real-world implantation effectively called test beds. These vulnerabilities arise from information communication technologies nature inherently.

Test beds are used to find productive solutions against cyberattacks. These test beds are used to test the information that came from the user side and neglects the cyberattack capabilities. And, they also give recommendations for security-orientated aspects [25]. They discuss an approach termed smart grid security categorization in this study, which is designed for complicated circumstances such as sophisticated metering infrastructure. These methodologies result in a standard classification in an organization. Analysis collected through these methodologies is transferred to these organizations to get a valid report. The Strategic and Governance Services Centre (SGSC) related to some risk analysis assigns a system of security class. These SGSC offers a path for decision makers about these security aspects also to maintain the correct functionalities to the proposed system [26].

2.2.3 ENERGY-EFFICIENT MANAGEMENT

This paper focuses on cost optimization from smart homes under price-based demand response program. The development of IoT smart devices enables to make a strategy via demand response programs to work efficiently. They also proposed certain technical terms regarding the management of power usage of IoT-enabled homes like wind-driven bacterial foraging (WBFA) and bacterial foraging optimization (BFO) algorithms. This WBFA strategy responds to demand response programs against this major problem, which is the limitation of consumers' knowledge in the demand side [27]. This research offers a Deer Hunting Optimization (DHO) and Crow Search Algorithm-Based Energy Management System (EMS) for distribution systems using an IoT platform and a hybrid approach (CSA). With the aid of IoT, the major goal of this concept is to regulate electricity transmission and monitor metered data to end users. With the support of IoT, this proposed approach allows service providers to be more flexible in their operations [28]. This paper mainly discusses elaborately about simulation tests and collected data from an electricity management system whose main objective is to minimize the high amount of load voltage values from the production from the prosumer. The above experimental research is based on Elastic Energy Management (EEM), which enables data from the source to be transferred with correct maintenance with flexible features. The data are collected from a message broker called *MQTT* protocol. Using this node, various searching procedures were used in the EEM [29].

2.2.4 EDGE AND FOG COMPUTING

In this paper, a new fog-enabled privacy-preserving data aggregation is introduced. This proposal differs from the traditional smart grid communication with false data injection attacks. Just like managing a large group of IoT systems is a challenging task, data aggregation for next generation SG is also an issue. FESDA is a fault-free equipment even if we make changes in the SMs. It also evaluates its performance through aggregation and decryption [30]. This invention covers MapReduce apparatus for handling huge amounts of data with flexibility at the distribution end. As a result, whenever a wireless IoT node covers sensor data sets, MapReduce runs the procedures and sends the results to the grid. This section has significantly decreased consumption in huge networks while also enhancing the smart grid. Since a wireless IoT edge causes congestion, it employs WIERBS technology to alleviate it, resulting in lower operational latency and improved performance [31]. This paper gives an insight about another major challenge in IoT, i.e., managing a large volume of data. To overcome this issue, edge computing is introduced. These data are processed at the edge of the IoT network, near to the embedded devices that gather the data. IoT even makes small concerns to be networked and interconnected to each other to follow the protocols of measuring and analyzing with the SG. Smart grid also consists of a large amount of sensor data. Edge computing is used to manage all the networks like these kinds of sectors in smart grid [32]. This paper carries out the concept of data processing before it is transmitted for deeper data mining intelligence. This presents an edge-node-aware framework, namely, distributed computing, which brings enterprise applications closer to data sources such as local edge servers. Since smart grid is an effective tool of energy management domain such as SMs renewable sources, etc., with the rise of transforming network structure toward green environment, edge node is adapted in a node-to-node manner [33]. This paper mainly talks about the demand side management in the smart grid with IoT and cloud computing. It initially talks about the traditional power grids and creates awareness about wastage of energy and money. This method creates a massive demand–supply disparity, resulting in exorbitant prices. They offer an application of IoT and cloud computing in this work, which provides a customer load profile that can be managed remotely by service businesses. This can be done through installing project circuit board, which can be accessed via web portal [34].

2.2.5 Applications, Fault Analysis, and Distributions

This paper reviews the knowledge-based analytics that maybe used to observe the Stuck at fault condition in advanced metering infrastructure meters, which stores the data in the form of charges. This corrects the abnormal variation and predicts what cell is going to be stuck in the AMI meter. Since the grid uses sensor data for analyzing and for monitoring the IoT devices, it has to go through the process wherein the AMI meter plays the main role in knowledge-based data processing strategy in smart grid [35]. This paper talks about the basic architecture of IoT with smart grid technology and is followed by the key technologies of IoT-SG. It is evident that in the past decade there is no established technology which aids in realizing smart electric power sector that has the capacity to compensate the production with the demand side of consumers. Smart grid becomes a vital part of transmission, which combines with IoT to monitor, control, and connect in the electrical system. It also discusses about the deployment of IoT technologies that increases the potential of smart grid development and will be able to overcome the fundamental challenges [36]. This paper discusses the study of application of big data and ML in the emergence of the next power system. Connectivity plays an important role in this new grid infrastructure, which is one of the domains of IoT. The integration of big data and IoT-SG makes efficient metering of load with proper estimation of cost. Again, the IoT with smart grid technology plays with its major challenge, i.e., security becomes a critical issue. Big data and ML still try to overcome the security issue [37].

2.2.6 Blockchain-Based IoT

This study describes a hybrid block chain approach for 5G networks, in which the service provider deploys both private and public broker platforms (MEC). We use block chain technology with an IoT device identification to make data easier to access and secure. This research looks at how IoT-SG devices are properly identified. This scenario also combines block chain technology with a 5G network to identify and report IoT fraud in the smart grid. In addition to this technology, they analyzed various consensus algorithms from the respective feasible hybrid block chain [38].

2.3 REAL-TIME APPLICATIONS OF IoT-ENABLED SMART GRID

2.3.1 IoT-Based Smart Applications

This paper introduces a special feature called smart plugs that enhances IoT for establishing hybrid connected system and also compromises the demand for existing response of end users. Since IoT-SG being a complicated process, we can't give preference for every device, instead of that we are using smart plugs. These smart plugs can't be installed as such, it has to prestored in the domains of microprocessor before using. This also discussed about the current plug technologies and limitations to overcome this hybrid monitoring system [39]. This paper designs the model and implementation of IoT-based smart energy meter as we know that SM uses advanced metering infrastructure, they enhance two-way communication between servers and consumers. Such kind of infrastructure makes flexible benefits to the power system easy and reliable. This also talks about cyberattacks to the power researchers. So, to avoid the basic carelessness and some security concerns it is mandatory to utilize SMs. This also integrates with IoT fabrication to connect with hardware technologies [40]. This paper gives an idea to defeat the challenges faced by the current smart grid through various measures. By implementing these IoT techniques, end users and operators are able to maintain the supply of resources, and reduce operation and maintenance cost by using smart appliances. As of now, cyber security being a vital challenge for the demand side, it can be performed through combining analog and digital aspects in this task [41]. Through the selection of mobile couriers, this article presents a future notion of smart grid in smart cities. This is commonly utilized in smart city "pervasive sensing paradigms" for sharing data applications. This perspective has broadened the IoT mission, which now includes radio frequency identification (RFID) sensors, readers, and other devices. As a consequence, we're providing smart cities with a hybrid data-gathering platform. This connects with the city's base station, data access points, and data couriers. This concept plays with the mobile functioning protocols to gather data with some limitations [42].

2.4   IoT-BASED SMART GRID ARCHITECTURE

This paper reviews the smart grid architecture by enabling adaptivity in IoT. Since we were using a huge amount of electric power in the previous decade, we blindly ignored the estimation of data. This paper researches how to integrate the state-of-art information with the existing adaptivity in the smart grid. This literally means to reconfigure IoT service parts without other service parts stopping working. Therefore, proactively adapting devices with this kind of context fluctuation becomes a concern [43]. The motive of this paper is to convey the challenges, issues, and opportunities for the development of SG. Smart grid is accelerating to meet the electricity demand providing a way toward eco-friendly era in a reliable manner. This two-way communication feature in SG being a more efficient property, they use an application of renewable energy resources, self-healing distribution intelligence, etc. This paper also gathered information about the upcoming opportunities for the development of a smart grid environment [44]. This article examines the IoT application, as well as addresses the necessary smart grid and data collecting criteria using the feature known as power Internet of Things. It's a smart service system that makes extensive use of modern information technology and artificial intelligence to enable human–computer interaction. Power Internet of Things exclusively integrates various infrastructure resources and updates the level of power system communication. It designs a full-service coverage of power Internet of Things and guides for the future planning of communication [45].

2.5   DETECTION FOR IoT-ENABLED SMART GRID SYSTEM

This study provides a probabilistic incremental clustering approach for analyzing energy consumption from current smart meters, which may be represented by load patterns retrieved from load data. This suggested technique performs a pattern reading trip on existing load patterns before switching to new load patterns from smart meters. Some of them are experimented with real data set and are evaluated by combining clustering, which outperforms this type of algorithm. Plus, this strategy displays the drift through its results [46]. This paper reviews the intrusion detection system (IDS) designed for smart grid that collects data from a power plant. Smart grid satisfies both services company and end consumers in a two-way communication model. Because the smart grid confronts security issues as a result of vulnerabilities in IoT systems, which produce security risks that affect smart applications, a rising demand for IDS tailored for IoT to combat IoT-related security assaults that exploit particular security weaknesses has emerged. Since IDS acts as a data check tool, it can also be used for ML and DL [47].

2.6   RECENT ADVANCEMENTS IN IoT-SMART GRID TECHNOLOGY

Signcryption with proxy re-encryption approach is used to meet the security needs of IoT connection in this article. Because the smart grid platform's IoT devices create a large quantity of data that must be handled and saved in the cloud server, it must be maintained and authorized independently. Since signcryption with schemes for SG is facing difficulty from more bandwidth, the paper proposes a light weight certificate based signcryption (CBSERE). This minimizes the commutation and communication cost along with increased security [48]. The goal of this study is to tackle the resource allocation problem of the smart grid's radio access network slice. Since this is term is not familiar in this technology, deep reinforcement learning is deployed. This also discusses the real-time application with the mapping of RAN to DRL. This also promotes 5G network which is and interconnected network that shows great potential with features of slicing. There is a picture clear virtualization between the industries to make it reliable and flexible. So, considering these service characteristics, a smart RAN slice is proposed to get simulation results through DRL (reducing the cost) [49]. They provide an overview of smart grid in this article, with its adaptable characteristics and many roles in the power distribution system. With day-by-day enhancement in technology, there will be a huge demand for electrical supply for its production. With the development of the distribution system, the rising demand is also increasing with the speed of maintaining the smart grid infrastructure. This paper discusses the broad scope of reliable and efficient deployment of smart grid technology in the power sector [50].

2.7   CONCLUSION

As of now, we all know that IoT with some emerging technologies like smart grid or renewable energy is being most wanted for consumers and for efficient processes. We have given a clear picture of how the end users can handle the device or smart appliances by knowing how they analyze the data and being aware of the price clustering and power demand. We have also adapted the technology that when one cycle of process of data observation and the resultant price was higher than expected, it automatically adjusts the usage and efficiency of the device or appliance that enabled IoT-SG to get the estimated price of the power that has been issued. Communication is an important enabler of smart grid systems. The smart grid is expected to use a wide range of communication technologies for future advancement. The use of communication technologies in the smart grid is influenced by application demands, geographic locations, surroundings, regulation, price, and other factors. The deployment of IoT technology at various levels of the smart grid has been covered in this chapter. It has been demonstrated that new types of applications and architectures, such as improved smart metering infrastructure, multi-agent systems, and fog computing, have been enabled by the technologies. These provide benefits like autonomy, scalability, mobility, and resiliency, among others. Security and efficiency are still issues to improve the efficacy of IoT; future research should focus on this.

REFERENCES

1. Al-Turjman, F. and Abu Jubbeh, M., 2019. IoT-enabled smart grid via SM: An overview. *Future Generation Computer Systems*, *96*, pp. 579–590.
2. Tightiz, L. and Yang, H., 2020. A comprehensive review on IoT protocols' features in smart grid communication. *Energies*, *13*(11), p. 2762.
3. Kimani, K., Oduol, V. and Langat, K., 2019. Cyber security challenges for IoT-based smart grid networks. *International Journal of Critical Infrastructure Protection*, *25*, pp. 36–49.
4. Ghasempour, A., 2019. Internet of things in smart grid: Architecture, applications, services, key technologies, and challenges. *Inventions*, *4*(1), p. 22.
5. Tanwar, S., Tyagi, S. and Kumar, S., 2018. The role of internet of things and smart grid for the development of a smart city. In *Intelligent communication and computational technologies* (pp. 23–33). Springer, Singapore.
6. Davoody-Beni, Z., Sheini-Shahvand, N., Shahinzadeh, H., Moazzami, M., Shaneh, M. and Gharehpetian, G.B., 2019, December. Application of IoT in smart grid: Challenges and solutions. In *2019 5th Iranian Conference on Signal Processing and Intelligent Systems (ICSPIS)* (pp. 1–8). IEEE.

7. Mugunthan, S. and Vijayakumar, T., 2019. Review on IoT based smart grid architecture implementations. *Journal of Electrical Engineering and Automation*, 1(1), pp. 12–20.

8. Khan, F., Siddiqui, M.A.B., Rehman, A.U., Khan, J., Asad, M.T.S.A. and Asad, A., 2020, February. IoT based power monitoring system for smart grid applications. In *2020 International Conference on Engineering and Emerging Technologies (ICEET)* (pp. 1–5). IEEE.

9. Pawar, P., 2019. Design and development of advanced smart energy management system integrated with IoT framework in smart grid environment. *Journal of Energy Storage*, 25, p.100846.

10. Gunduz, M.Z. and Das, R., 2020. Cyber-security on smart grid: Threats and potential solutions. *Computer Networks*, 169, p.107094.

11. Barman, B.K., Yadav, S.N., Kumar, S. and Gope, S., 2018. IOT based smart energy meter for efficient energy utilization in smart grid. In *2018 2nd International Conference on Power, Energy and Environment: Towards Smart Technology (ICEPE)* (pp. 1–5). IEEE.

12. Bagdadee, A.H., Hoque, M.Z. and Zhang, L., 2020. IoT based wireless sensor network for power quality control in smart grid. *Procedia Computer Science*, 167, pp. 1148–1160.

13. Bera, B., Saha, S., Das, A.K. and Vasilakos, A.V., 2020. Designing blockchain-based access control protocol in iot-enabled smart-grid system. *IEEE Internet of Things Journal*, 8(7), pp. 5744–5761.

14. Chhaya, L., Sharma, P., Kumar, A. and Bhagwatikar, G., 2018. IoT-based implementation of field area network using smart grid communication infrastructure. *Smart Cities*, 1(1), pp. 176–189.

15. Babar, M., Tariq, M.U. and Jan, M.A., 2020. Secure and resilient demand side management engine using machine learning for IoT-enabled smart grid. *Sustainable Cities and Society*, 62, p.102370.

16. Reka, S.S. and Dragicevic, T., 2018. Future effectual role of energy delivery: A comprehensive review of Internet of Things and smart grid. *Renewable and Sustainable Energy Reviews*, 91, pp. 90–108.

17. Saleem, Y., Crespi, N., Rehmani, M.H. and Copeland, R., 2019. Internet of things-aided smart grid: technologies, architectures, applications, prototypes, and future research directions. *IEEE Access*, 7, pp. 62962–63003.

18. Hussain, M. and Beg, M.M., 2019. Fog computing for internet of things (IoT)-aided smart grid architectures. *Big Data and Cognitive Computing*, 3(1), p.8.

19. Wang, P., Liu, S., Ye, F. and Chen, X., 2018. A fog-based architecture and programming model for iot applications in the smart grid. *arXiv preprint arXiv:1804.01239*.

20. Wang, Z., Liu, Y., Ma, Z., Liu, X. and Ma, J., 2020. LiPSG: lightweight privacy-preserving Q-learning-based energy management for the IoT-enabled smart grid. *IEEE Internet of Things Journal*, 7(5), pp. 3935–3947.

21. Ozger, M., Cetinkaya, O. and Akan, O.B., 2018. Energy harvesting cognitive radio networking for IoT-enabled smart grid. *Mobile Networks and Applications*, 23(4), pp. 956–966.

22. Wang, C., Zhu, Y., Shi, W., Chang, V., Vijayakumar, P., Liu, B., Mao, Y., Wang, J. and Fan, Y., 2018. A dependable time series analytic framework for cyber-physical systems of IoT-based smart grid. *ACM Transactions on Cyber-Physical Systems*, 3(1), pp. 1–18.

23. Yılmaz, Y. and Uludag, S., 2021. Timely detection and mitigation of IoT-based cyberattacks in the smart grid. *Journal of the Franklin Institute*, 358(1), pp. 172–192.

24. Kaur, M. and Kalra, S., 2018. Security in IoT-based smart grid through quantum key distribution. In *Advances in computer and computational sciences* (pp. 523–530). Springer, Singapore.

25. Gunduz, M.Z. and Das, R., 2018, March. A comparison of cyber-security oriented testbeds for IoT-based smart grids. In *2018 6th International Symposium on Digital Forensic and Security (ISDFS)* (pp. 1–6). IEEE.

26. Shrestha, M., Johansen, C., Noll, J. and Roverso, D., 2020. A methodology for security classification applied to smart grid infrastructures. *International Journal of Critical Infrastructure Protection*, 28, p.100342.

27. Hafeez, G., Wadud, Z., Khan, I.U., Khan, I., Shafiq, Z., Usman, M. and Khan, M.U.A., 2020. Efficient energy management of IoT-enabled smart homes under price-based demand response program in smart grid. *Sensors*, 20(11), p.3155.

28. Krishnan, P.R. and Jacob, J., 2022. An IOT based efficient energy management in smart grid using DHOCSA technique. *Sustainable Cities and Society*, 79, p.103727.

29. Powroźnik, P., Szcześniak, P. and Piotrowski, K., 2021. Elastic energy management algorithm using IoT technology for devices with smart appliance functionality for applications in smart-grid. *Energies*, 15(-1), p.109

30. Manu, D., Shorabh, S.G., Swathika, O.G., Umashankar, S. and Tejaswi, P., 2022, May. Design and realization of smart energy management system for Standalone PV system. In *IOP Conference Series: Earth and Environmental Science* (Vol. 1026, No. 1, p. 012027). IOP Publishing.

31. Swathika, O.G., Karthikeyan, K., Subramaniam, U., Hemapala, K.U. and Bhaskar, S.M., 2022. Energy efficient outdoor lighting system design: case study of IT campus. In *IOP Conference Series: Earth and Environmental Science* (Vol. 1026, No. 1, p. 012029). IOP Publishing.

32. Sujeeth, S. and Swathika, O.G., 2018, January. IoT based automated protection and control of DC microgrids. In *2018 2nd International Conference on Inventive Systems and Control (ICISC)* (pp. 1422–1426). IEEE.

33. Patel, A., Swathika, O.V., Subramaniam, U., Babu, T.S., Tripathi, A., Nag, S., Karthick, A. and Muhibbullah, M., 2022. A practical approach for predicting power in a small-scale off-grid photovoltaic system using machine learning algorithms. *International Journal of Photoenergy, 2022*. https://doi.org/10.1155/2022/9194537

34. Odiyur Vathanam, G.S., Kalyanasundaram, K., Elavarasan, R.M., Hussain Khahro, S., Subramaniam, U., Pugazhendhi, R., Ramesh, M. and Gopalakrishnan, R.M., 2021. A review on effective use of daylight harvesting using intelligent lighting control systems for sustainable office buildings in India. *Sustainability, 13*(9), p.4973.

35. Swathika, O.V. and Hemapala, K.T.M.U., 2019. IOT based energy management system for standalone PV systems. *Journal of Electrical Engineering & Technology, 14*(5), pp. 1811–1821.

36. Swathika, O.V. and Hemapala, K.T.M.U., 2019, January. IOT-based adaptive protection of microgrid. In *International Conference on Artificial Intelligence, Smart Grid and Smart City Applications* (pp. 123–130). Springer, Cham.

37. Kumar, G.N. and Swathika, O.G., 2022. 19 AI Applications to. *Smart Buildings Digitalization: IoT and Energy Efficient Smart Buildings Architecture and Applications*, p. 283.

38. Swathika, O.G., 2022. 5 IoT-Based Smart. *Smart Buildings Digitalization: IoT and Energy Efficient Smart Buildings Architecture and Applications*, p. 57.

39. Lal, P., Ananthakrishnan, V., Swathika, O.G., Gutha, N.K. and Hency, V.B. 2022. Chapter 14 IoT-Based Smart Health. *Smart Buildings Digitalization: Case Studies on Data Centers and Automation*. 2022 Feb 23:149.

40. Chowdhury, S., Saha, K.D., Sarkar, C.M. and Swathika, O.G. 2022. IoT-based data collection platform for smart buildings. In *Smart Buildings Digitalization* (pp. 71–79). CRC Press.

41. Yang, Q., 2019. Internet of things application in smart grid: A brief overview of challenges, opportunities, and future trends. In *Smart Power Distribution Systems* (pp. 267–283). Elsevier.

42. Al-Turjman, F., 2018. Mobile couriers' selection for the smart-grid in smart-cities' pervasive sensing. *Future Generation Computer Systems, 82*, pp. 327–341.

43. Petrović, N. and Kocić, Đ., 2019. Enabling adaptivity in IoT-based smart grid architecture. In *ICEST 2019* (pp. 1–4).

44. Salkuti, S.R., 2020. Challenges, issues and opportunities for the development of smart grid. *International Journal of Electrical and Computer Engineering (IJECE), 10*(2), pp. 1179–1186

45. Wang, Q. and Wang, Y.G., 2018, October. Research on power Internet of Things architecture for smart grid demand. In *2018 2nd IEEE Conference on Energy Internet and Energy System Integration (EI2)* (pp. 1–9). IEEE.

46. Jiang, Z., Lin, R. and Yang, F., 2021. An incremental clustering algorithm with pattern drift detection for IoT-enabled smart grid system. *Sensors, 21*(19), p.6466.

47. Efstathopoulos, G., Grammatikis, P.R., Sarigiannidis, P., Argyriou, V., Sarigiannidis, A., Stamatakis, K., Angelopoulos, M.K. and Athanasopoulos, S.K., 2019, September. Operational data based intrusion detection system for smart grid. In *2019 IEEE 24th International Workshop on Computer Aided Modeling and Design of Communication Links and Networks (CAMAD)* (pp. 1–6). IEEE.

48. Hussain, S., Ullah, I., Khattak, H., Adnan, M., Kumari, S., Ullah, S.S., Khan, M.A. and Khattak, S.J., 2020. A lightweight and formally secure certificate based signcryption with proxy re-encryption (CBSRE) for Internet of Things enabled smart grid. *IEEE Access, 8*, pp. 93230–93248.

49. Meng, S., Wang, Z., Ding, H., Wu, S., Li, X., Zhao, P., Zhu, C. and Wang, X., 2019, October. RAN Slice Strategy Based on Deep Reinforcement Learning for Smart Grid. In *2019 Computing, Communications and IoT Applications (ComComAp)* (pp. 6–11). IEEE.

50. Butt, O.M., Zulqarnain, M. and Butt, T.M., 2021. Recent advancement in smart grid technology: Future prospects in the electrical power network. *Ain Shams Engineering Journal, 12*(1), pp. 687–695.

3 Securing Smart Power Grids Against Cyber-Attacks

Sanjeevikumar Padmanaban
University of South-Eastern Norway

Mostafa Azimi Nasab, Tina Samavat,
Morteza Azimi Nasab, and Mohammad Zand
CTiF Global Capsule

CONTENTS

ABSTRACT

Creating an intelligent electricity network is based on data exchange in different parts of the network at the production, transmission, and distribution level, and its vulnerability is increasing day by day. This network needs a suitable, safe, and reliable structure to meet the goals of intelligence. There are two types of data in smart grids: physical measurements and variable states. This information is collected every 2–4 seconds and used to estimate the status of the network. Current systems can only detect destructive data due to incorrect measurements of measuring instruments and cannot detect malicious data injection by attackers. This research has investigated and identified this attack in the smart grid mode vector estimation. In both cases, the attack vector injects inaccurate data at random, and attackers seek to enter any data to create an error in estimating the network state vector without the target being a specific component of the state vector. In this type of attack, from the beginning, it seeks to form a suitable data vector so that it is considered to create its error in a specific component of the network mode vector.

DOI: 10.1201/9781003331117-3

3.1 INTRODUCTION

Today, electricity is the main source of energy consumption in the world. In the early days, electricity was generated and consumed locally, i.e., small power plants were built in each region for fulfilling energy consumption needs of that region. With the industrialization of countries, the electricity demand increased sharply. The construction of small power plants no longer seemed logical, because the construction of large power plants and the transfer of electricity to distant places were not economically viable. The solution to overcoming this concern was introducing the concept of smart grids. Just as it is impossible to imagine a world without e-mail technology today, there will soon be a time when human life would not be conceivable without a smart grid. In such a system, communication links for transferring data and commands are vital; therefore, having uninterrupted communication is required for the reliable operation of electricity distribution and transmission networks. In addition to managing and protecting the network, real-time communication is also needed for an uninterrupted power supply to customers. As the population grows and energy demand expands, the electrical and low efficiency of traditional power distribution networks in the face of the burden imposed by customers requires intelligent energy management. The conversion of traditional energy networks to smart grids has led to a revolution in the energy industry [1].

The basic concept of smart grids is entwined with improving demand-side management and energy efficiency using advanced metering infrastructure and building a flexible and reliable power grid. Electrical energy flows from generating stations to consumers through a one-way hierarchical flow. This one-way, uncontrolled flow of electricity poses several challenges for traditional networks and their operators. Power outages can have irreparable economic and social effects. For example, the 2003 power outage in North America, which took 4 days, was caused by a relatively small fault and left about $6 billion to $13 billion impairments. Smart grids generally include power, information, and communication infrastructure, which require exchanging information and two-way telecommunications across the entire network. By using intelligent technology to make fundamental changes for making timely decisions in critical situations in production, transmission, distribution, and energy consumption, unwanted blackouts can be prevented, and in addition to providing customers with secure and stable electricity access, incurring additional costs can be avoided [2].

A smart grid brings many opportunities and challenges to the electricity industry. These challenges include high operating costs and lack of clear strategy, modeling complexity, data security and privacy, sufficient knowledge, and experience to use smart equipment [3].

The security challenge in smart grids, due to the variety of components, different ways of sending data, and the definition of different layers for these networks, is very diverse and complex. Despite the significant advantages of connecting further sub-sections of smart grids, these connections can also make smart grids more vulnerable to malicious security attacks. Therefore, it is clear that the security of smart grids is one of the most important technical challenges in advancing the smartening of the electricity industry. Large-scale coverage of smart grids makes it possible to manage operations remotely, but the first concern is security against hackers, misuse, and malicious activities [4]. One of the attacks that have caught the attention of many researchers is the False Date Injection Attack (FDIA) malware injection because traditional networks are unable to detect this attack, so attackers can distract the control center from making the right decision. This type of attack in smart grids has been investigated and identified in reference [5].

3.1.1 History of Smart Electricity Networks

The power grid structure, which has been delivering power to numerous consumers for years, is evolving rapidly, with a population growth of 25% since 1982. Traditional networks are

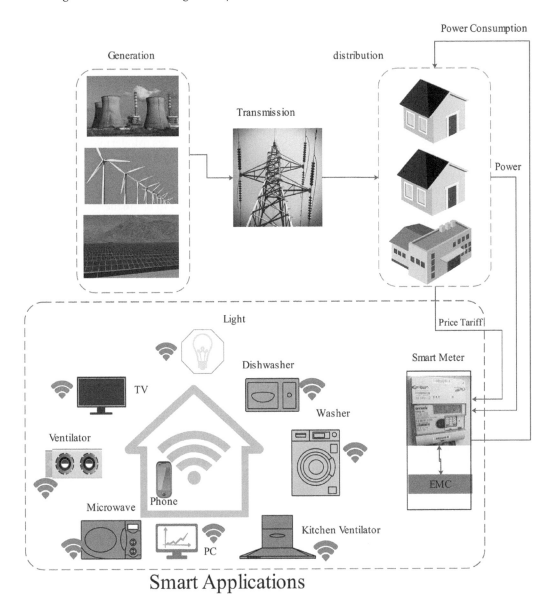

FIGURE 3.1 Overview of a smart grid.

insufficient to meet this increasing demand, and constructing new generators to add to our power network is not logical. Therefore, smart grids are considered a new generation of power systems with advanced computing and communication technologies. Figure 3.1 shows an overview of an intelligent power grid. The purpose of a smart grid is to use telecommunications, information technology, and data science in the current grid to be more flexible, efficient, and capable. This rapid growth in size, scale, and complexity also has limitations. These constraints pose both national challenges, such as the security of the power system, and global challenges, such as climate change, which are expected to be highly controversial in the future. Although the grid is about 99.97% reliable today, power outages in a country like the United States still cost at least $150 billion a year (about $500 per capita) (only one outage). Electricity caused $4 billion to $10 billion in damage to Canada and the United States in 2003 [6]. In December 2007, the then President of the United States signed the Energy Independence and Security Act, known as the EIAS, which was formed to monitor the electricity power section, the third part of which was called smart grids [6]. The world's first smart grid was introduced in March 2008, and the city of Boulder, Colorado, became the first smart grid city. Over the years, China has become the largest and most important global smart grid market. The Consumer Electronics Control (CEC) recently reported that 26.53% of electricity generated in Japan in 2010 was from clean energy sources [7]. It is estimated that by 2030, $378 billion will be spent worldwide to build smart grids. Most of the cost will be contributed by ten countries. The main payers are the United States, Brazil, India, China, and several European countries. Modernizing transmission and distribution is also expected to provide the largest investment in the energy structure in the next 20 years. Designers' goal is to use smart technology around the three main axes of subscribers, equipment, and communications. Given the above, it seems that the use of smart grids will be inevitable in the future. Therefore, this network needs to be further evaluated and its challenges and threats to be addressed.

3.1.2 COMPARISON OF CURRENT ELECTRICITY NETWORKS WITH SMART ELECTRICITY NETWORKS

The electricity grid has remained unchanged for many years. The existing electricity grid has resulted from the expansion of urbanization and rapid development of infrastructures globally during the past decades and inappropriate adoption of the grid by a new generation of consumers. Although power companies are located in different regions, they typically use the same technology. However, the growth of the electricity system has been affected by economic, political, and geographical issues that are unique to each company. Despite such differences, the overall structure of the power system is the same [8]. In these networks, electricity is generated through power plants (such as coal, gas, and nuclear) and transferred to the consumer. As shown in Figure 3.2, the power plant is at the highest level, and the consumer is at the lowest level of a power grid. There is no two-way path for electricity flow, information exchange, and decision-making in the traditional power grid. Environmental

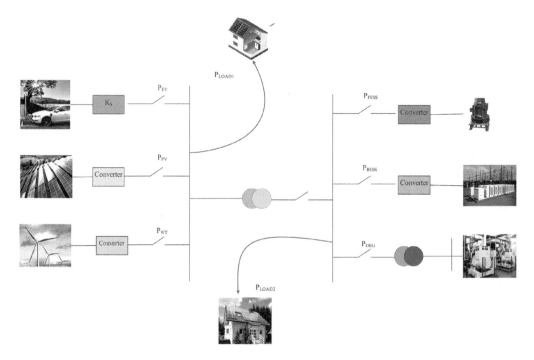

FIGURE 3.2   The structure of traditional networks.

pollutants and energy shortages are also among the problems of these networks. In addition, an unprecedented increase in power demand, delays in investing in power system infrastructure, reduced system stability, and the absence of sufficient security margins, any unforeseen or unusual demand in the distribution network that causes errors can lead to global blackouts [9].

Traditional power grids have the following disadvantages:

- The inefficiency of the electricity network in managing maximum demand periods
- The inability of the network to exchange reliable information
- Limited network capability in the use of distributed energy sources
- Network vulnerability to blackouts and power quality disruption
- Network vulnerabilities due to natural disasters.

Smart grids have been proposed to solve current power problems and provide better and more efficient management. One of the main goals of smart grids is to provide reliable electricity and meet customers' growing needs with minimal damage to the environment. Figure 3.3 shows the structure of an intelligent power grid. These networks include advanced monitoring, automation, production, distribution, and transmission control.

The main characteristics of smart networks include the following:

- Conscious and active participation of consumers in smart electricity networks
- Modify production and storage (provide the required power quality)
- Flexibility in the face of natural disasters and disasters
- New products, new services, new markets.

Figure 3.4 shows an intelligent power grid next to a current hierarchical grid. The smart grid consists of several small networks, each with production resources, energy storage, and control centers.

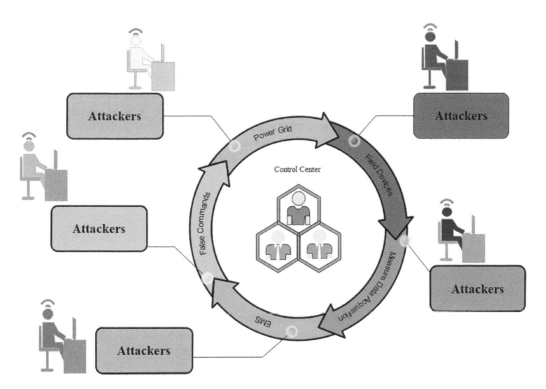

FIGURE 3.3 Intelligent network structure.

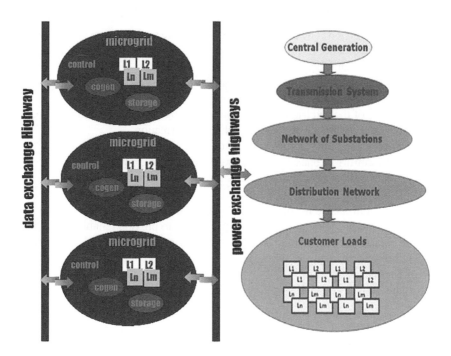

FIGURE 3.4 Traditional network compared to a hierarchical smart grid.

In addition to connecting to the electricity network, these networks are connected through telecommunication lines.

3.2 NECESSARY TECHNOLOGY FOR SMART GRID

Intelligence can be implemented in all three stages of production, transmission, and distribution. On the other hand, there are two views on these changes: (a) the design and implementation of an intelligent power grid, regardless of the current structure, and (b) smartening alongside the existing pieces of equipment. The first view is speedy and provides humanity with an entirely new and advanced network, but the second view is more suitable for places with a reliable and up-to-date power grid. In order to make the electricity network smarter, it is necessary to apply technologies in different parts of the current electricity network [10,11]. These technologies include the following:

- Advanced smart meters
- Extensive and decentralized energy sources
- Energy savers
- Extensive automation
- Broadband telecommunication network
- Low-cost telecommunication systems
- Smart switching when error detection.

Using the proposed technologies is expected to increase the security and efficiency of the electricity network and reduce the cost of construction, design, and maintenance [12]. Smartening in the production, transmission, and distribution sectors is briefly introduced.

- **Production:** In the issue of electricity intelligence at the production level, scattered and green resources are discussed. Scattered sources are small generators or energy savers.

In the current electricity network, the presence of these sources with low penetration percentage does not affect the design of the network. However, in an intelligent electricity network, these resources will be used extensively and flexibly, removing power generation from centralized mode. Small generation sources such as wind turbines, solar cells, and other sources that do not use fossil fuels to generate electricity are in this category. These resources are easy to install and cost very little to set up large power plants. On the other hand, these sources do not introduce polluted carbon compounds into the environment. Nevertheless, the amount of electricity generation in distributed sources is variable and depends on factors such as weather conditions and installation location.

- **Electricity distribution:** This area is responsible for transferring electricity to consumers and links for communication to control and manage the grid [13]. This area can be seen in Figure 3.5.
- **Control and monitoring**: The structure and technology of the monitoring and control system in the current electricity network date back to 1960. In this structure, called SCADA, the control is performed centrally, and information is collected by sensors in different parts and sent to the control center. The upper layers of this structure do necessary processing and applications, and the information collected is local and limited and arrives at the center with great delay [14,15]. From the mid-1980s, the Global Positioning System (GPS) with all its possibilities was released for civilian use, and subsequently, the horizon of synchronization of measurement data across the power grid was clarified. The possibility of synchronizing measurements and the symmetric component measurement method introduced the concept of simultaneous measurement technology and phasor measurement unit (PMU). At the same time, the technical specifications of these devices were standardized with the

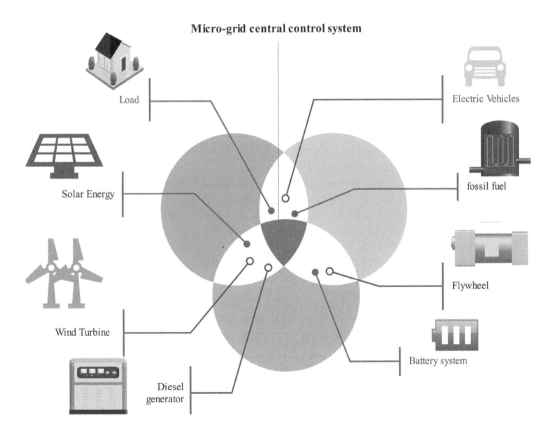

FIGURE 3.5 Intelligent power grid – distribution area.

publication of the IEEE standard in 1995. Then, in 2005, the second edition of the PMU standard was published by the IEEE [16]. They are accompanied by very precise timing at the same measurement location, so delays in sending this data are no longer critical to their usability. Having a timestamp of measured and sent data allows the data of a particular moment to be used together, thus providing a true and accurate picture of the network. In a smart grid, PMUs measure phase and voltage accurately and simultaneously using GPS with primary processing, and information are sent and analyzed to control centers using Wide Area Measurement System (WAMS). Special protocols have been defined for this network, the most important of which is CEI 61850. In 1990, monitoring and the data transmission network evolved. In line with these issues, in 2000, the WAMS network was implemented in the US pilot projects. Other countries are moving in this direction quickly. China had a comprehensive implementation plan for WAMS that was completed by 2021 [17]. Today, the tendency to use phasor measurement technology is at its peak, as the need for an accurate estimate of the state of the power grid is an inevitable necessity for improving the power grid's performance and dealing with catastrophic events. The first step in different countries is to get acquainted with the PMU and how it works in providing a continuous picture of the network mode. PMUs are installed to achieve full visibility in the power grid to perform a variety of monitoring, control, and protection functions of the system optimally in a control center, using simultaneous measurements of the power grid range [18].

3.3   SECURITY THREATS IN SMART ELECTRICITY NETWORKS

Today, electricity networks are considered vital highways for sustainable development in any country. Proper and cheap supply and distribution of energy to consumers, which are the country's major industries, can increase national production.

In the meantime, security, due to its great importance in the energy industry, is one of the vital issues. This section tries to introduce the technology infrastructure in the power grid and its security challenges to be examined. The following are three main levels of cyber security that are shown in Figure 3.6. The Cyber Security Group at the National Institute of Standards and Technology (NIST) has published a comprehensive approach to the issue of security in smart power systems [19].

- **Accessibility:** Ensuring availability at the desired time and using the required information is one of the most important issues in the smart grid; because if the ability to use information is lost, it may disrupt energy distribution.
- **Accuracy:** Protection against manipulation or destruction of information means the ability not to deny and verify its identity. Loss of accuracy, which may be due to unauthorized changes or destruction of information, can lead to poor management decisions in the future.
- **Confidentiality:** Maintaining permissible restrictions on access to and disclosure of information to protect the privacy of individuals and private information. This is especially important to prevent unauthorized disclosure of information not permitted to the general public.

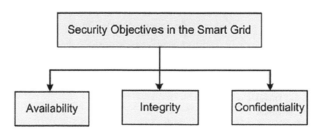

FIGURE 3.6   Various security components in the smart grid.

Accessibility, accuracy, and confidentiality, as mentioned, are the three issues at the high level of security in smart grids. In addition to such general issues, the NIST report sets out specific security requirements for an intelligent network, including cyber security and physical security. Specifically, in cyber security, it includes details of the problems and requirements related to security in the smart energy network regarding information and network. The field of physical security includes the security of physical equipment and environmental protection and security policies related to employees and contractors. Since in this review, our focus is on cyber security, so we state the requirements of cyber security below:

I. **Attack and Defensive Operations Detection:**
   Compared to a traditional power system, an intelligent energy grid is an open, distributed communication network over a wide geographical area. It is impossible to ensure that all nodes in this vast network are resistant to attack. Therefore, the communication network needs to have its traffic constantly monitored, tested, and measured to detect unusual activities related to the attacks as soon as possible. In addition, the network must have self-repair capability to operate accurately after attacks. Given the importance of power infrastructure, offensive operations are a basic requirement to maintain network accessibility.

II. **Identification, Authentication, and Access Control:**
   The intelligent system communication network includes millions of electrical devices and consumers. Identification and authentication are key processes for authenticating an instrument or consumer as a prerequisite for granting access to resources in an intelligent energy system information network. The purpose of access control is to ensure that resources are accessible to the correct user who has been correctly identified. Access control must be strict to prevent unauthorized persons from accessing sensitive information and controlling infrastructure. To achieve these goals, each node in the smart grid must have basic encryption systems such as symmetric and asymmetric systems for authentication and data encryption.

III. **Secure and Efficient Communication Protocols:**
   Unlike traditional communication systems, communications in an intelligent power grid have time and security constraints; especially in the distribution and transmission, these two constraints sometimes even interfere with each other. If it is impossible to achieve both these requirements in some cases completely and simultaneously, an optimal balance must be struck between information security and communication efficiency in the design of smart grids.

In communication networks, security attacks can be divided into two types:

1. **Abusive users:** These users are users who try to obtain more resources on the network than legitimate users, in violation of network rules.
2. **Malicious users:** In contrast to abusive users, malicious users have no intention of self-interest but tend to obtain, modify, or disrupt the information contained in the network.

Both groups of users pose network security challenges. In an intelligent energy system, destructive behavior is more important because malicious attacks can cause catastrophic damage to power supplies and widespread power outages, which is a very sensitive issue in an intelligent energy system. It is not possible to list all types of attacks due to the vastness and complexity of the network in intelligent systems, so malicious attacks are divided into three main categories according to accessibility, accuracy, and confidentiality:

1. Attacks that target accessibility deny response (Denial of Service—DOS) and try to delay, stop, or disrupt the communication of an intelligent system.
2. Attacks that target accuracy and attempt to intentionally and illegally alter or disrupt the data in the system.

3. Attacks aimed at secrecy and attempted to obtain unauthorized data from network resources in an intelligent system.

3.4 DATA ATTACK ON SMART POWER GRIDS

One of the issues in smart grids is the state estimation in the control center and the error detection in this center. Estimating the situation using the data sent to the control center helps have a real-time view of the network in the Energy Management Center. The most important tasks of the control center are as follows [20]:

1. **Processing and analysis of observations:** Before the state estimating is done, the collected data are analyzed to see if there is a unique estimation for the system state in the control center.
2. **State estimation:** Using real-time data, they estimate the voltage and phase of each bus.
3. **Improper bad data processing:** In this step, inappropriate data are detected and deleted from the data.

One of the cases that creates inappropriate data in the network is called data attacks, which will cause an error in estimating the state vector in the control center. It is necessary to detect a data attack and take the necessary measures quickly to prevent the spread of this error in the network. Data are attacked by an attacker by injecting improper data into the network. In fact, by taking control of the measuring instruments, the attacker injects his data into the network in such a way that he achieves his goals by making an error in estimating the system state vector. False data attacks can be divided into the following two categories based on the attacker's knowledge or lack of knowledge of the system structure [21]:

1. **Undetectable attack:** In this type of attack, the attacker, knowing the structure of the system and knowing the system parameter called the load measurement matrix (H), adjusts his data in a way that is undetectable using detection methods. The data attack in this review falls into this category.
2. **Detectable attack:** In this type of attack, the attacker, without knowing the H matrix, forms his data vector. Because of the structure of injected data, it can be detected easier.
   - Another category for attacking data according to the attacker's target is as follows [21]:
   - Random data injection attack: In this type of attack, the attacker seeks to enter any type of data to create an error in estimating the network state vector without having any specific component of the state vector as the target.
   - Targeted incorrect data injection attack: In this type of attack, the attacker from the beginning seeks to form a suitable data vector in such a way that in a specific component of the network mode vector, the desired error is received.

From this point of view, the attacks considered in this research are of the targeted type. The DC load distribution model detects data attacks in smart power grids. The load distribution model used in the control center estimator is a set of equations that show the power distribution across all transmission lines. The AC load distribution model includes nonlinear equations from which active and reactive power are obtained. This model increases the computational volume and complexity, so the linear approximation of this model, such as the DC load distribution model, is used. Although this model is less accurate, it greatly reduces the computational volume. Therefore, we use this model in this research.

3.5 CONSERVATION-BASED DESIGNS

Since a power system typically covers a large geographical area, more than 1 million square kilometers, the network operator can select important measuring devices from all devices for protection, such

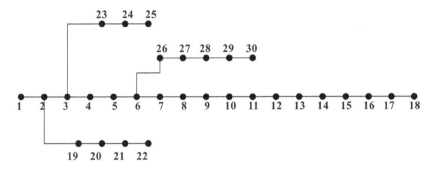

FIGURE 3.7 30-sample networks with measuring devices.

as secret methods, protection encryption, permanent monitoring, or disconnection from the Internet [22]. Measuring devices in DC estimation, power systems are divided into two types: reactive injection in busbars and active power flow in lines. As mentioned before, one device is needed to calculate the active power in each bus, and two devices are needed to calculate the active power in each line. For example, consider a grid consisting of 30 busbars, as shown in Figure 3.7, in which busbar 9 is connected to busbars 6, 10, and 11. If an attacker wants to change the 9-state variable, it is necessary to adjust the measuring devices for injection power P6, P9, P10, and P11 according to the type of manipulation [23]. On the other hand, if the system operator protects the P9 injection power computing device, the attacker has no chance of manipulating and attacking the 6, 9, 10, and 11 bus mode variables. Nevertheless, if the measuring device on the P6 and P9 transmission lines is protected, only the busbar variables 9 and 6 can be manipulated. According to the above example, it can be understood that the devices for measuring the power of injection in the busbars are more important than those for measuring the power flowing in the lines. From the point of view of state estimation, injection power measurements play an important role in determining the specific state variable. In contrast, the measured power in the measurement lines is added and only improves state estimation accuracy [24]. In these sources, several schemes have been proposed to deal with the occurrence of false data injection attacks, including the protection of a set of basic measurements and PMU-based protection.

3.5.1 Protection of a Set of Basic Measurements

Basic measurements are at least a set of measurements in which the corresponding rows in the lattice Jacobin matrix used to estimate state variables are independently linear. Reference [25] shows that with the knowledge of K which is the number of measurements that attackers can manipulate, m which is the number of measurements, and n that indicates state variables, there is always a successful attack vector that can inject false information into the power grid without identification. On the other hand, the attack will be detectable if the number of measurements that the attacker can manipulate is less than 1 mn. Reference [26] uses graphical methods to examine the protection mechanisms against FDIA attacks. Protecting selected measurements indicates that no attack can be launched to affect any set of state variables. Reference [27] divides the power grid into several sub-grids. It uses an asymmetric decoding method called McAllis to protect the measured values while transferring them to the control center. The ability to calculate limited units of measurement prevents complex cryptographic operations. One of the disadvantages of this algorithm is the large cryptographic key size and it is important to suggest an improved and lighter version of this cryptography with a smaller key.

3.5.2 PMU-Based Protection

In a smart grid, PMUs play an important role in measuring the phase, voltage, and current of different parts of the grid accurately and simultaneously using GPS. It is very difficult for attackers to gather information collected by PMUs. However, given the high cost of large-scale installation and the complexity of synchronizing PMUs with each other, it is important to find the best places to place PMUs on the network so that the number of PMUs can be determined. In recent years, the issue of optimal placement of these units in power grids to full visibility of the system has become very important.

Chen and Zhu [24] proposed an algorithm for placing PMUs in an intelligent power grid and introduced it as an integer programming problem. They showed that additional PMUs could help improve the ability to detect bad data in a given system. Chen and Zhu [24] proposed an algorithm for the convenient placement of PMUs in a power grid to find suitable locations in the shortest possible time. The proposed algorithm has little complexity and applies to different types of PMU measurements. This reference also shows that if PMUs protect the bus boundaries, an attacker cannot inject false information without changing other network measurements. References [24] and [28] indicate that by optimally placing PMUs, the power system can effectively protect itself against data injection attacks. However, Khalili et al. [29] proposed a cyber-attack by spying on GPS. By sending a fake signal to the GPS receiver, attackers could easily launch an attack without accessing the communication network. This attack plan could lead to an initial failure of the PMU system. Roomi et al. [30] showed that an attacker could execute attacks by cutting a line only by attacking a set of critical measurements, even if the PMUs were stationed in the system.

3.5.3 DIAGNOSIS-BASED DESIGNS

Many methods have been proposed to detect intrusion and detect fake data injection attack. Table 3.1 summarizes several tasks performed along with the advantages and disadvantages of each method.

3.5.4 DETECTION OF ATTACKS BASED ON STATE ESTIMATION METHODS

Another area of research is the attempt to improve methods for detecting bad data based on state estimation to detect false data injection attacks. Reference [17] detects DOS and FDIA attacks online and models the system as a discrete linear dynamic system. In this paper, the CUSUM algorithm is used to detect cyber-attacks quickly. In Reference [27], data prediction methods are used to identify and determine abnormal data. In this way, the deviation of predicted and measured values increases whenever an attack occurs, which identifies the attack. In addition, the abnormal data are

TABLE 3.1

Advantages and Disadvantages of False Data Injection Attack Detection Methods

Diagnosis Method	Advantages	Disadvantages	Reference
L-norm	Incorrect data in terms of the non-ideal recognizes it effectively.	Threshold settings affect detection accuracy.	[3]
Adaptive partition	High detection accuracy.	Unable to identify incorrect data under ideal conditions.	[22]
Generalized Likelihood Ratio	False data in terms of ideal and non-ideal identifies.	High computational complexity.	[25]
Adaptive CUSUM1	false data with ideal conditions and the non-ideal identifies with high detection accuracy.	High time consumption for large systems.	[27]
Euclidean distance metric	Ability to identify false data in ideal and non-ideal conditions.	Abnormal fluctuations due to normal data cause significant changes as a result of this algorithm.	[14]
Principal component analysis	High detection accuracy.	Loss of communication data leads to errors are detected in the results.	[5]

replaced with the predicted data, and the calibration process takes place. Unlike previous studies in which the fixed and diagonal network topology matrix is considered, in reference [28], this matrix is non-diagonal and variable with time. Non-diagonal means that the predicted value of one node depends on the values of other nodes, and variable with time means that whenever a new measurement is made, the values of this matrix are updated.

Reference [22] also examines the relationship between two physical parameters of the power grid to detect an FDIA attack. The first parameter is Controller to the Static Var Compensator, which can be used to analyze the voltage fluctuations of the nodes in the power grid. The second parameter is the New Voltage Stability Index (NVSI) voltage index that examines the voltage stability of the node, which can quantify the effect of FDIA on system measurements. Accordingly, a state forecasting method is proposed to predict and detect FDIA, and simulations in the IEEE system with 39 and 118 busbars for the efficiency and performance of the proposed design are evaluated.

In reference [29], the problems of estimating the state and detecting incorrect data injection in smart networks are also investigated. An autoregressive detector in the presence of colored Gaussian noise during state estimation is provided to detect random data in the smart grid. By modeling noise with this detector, it estimates the status of power transmission networks. This study also evaluates and compares the performance of the proposed state estimator and detector on the IEEE system with 30 busbars with normal Gaussian noise-based detectors and shows that normal Gaussian noise can be a special case of the proposed approach. This article is widely used in processing power signals. Reference [27] also examines the effect of FDIA on an intelligent network. It analyzes the effect of FDIA on nonlinear state estimation and finally compares linear and nonlinear state estimation and the error propagation rate for different attack vectors.

3.5.5 Attack Detection Using Machine Learning Algorithms and Neural Networks

There are several ways to identify FDIA in neural networks and machine learning. The FDIA detection problem can be considered a binary classification problem from a machine learning perspective. In this approach, machine learning algorithms are used to classify healthy measurements or measurements that have been attacked. For example, the article [30] uses well-known (supervised and semi-monitored) algorithms with high-level capabilities to model the attack detection problem. The proposed algorithms in this article have been investigated in different IEEE systems. The relationship between the statistical and geometric properties of attack vectors used in attack execution scenarios and learning algorithms for attack detection using statistical learning methods has been analyzed. It shows that the performance of machine learning algorithms for detecting attacks is better than that of the algorithms proposed for attack detection by state estimation methods.

3.5.6 Other FDIA Defense Strategies

Several other methods have been proposed to defend against FDIAs. Based on the fact that knowing network topology information is a key pioneer for an attacker, Zhu et al. [31] proposed a step-by-step verification scheme to help the base station detect false data packets. Chajon et al. [32] proposed a distance-based method (Kullback–Leibler Distance, KLD) by tracking the dynamics of measurement changes to detect FDIAs in the AC mode estimation model. The key idea in this paper is to dynamically measure by calculating the distance between adjacent steps using KLD: Under normal circumstances, the KLD is very small, but at the moment that incorrect information enters the network, the KLD becomes larger than normal, which allows attacks to be detected.

3.6 MODE ESTIMATION IN SMART GRIDS

The idea of state estimation based on least-squares has existed since the early 19th century. Major advances in its application to aerospace have occurred in the 20th century. In this method, a value is assigned to an unknown state variable of a system according to a special criterion performed using measurements from that system. In particular, the actual values of the state variables are estimated. A common and familiar criterion is to minimize the sum of the squares of the difference between the estimated and actual values [32].

In the case of state estimation in power systems, if the system model is correct and the measurements are accurate, it can be said that the state estimation results are also correct. However, the measuring transducers may be incorrectly connected or defective, or their calibration not true in a way that does not provide accurate measurements. In this case, the data collected from the controlled system to estimate the state will no longer have Gaussian errors. Of course, the efficiency of state estimators will be overshadowed by such data and will not have desirable outputs. Therefore, if the measurement is inadequate or has a large error, such data should be disclosed and deleted or replaced before the calculation calculations, thus ensuring the accuracy and efficiency of the estimators.

Direct use of the measured data to estimate the state is illogical. Due to the physical nature of the measurement operation, the values sent to the state estimator will not be free of large and unavoidable errors. The most common method of detecting inappropriate data in estimating the state of power systems in ancient times was a statistical method. Using this method requires having enough information about the statistical properties of measurement errors. This method of detecting inappropriate data is theoretically simple, but it is time-consuming and uses iterative computational algorithms in practice. It is clear that this method will not be fast enough for real-time control of complex and very large power systems. Recent advances in signal processing and control have led to conventional methods in these theories being replaced by techniques based on artificial intelligence. Analyzing power systems is one of the branches in which the neural network technique is widely used. The analysis of power systems leads to very complex and extensive differential and dynamic equations; because power systems consist of many parts with dynamic behavior and very complex equations, which leads to more complex analysis. Also, in many cases, studying and generalizing the power system in a very short time are required. To achieve this goal employing analytical and mathematical methods are not desirable due to being time-consuming [33].

In a power system, state variables include voltage values and relative phase angles at system nodes. Measurements are needed to estimate system performance in real-time, both for reliability control and constraints on the economic distribution of the load. An estimator input contains incomplete measurements of voltage and power values. Since there is an error in the measured values and measurements, equipment and techniques may be additional, and the estimator is designed to estimate the voltage values and phase angles best. The estimator output information is then used in system control centers to study the economic distribution of the load, taking into account system reliability and system control.

3.7 BAD DATA

Any false information injected into the estimator is called bad data. One of the main purposes of a state estimator is to detect bad data and remove it if possible. Measurements may include errors for a variety of reasons. Random errors usually occur due to the limited accuracy of measuring instruments and communication devices. If there are enough measurements, such errors are filtered by the state estimator. How this is filtered depends on the method used to estimate the state. Large measurement errors can also occur in measuring devices affected by other factors or damaged or have incorrect connections. Communication system failures and noise caused by unexpected interference can also lead to large deviations and errors in recording measurements. In addition, a state estimator may be deceived and misled by incorrect topology information, known as bad data in the state estimator. The management of such cases is complex, and the correction of topological errors is debatable in a separate topic and is beyond the scope of this study. Some common tests can identify and remove some bad data before estimating the status. Negative voltage measurements several times larger or smaller than expected or the large difference between the input and output current of a node inside the substation are examples of such bad data. Unfortunately, not all bad data types can be easily detected by such methods, so state estimators must be equipped with more advanced features to detect and identify any bad data.

Bad data may appear differently depending on the location and the number of measurement errors. This data can be divided into two general categories:

- Bad single data: Only one measurement in the whole system has a large error.
- Multiple bad data: Have more than one measurement error in the system. Multiple bad data are classified into three groups:
  - Multiple non-interactive bad data do not affect each other: Bad data have a poor correlation with the rest of the measurements.
  - Uncoordinated multiple interactive bad data: Uncoordinated bad data in strongly correlated measurements.
  - Coordinated bad interactive data: Bad data are consistent in measurements that have strong correlations [22].

3.7.1 Bad Data Types in the Power System

Bad data in a system can be caused by the following factors [12]:

1. **Insufficient data due to topological errors:** SE performs its operation with the electrical model obtained by the Transaction processor (TP); in other words, TP converts the detailed model of the switch/bus section into a more compact and useful bus/branch model. In this process, some measurements should be ignored (such as load distribution by circuit breakers (CBs)). A CB contains one or more logic states with multiple isolated switches, often the correct condition of all CBs in a known system. However, sometimes the status of a particular CB may be wrong. This happens when some remote switches have not been measured or operated remotely and are defective. Other reasons include a breaker being repaired and not reported by the repair team, a line or transformer being cut, a bus detachment, or a mechanical fault in signaling devices. In such cases, the TP encounters a CB whose status is unknown. In such cases, the TP must decide on a situation similar to the CB status previously recorded for the same breaker or using measurement-related values. Therefore, the possibility of choosing the wrong situation for CB is inevitable. When this happens, the bass/branch model produced by TP is wrong and leads to a topology error. Therefore, as its name implies, this error is related to the physical condition of the system. Unlike parametric errors, most of these errors remain unknown as long as they do not exceed the threshold value. Topological errors typically cause the SE to become significantly problematic. As a result of these topological errors, the misdiagnosis process is disrupted, and some correct and analog data are also misdiagnosed as bad data. Therefore, due to misdiagnosis, this data is deleted from the data set (in which case it should not be deleted), and again the state estimation algorithm is implemented. As a result, the estimator presents an incorrect estimation or diverges. A type of data can also be presented in a specific way. However, in the first place, the topological error must be controlled, so it is necessary to use an advanced method to detect and identify these bad errors [24].
2. **Insufficient data due to measurement error:** Errors due to limited accuracy of measuring devices, failure of devices, and disconnection of communication lines are among the measurement errors.
3. **Insufficient data due to cyber-attacks that occur in smart grids:** This condition occurs only in smart grids, but the previous two modes, whether it is a smart grid or a traditional grid, can occur. It should also be noted that the first factor is a separate category and is not considered in issues related to bad data processing, so we will not discuss this type of bad data and its detection methods. It is clear that in smart grids, all devices have their

own IP so that consumers may have access to their power consumption information and control it via the web. This two-way path provides the possibility of vulnerability, or the attacker may infiltrate the network through unpredictability and change the load through access to network software to destabilize the network. Therefore, due to the expansion of smart grids, it is necessary to conduct studies in this field to provide solutions to detect and identify this data.

4. **Reinforced learning:** This type of learning is very similar to the unsupervised type, and the similarity is that the data used for learning are not labeled; however, when asked about the data, the result will be graded. A good example of this type of learning is playing. If the car wins the game, then it uses the result to strengthen its future moves during the game.

3.7.2 Machine Learning Performance

Engineers working in machine learning systems use various techniques for this purpose. As mentioned earlier, many of these techniques involve exploring data and statistics. For example, suppose you have a body of information that defines the characteristics of different types of coins (including weight and radius). In that case, you can use statistical techniques such as the "nearest neighbor" algorithm to classify coins that have not been seen before. The "nearest neighbor" algorithm performs that which seeks to classify the nearest neighbor to the coin and then assigns the same classification to the new coin. The number of neighbors used to make this decision is known as "k". Countless others try to do the same thing using different methods.

Figure 3.8 shows that the image at the top left is related to the available data set. These data are classified into two groups, blue and red, and are completely hypothetical. As can be seen, there are some definite groupings in these images. Everything in the top left corner of the image is in the

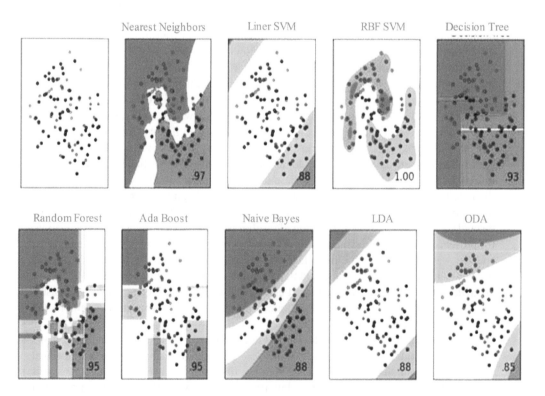

FIGURE 3.8 Different machine learning techniques for data classification.

red category, and everything in the bottom right of the image is in the blue group. There is also some overlap in the middle of the image. If new data is placed in the middle, whether the sample belongs to the red or blue group arises. Other images show different algorithms and how to group new samples. If the new data is in a white area, it cannot be classified using these methods. The numbers in the lower right corner of the photos also show the accuracy of the classification. Several methods for identifying FDIA in machine acquisition are available for the binary classification problem described above. The following are two of these methods that have been developed in this study with Python software.

- Proposed nearest neighbor algorithm in detecting false data injection attack

    The nearest neighbor algorithm is widely used in data mining, machine learning, and pattern recognition (Figure 3.9). The simplest case in this algorithm is when it is 1k. In this case, the algorithm is known as the nearest neighbor. According to Figure 3.10, to classify a record with an unspecified category, the distance of the new record from all training records is first calculated to identify the nearest neighbors. The distance between points can be calculated from Euclidean distance methods, Hemming distance, Manhattan distance, and Minkowski distance.

    Selecting the value of k in this classification method is very important. If the value of k is selected too small, the algorithm becomes sensitive to noise, and if the value of k is selected too large, records from other categories may be placed among the nearest neighbors.
- The proposed decision tree algorithm in detecting false data injection attack

    The decision tree algorithm is a supervised algorithm used as a support vector machine for classification. Suppose y, which is the predicted response or the output of the target, is a classification variable or an array of characters or strings. In that case, the decision tree will perform the classification operation. Attach to these tags. The average number of middle nodes per tree surface is called the average width of the tree, and the average number of layers from root to end nodes is the average depth. If it is assumed that each depth has two edges and examines only one property, then we have:
- The first depth of a feature
- The second depth of the two features
- Third depth of the four features.

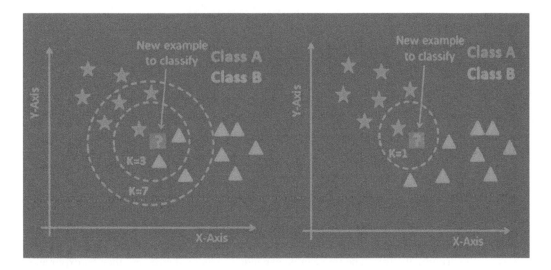

FIGURE 3.9 Example of the nearest neighbor algorithm.

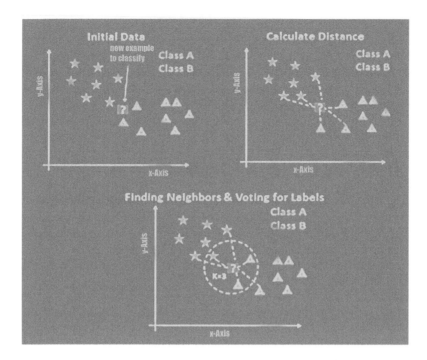

FIGURE 3.10   The nearest neighbor algorithm process.

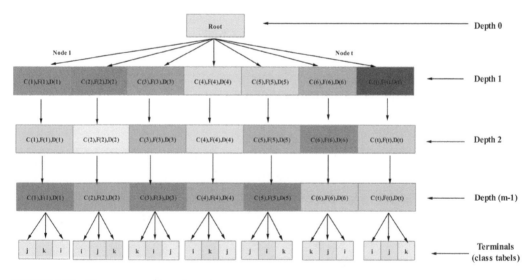

FIGURE 3.11   The process of learning a tree.

Finally, the data can be divided into four modes based on seven characteristics at three depths. Figure 3.11 shows the learning process of a tree. Models are made quickly. If the tree is allowed to grow without restriction, it will take a long time, unintelligent. The size of the trees can be controlled by stopping rules. A common rule of thumb is to limit the depth of tree growth. Another way is to stop pruning the tree. The tree can be expanded to its final size, and then by exploratory methods or user intervention, the tree is reduced to the smallest size so that accuracy is not lost.

3.8   SUMMARY

Unlike many research areas in artificial intelligence, machine learning cannot be considered an intangible goal; in fact, machine learning is a reality that is currently being used to improve the services used by humans. In many ways, it can be considered a forgotten star that works behind the scenes and does its best to find the answers we are looking for. This technology also helps to save operating costs and improve the speed of data analysis. For example, in the oil and petrochemical industry, using machine learning, the operational data of all drills are measured. By analyzing the data, algorithms are set that have the most results and optimal extraction in subsequent drilling. In this chapter, after modeling an intelligent network, the relationships in estimating the network state were explained, and the types of bad data that could cause problems were expressed. Then how to design a false data injection attack vector was discussed in detail. After stating the relationships required to extract the measurement data, the proposed algorithms for detecting false data injection attacks in smart grids were proposed.

REFERENCES

1. Asadi, A. H. K., Jahangiri, A., Zand, M., Eskandari, M., Nasab, M. A., & Meyar-Naimi, H. (2022, January). Optimal Design of High Density HTS-SMES Step-Shaped Cross-Sectional Solenoid to Mechanical Stress Reduction. In *2022 International Conference on Protection and Automation of Power Systems (IPAPS)* (Vol. 16, pp. 1–6). IEEE.
2. Zand, M., Nasab, M. A., Hatami, A., Kargar, M., & Chamorro, H. R. (2020) Using Adaptive Fuzzy Logic for Intelligent Energy Management in Hybrid Vehicles. In *2020 28th ICEE* (pp. 1–7). doi: 10.1109/ICEE50131.2020.9260941
3. Ahmadi-Nezamabad, H., Zand, M., Alizadeh, A., Vosoogh, M., & Nojavan, S. (2019). Multi-objective optimization based robust scheduling of electric vehicles aggregator. *Sustainable Cities and Society*, 47, 101494.
4. Zand, M., Nasab, M. A., Sanjeevikumar, P., Maroti, P. K., & Holm-Nielsen, J. B. (2020). Energy management strategy for solid-state transformer-based solar charging station for electric vehicles in smart grids. *IET Renewable Power Generation*. doi: 10.1049/iet-rpg.2020.0399
5. Nasab, M. A., Zand, M., Hatami, A., Nikoukar, F., Padmanaban, S., & Kimiai, A. H. (2022, January). A Hybrid Scheme for Fault Locating for Transmission Lines with TCSC. In *2022 International Conference on Protection and Automation of Power Systems (IPAPS)* (Vol. 16, pp. 1–10). IEEE.
6. Ghasemi, M., Akbari, E., Zand, M., Hadipour, M., Ghavidel, S., & Li, L. (2019). An efficient modified HPSO-TVAC-based dynamic economic dispatch of generating units. *Electric Power Components and Systems*, 47(19–20), 1826–1840.
7. Phiri, L., & Tembo, S. (2022) Cyberphysical security analysis of digital control systems in hydro electric power grids. *Computer Science and Engineering*, 12, 15–29.
8. Nasri, S., Nowdeh, S. A., Davoudkhani, I. F., Moghaddam, M. J. H., Kalam, A., Shahrokhi, S., & Zand, M. (2021). Maximum Power Point Tracking of Photovoltaic Renewable Energy System Using a New Method based on Turbulent Flow of Water-based Optimization (TFWO) under Partial Shading Conditions. In *Fundamentals and Innovations in Solar Energy* (pp. 285–310). Springer, Singapore.
9. Rohani, A., Joorabian, M., Abasi, M., & Zand, M. (2019). Three-phase amplitude adaptive notch filter control design of DSTATCOM under unbalanced/distorted utility voltage conditions. *Journal of Intelligent & Fuzzy Systems*, 37(1), 847–865.
10. Zand, M., Nasab, M. A., Neghabi, O., Khalili, M., & Goli, A. (2019). Fault Locating Transmission Lines with Thyristor-Controlled Series Capacitors by Fuzzy Logic Method. In *2020 14th International Conference on Protection and Automation of Power Systems (IPAPS)* (pp. 62–70). Tehran. doi:10.1109/IPAPS49326.2019.9069389
11. Zand, Z., Hayati, M., & Karimi, G. (2020). Short-Channel Effects Improvement of Carbon Nanotube Field Effect Transistors. In *2020 28th Iranian Conference on Electrical Engineering (ICEE)* (pp. 1–6). Tabriz. doi:10.1109/ICEE50131.2020.9260850
12. Tightiz, L., Nasab, M. A., Yang, H., & Addeh, A. (2020). An intelligent system based on optimized ANFIS and association rules for power transformer fault diagnosis. *ISA Transactions*, 103, 63–74.
13. Dashtaki, M. A., Nafisi, H., Pouresmaeil, E., & Khorsandi, A. (2020). Virtual Inertia Implementation in Dual Two-Level Voltage Source Inverters. In *2020 11th Power Electronics, Drive Systems, and Technologies Conference, PEDSTC 2020.* IEEE.

14. Zand, M., Neghabi, O., Nasab, M. A., Eskandari, M., & Abedini, M. (2020, December). A Hybrid Scheme for Fault Locating in Transmission Lines Compensated by the TCSC. In *2020 15th International Conference on Protection and Automation of Power Systems (IPAPS)* (pp. 130–135). IEEE.

15. Zand, M., Azimi Nasab, M., Khoobani, M., Jahangiri, A., Hossein Hosseinian, S., & Hossein Kimiai, A. (2021). *Robust Speed Control for Induction Motor Drives Using STSM Control. In 2021 12th (PEDSTC)* (pp. 1–6). doi: 10.1109/PEDSTC52094.2021.9405912

16. Sanjeevikumar, P., Zand, M., Nasab, M. A., Hanif, M. A., & Bhaskar, M. S. (2021). Spider Community Optimization Algorithm to Determine UPFC Optimal Size and Location for Improve Dynamic Stability. In *2021 IEEE 12th Energy Conversion Congress & Exposition - Asia (ECCE-Asia)* (pp. 2318–2323). IEEE. doi: 10.1109/ECCE-Asia49820.2021.9479149

17. Choeum, D., & Choi, D.H. (2021). Trilevel smart meter hardening strategy for mitigating cyber-attacks against Volt/VAR optimization in smart power distribution systems. *Applied Energy, 304,* 117710.

18. Borenius, S., Gopalakrishnan, P., Bertling Tjernberg, L., & Kantola, R. (2022). Expert-guided security risk assessment of evolving power grids. *Energies, 15*(9), 3237.

19. Azimi Nasab, M., Zand, M., Eskandari, M., Sanjeevikumar, P., & Siano, P. (2021). Optimal planning of electrical appliance of residential units in a smart home network using cloud services. *Smart Cities, 4,* 1173–1195.

20. Nasab, M. A., Zand, M., Padmanaban, S., Bhaskar, M. S., & Guerrero, J. M. (2022). An efficient, robust optimization model for the unit commitment considering renewable uncertainty and pumped-storage hydropower. *Computers and Electrical Engineering, 100,* 107846.

21. Nasab, M. A., Zand, M., Padmanaban, S., Dragicevic, T., & Khan, B. (2021). Simultaneous long-term planning of flexible electric vehicle photovoltaic charging stations in terms of load response and technical and economic indicators. *World Electric Vehicle Journal,* 190, doi: 10.3390/wevj12040190

22. Chen, J., & Zhu, Q. (2022). A system-of-systems approach to strategic cyber-defense and robust switching control design for cyber-physical wind energy systems. In *Security and Resilience of Control Systems* (pp. 177–202). Springer.

23. Zand, M., Nasab, M. A., Padmanaban, S., & Khoobani, M. (2022). Big data for SMART sensor and intelligent electronic devices–Building application. In *Smart Buildings Digitalization* (pp. 11–28). CRC Press.

24. Chen, J., & Zhu, Q. (2022). A system-of-systems approach to strategic cyber-defense and robust switching control design for cyber-physical wind energy systems. In *Security and Resilience of Control Systems* (pp. 177–202). Springer.

25. Stright, J., Cheetham, P., & Konstantinou, C. (2022). Defensive cost–benefit analysis of smart grid digital functionalities. *International Journal of Critical Infrastructure Protection, 36,* 100489.

26. Dashtaki, M. A., Nafisi, H., Khorsandi, A., Hojabri, M., & Pouresmaeil, E. (2021). Dual two-level voltage source inverter virtual inertia emulation: A comparative study. *Energies, 14*(4), 1106.

27. Padmanaban, S., Khalili, M., Nasab, M. A., Zand, M., Shamim, A. G., & Khan, B. (2022). Determination of power transformers health index using parameters affecting the transformer's life. *IETE Journal of Research,* 1–22. doi: 10.1080/03772063.2022.2048714

28. Sanjeevikumar, P., Samavat, T., Nasab, M. A., Zand, M., & Khoobani, M. (2022). Machine learning-based hybrid demand-side controller for renewable energy management. In *Sustainable Developments by Artificial Intelligence and Machine Learning for Renewable Energies* (pp. 291–307). Elsevier.

29. Khalili, M., Ali Dashtaki, M., Nasab, M. A., Reza Hanif, H., Padmanaban, S., & Khan, B. (2022). Optimal instantaneous prediction of voltage instability due to transient faults in power networks taking into account the dynamic effect of generators. *Cogent Engineering, 9*(1), 2072568.

30. Roomi, M. M., Ong, W. S., Hussain, S. S., & Mashima, D. (2022). IEC 61850 compatible OpenPLC for cyber attack case studies on smart substation systems. *IEEE Access, 10,* 9164–9173.

31. Chen, M., Wang, Y., & Zhu, X. (2022). Few-shot website fingerprinting attack with meta-bias learning. *Pattern Recognition,* 108739.

32. Su, Q., Wang, H., Sun, C., Li, B., & Li, J. (2022). Cyber-attacks against cyber-physical power systems security: State estimation, attacks reconstruction and defense strategy. *Applied Mathematics and Computation, 413,* 126639.

33. Yohanandhan, R. V., Elavarasan, R. M., Pugazhendhi, R., Premkumar, M., Mihet-Popa, L., & Terzija, V. (2022). A holistic review on Cyber-Physical Power System (CPPS) testbeds for secure and sustainable electric power grid–Part-I: Background on CPPS and necessity of CPPS testbeds. *International Journal of Electrical Power & Energy Systems, 136,* 107718.

4 Design and Modelling of a Stability Enhancement System for Wind Energy Conversion System

Ruhul Amin Choudhury and Mandeep Singh
Lovely Professional University

CONTENTS

ABSTRACT

With steep growth in urbanization and industrialization, there is huge rise in energy demand, which is mostly compensated by fossil fuel, the largest contributor to global warming and thus climate change. With advancement in technology, Renewable Energy sources become most popular alternative sources of generating energy which can solve the aforementioned problem. Wind energy is one of the most popular and one of the rapidly increasing sources of electrical energy produced by wind turbines. But, due to the intermittent nature of the Wind as it is highly influenced by the change, it creates a barrier in complete dependency for energy. Though wind energy has shown tremendous growth in recent years, it still faces challenges in terms of grid integration, stability, wind nature and location for the setup. In order to incorporate wind turbine into power grid, many advanced control system have been proposed and implemented so that the overall stability and efficiency of the system can be enhanced. The generation of power through wind turbine is totally dependent on the speed of the wing and slight variance in the speed will impact the output power. This power can't be further fed to grid as the grid voltage must be of constant amplitude and frequency, and thus many control strategies have been proposed and implemented. In this chapter, we have discussed various components and parameters contributing to wind energy generation and proposed a novel control system to maintain the stability of the wind turbine-connected grid system.

DOI: 10.1201/9781003331117-4

4.1 INTRODUCTION

Energy is playing a vital role in recent industrialization and urbanization. The steep increase in energy demand due to the rapid advancement in all sectors has led to crisis in energy generation. To meet this energy demand, all sectors are mostly dependent on fossil fuels which exhibit high negative impact on the atmosphere, and also the major dependency on fossil fuel leads to the depletion of the fossil source too. So to compensate or reduce the dependency on fossil fuel, the energy sector is shifting its focus towards the renewable sources of energy such as solar, wind and biomass. Utilization of energy is greatly influenced by affordability, availability and environment-friendliness. With recent advancement in technology and material science, renewable energy sources have found a strong ground to replace fossil fuel; besides this, the non-conventional sources such as solar and wind are readily and abundantly available in nature. But the intermittent nature of wind and solar creates a strong boundary for full-scale dependency or application in energy sector. Both the solar and wind are highly influenced by climate change and thus cause a serious concern in grid integration. Among the many renewable sources, wind energy has shown great potential and has increased its implementation in energy sector in multi-fold level. Though wind energy has exhibited a tremendous growth in recent years, it is still facing a strong challenge while incorporating with the grid system, wind nature, location of the setup, etc. Many advanced control system has been proposed and implemented, such as modern generator and power converter to increase the overall efficiency of the system and to incorporate wind energy into the grid system. The energy generation via wind turbine is highly dominated by the nature and speed of the wind and thus a slight variation in speed will exhibit its effects on output power. As grid voltage always must be of constant amplitude and frequency, thus the unstable output of wind energy becomes incompatible with the grid and in that reference various control strategies have been proposed and implemented. In another aspect, to generate electricity from wind energy, the wind turbine plays a pivotal role. Along with wind turbine, there are other parts which in total forms the system and these are wind generator, power unit and wind energy converter. Wind turbines are basically classified into two types based on their axis of rotation, namely horizontal-axis turbine and vertical-axis turbine. In the present scenario, mostly horizontal axis is used [1].

4.1.1 HORIZONTAL-AXIS WIND TURBINES

In horizontal-axis wind turbines, the rotor shaft coupled with blades and with electric generator is placed at the roof of the axis. A gearbox is also present, which drives the electric generator during the slow turns of the blades. In general, the turbine is placed pointing in a upwind manner to compensate the turbulence produced by the tower. The tower blades are placed at a suitable distance from the pole or sometimes in a tilted fashion to avoid any conflict with other parts during high wind situation. The blades are also made firm to keep the edges from being pushed due to high winds.

Though there is significant problem of turbulence, still downwind machines comes into market and have their own position and significance. During high wind force, the blades of the downwind machine actually get slightly bent to reduce swept area, which in turn reduces wind resistance. The excess amount of turbulence sometimes causes severe damage and leads to system failure.

4.1.2 VERTICAL-AXIS WIND TURBINES

In this type of wind turbine, the rotor is not need to be placed at the top facing the wind direction; rather it is placed horizontally, and this makes its construction more stable as it does not need the tower to support the rotor and requires less maintenance. This design also brings another advantage as it can easily cope with wind direction, which means it can utilize winds from varying directions. The sole disadvantage is that it produces pulsating torque.

4.1.3 POWER SYSTEM STABILIZATION

The wind energy generation system must have the potential to reduce fluctuation in power, which occurs due to transient fault. Before the stability of the electrical system gets disturbed, the wind turbine must reduce the effect.

In our proposed work, the following aspects are considered:

- Fast attaining of stability after transient fault
- Generation of controlled active power to suppress the system oscillation
- Controlling the somewhat sluggish pitch system to perfection
- Reducing the oscillation due to transient by oscillating power extraction
- To investigate the impact of wind turbine control system on total stability.

The wind turbines are basically classified as a variable speed, variable pitch turbine. These are driven in conventional ways, as it is composed of gearbox and fully scaled synchronous generator. Though, in majority of situations, it is noticed that variable speed drive preferably works with doubly fed induction motor, still full-scaled synchronous generator are preferred, as they gained immense popularity due to some of their advantages over the former, and various research studies are also carried out in this field. Various control units and methods are proposed and implemented to make these turbines more enable to tackle the transient fault and to prevent the loss of stabilization in the system. Different controlling algorithm is proposed in order to suit various fault issues and is made utilized to handle that fault. A realistic power model is analysed by simulating different transient faults and also the mutual impacts of both power system and wind turbines that are taken into consideration [3]

4.1.4 GRID-CONNECTED REQUIREMENTS

The utility or generating energy through wind has not created that much of hype in early days energy situation. Therefore, there is no requirement for grid connection with the wind turbines. But as the time passed by, more works shifted towards the generation of wind energy, which leads to involvement of grid connection with the wind turbines. During the 1990s, these connection rules are presented and implemented in various countries such as Germany and Denmark. This harmonization is later accepted by national network association and wind energy association for the upliftment of the wind turbine manufacturers and wind farm owners [4].

a. **Active Power Control:**
There are generally two basic reasons to control the active power: one is to control the increase in frequency during steady condition and the second reason is to attain voltage and frequency stability during transient fault. However, it cannot be distinguished between active power flow during normal condition and transient condition from the analysis of ground cover ratio (GCR). The analysis of power flow is generally required during the transient fault condition in order to check if the power flow can be reduced when fault occurs so that the wind turbine speed can be prevented from increasing. If the wind turbine is directly connected to the grid-connected generator, then the reactive power demand is less, which effectively reduces the power demand after the fault is cleared so that the voltage dips due to fault can be eliminated and the system can regain its voltage stability. Another concern is the rate of increase in power flow when the fault is cleared. This analysis is required in order to prevent power surges to avoid interruption of energy generation. The analysis of power flow is required in all GCR. The consideration of factors for analysing greatly depends on various factors such as short circuit power. The low short circuit power

results in high power flow after the fault is cleared; that's why more control of power is necessary in order to keep the system in stable condition [7].

b. **Frequency Operating Range**

   The wind turbine must tolerate the frequency deviation. In many grid-connected systems, primary and secondary frequency control are utilized. The control of frequency is required during transient duration, as at that time there is much fluctuation in system frequency, which can easily lead the system to an unstable condition. The frequency tolerance of the wind turbine must be as high as possible, because if the turbine cannot be able to handle the fluctuation, then it leads to discontinuity in energy generation but excessive frequency range also effects the working of wind turbine because in some cases it is seen that the speed of the turbines greatly depends on the grid frequency. Variable speed turbines with doubly fed induction generators are mostly grid-independent, but variable speed turbines with full-scale converters are completely independent of grid frequency [6].

c. **Reactive Power Compensation and Controlling of Voltage**

   In both utility and customer side, each equipment possess a specified voltage rating with in which it can operate easily or else it will lead to system or equipment failure. So the control of reactive power demand and voltage control are necessary to keep the voltage within its limit. In some grid connections, the wind turbines also contribute to voltage control. The reactive power demand are compared in terms of power factor. In the figure it is noted that lagging power factor denotes reactive power demand and the leading power denotes the reactive power absorption [9].

d. **Requirements for Transient Fault and Voltage Operating Range:**

   In this section, the working voltage is related with trip times. The correlation must be considered in order to synchronize the wind turbine with the grid. The correlation between active and reactive power is not taken into consideration. Along with voltage operating limits, the transient fault ride through is also mentioned. The voltage and current are matched in order to compensate power flow as before when the fault does not occur. The above information reveals that the wind energy system must be designed for larger currents than the rated current.

4.2   MODELLING OF WIND TURBINE

The mechanical power executed by the wind turbine is generally depicted by

$$P_m = 1/2 C_p (\beta, \gamma) \rho \pi R^2 V_{\text{wind}}^3, \tag{4.1}$$

where

  $C_{p=}$ coefficient of rotor power
  β=pitch angle
  ρ=density of air
  R=radius of turbine blade
  γ=speed ratio.

The mechanical torque of the wind turbine is given by

$$T_m = \frac{P_m}{\omega_m}, \tag{4.2}$$

where

  ω_m=mechanical angular velocity.

The blade pitch angle and blade tip speed ratio, which are given below, determine the turbine's rotor power coefficient.

$$C_p = \left(0.44 - 0.0167\beta\right)\sin\frac{\pi(\gamma - 2)}{13 - 0.3\beta} - 0.00184(\gamma - 2)\beta \qquad (4.3)$$

The generating power of the wind turbine is directly influenced by the coefficient of rotor power at a particular speed of wind. The manufacturer of the wind turbine specifies its parameters, and the geographical location of the setup decides the air density. With increase in the rotational speed ω_m the pitch angle reference also increases and thus reduces the overall torque of the turbine. Therefore, it is said that pitch angle controls the torque of the turbine [2].

With variation in the wind speed the rotational speed also gets varied to achieve utmost output power. The mechanical torque is driven by the pitch angle to withdraw utmost power from the prevailing wind. Wind power exhibits great barrier with frequent changes in wind speed, and thus, to compensate that variation, an advanced controller is required to maintain the overall flow of the power and thus to increase the overall efficiency of the system. Many research studies have been carried out to design and develop a controller capable to withstand unpredictable wind nature during operation.

4.3   PROPOSED RESEARCH WORK

The aim of the research work is to improve the current and voltage profile and by the use of the reactive power efficiently, which will finally optimize the output of the system. But before this, several problems occur like harmonic distortion, and this will also be improved by the combination of the FUZZY+PI.

As the consumption of the power is increasing every day, we have to find the alternate sources of energy; as we all know, depletion of fossil fuels is increasing rapidly, and hence the question of what would be the next source of power becomes important. So wind power could be the new source after finishing of the fossil fuel. So merging of the different types of plant taken aids in increasing the efficiency. But in the merging of the plants a lot of problems occur and synchronization is really hard which is due to alternator and turbine.

4.3.1   FACTS DEVICES

The standard definition of FACTS per IEEE is: "Alternating current transmission systems incorporating power electronics based and other static controllers to enhance controllability and power transfer capability". The increase in effectiveness in electrical network is largely due to the increase in improvement in semiconductor device and the use of various static converters. This advancement in semiconductor devices also affects the controllability and transferability to a greater extent. Recent development in energy electronics introduces the usage of record controllers in power system. The FACTS devices also give an idea regarding the various costs involved in reactive power sources to compensate the lagging effect. The main disadvantage of FACTS devices is that it is expensive. The FACTS devices provide smooth and faster response. In general, it is based on thyristor or gate turn-off thyristor.

4.3.2   DIFFERENT METHODOLOGIES

Different technologies can be used to optimize the true and reactive power, which could enhance the voltage and current profile; there are various technologies, which can be used, like Fuzzy, Artificial Neural Network (ANN), Supercapacitor, and Maximum Power Point Tracking.

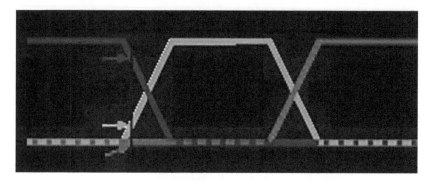

FIGURE 4.1 Representation diagram of fuzzy logic.

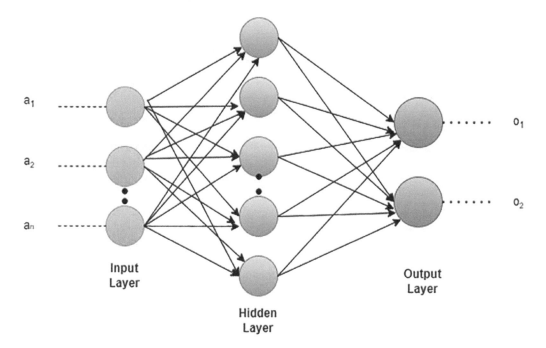

FIGURE 4.2 Representation diagram of ANN.

a. **Fuzzy Logic:**

It consists of different types of logics and the table is made from the logics like 0 and 1. Fuzzy logic contains different logics, which could be partially false or partially true. Additionally, if bilingual or multilingual variables are used, then it can be cosmos different function. We regularly use on daily basis in our homes and locals (Figure 4.1).

b. **Artificial Neural Network:**

Next, we will start from the ANN and this is not used on a great extent. Maximum time ANN is used for this application. It is based on the learning of the machine, which can store up to a greater extent. But it is hard to operate on populated inputted (Figure 4.2).

c. **Super/Ultra-Capacitor**

An ultra-capacitor is different from the conventional capacitor in different ways: The area of the plates is quite larger and the gap between the plates is also large as compared to the normal capacitor. The number of plates in it is same as those in the normal capacitor. The manufacturing material of ultra-capacitor is charcoal, which enhances the effective area of the plates, and upon the plates, the quantity of the charge is much more in the ultra-capacitor. The dielectric material made from thick sheet of mica does the separation of plates. When this capacitor is charged, then the plates carry positive and negative charges, producing an electric field, which stores energy.

4.4 IMPLEMENTED METHODOLOGY

Due to accessibility, wind energy becomes the most prominent source of energy generation. So in order to overcome various barriers during integration of wind with grid and make the system stable, a simple static pulse technique is used. There are many non-linear factors present in the wind system, so most of the control logic finds it difficult to adjust and solve the stability issue. In this case, fuzzy fits better as this logic depends on both crisp and average values. Earlier this logic has not been used much in wind turbine control. One of the prominent reasons is that the control unit of wind turbine depends on small signal analysis and it can easily be handled by the Proportional Integral (PI) and Proportional Integral Derivative (PID) controller. But as discussed above we are focusing on the large voltage sag due to fault, and this analysis possess lots of nonlinearities, which can easily be handled by the fuzzy controller, because in a system if the problem cannot be presented in mathematical form and possesses nonlinearity, such a problem can be resolved using fuzzy logic. Also, on the contrary, the tuning of the parameters with traditional controllers like PI or PID requires the analysis and variance in mathematical equation. For the design of a fuzzy logic controller, such a description is not necessary, as it is fully based on system behaviour. This knowledge is presented in the form of series rules, which are utilized to attain the system output from system input. The designing of a fuzzy controller generally depends on three parameters, namely input for the logic and creating fuzzy sets for those inputs; defining fuzzy rule; designing a controller to convert the results into an output signal, which is commonly called defuzzification. The derivatives are being generated from the set of rules, which are generally grid frequency and voltage. These rules are accounted as a set of crisp values and their importance can be presented as membership function. A crisp pitch angle is generated through defuzzification process. The incorporation of fuzzy pitch angle controller with the wind turbine is depicted in Figure 4.3.

4.5 IMPLEMENTED FUZZY RULE

See Figure 4.4.

```
Name='fc_buck' Type='mamdani' Version=2.0 NumInputs=2 NumOutputs=1
NumRules=37 AndMethod='min' OrMethod='max' ImpMethod='min'
AggMethod='max'
DefuzzMethod='centroid'
```

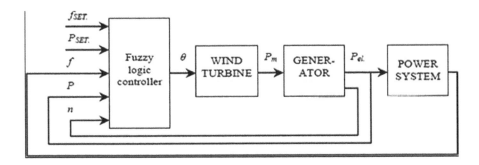

FIGURE 4.3 Fuzzy logic block diagram of proposed model.

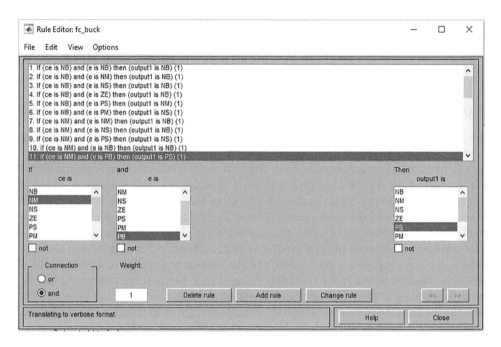

FIGURE 4.4   A snapshot of the fuzzy rule implementation in the model.

```
[Input1] Name='ce' Range=[-15 15] NumMFs=7
MF1='NB':'trapmf',[-19.2 -15.15 -11.25 -7.5]
MF2='NM':'trimf',[-11.25 -7.5 -3.75]
MF3='NS':'trimf',[-7.5 -3.75 0]
MF4='ZE':'trimf',[-3.75 0 3.75]
MF5='PS':'trimf',[0 3.75 7.5]
MF6='PM':'trimf',[3.75 7.5 11.25]
MF7='PB':'trapmf',[7.5 11.25 15.45 19.35]

[Input2] Name='e' Range=[-15 15] NumMFs=7
MF1='NB':'trapmf',[-19.5 -15.45 -11.25 -7.5]
MF2='NM':'trimf',[-11.25 -7.5 -3.75]
MF3='NS':'trimf',[-7.5 -3.75 0]
MF4='ZE':'trimf',[-3.75 0 3.75]
MF5='PS':'trimf',[0 3.75 7.5]
MF6='PM':'trimf',[3.75 7.5 11.25]
MF7='PB':'trapmf',[7.5 11.25 17.85 17.85]

[Output1] Name='output1' Range=[-10 10] NumMFs=7
MF1='NB':'trimf',[-13.33 -10 -6.666]
MF2='NM':'trimf',[-10 -6.666 -3.334]
MF3='NS':'trimf',[-6.666 -3.334 0]
MF4='ZE':'trimf',[-3.334 0 3.334]
MF5='PS':'trimf',[0 3.334 6.666]
MF6='PM':'trimf',[3.334 6.666 10]
MF7='PB':'trimf',[6.666 10 13.34]

[Rules]

1     1,    1       (1) : 1
1     2,    1       (1) : 1
```

```
1       3,      1       (1)  :   1
1       4,      1       (1)  :   1
1       5,      2       (1)  :   1
1       6,      3       (1)  :   1
2       2,      1       (1)  :   1
2       3,      1       (1)  :   1
2       5,      3       (1)  :   1
2       1,      1       (1)  :   1
2       7,      5       (1)  :   1
3       1,      1       (1)  :   1
3       2,      1       (1)  :   1
3       3,      2       (1)  :   1
3       6,      5       (1)  :   1
3       7,      6       (1)  :   1
4       1,      1       (1)  :   1
4       2,      2       (1)  :   1
4       3,      3       (1)  :   1
4       5,      5       (1)  :   1
4       6,      6       (1)  :   1
4       7,      7       (1)  :   1
5       1,      2       (1)  :   1
5       2,      3       (1)  :   1
5       5,      6       (1)  :   1
5       6,      7       (1)  :   1
5       7,      7       (1)  :   1
6       1,      3       (1)  :   1
6       3,      5       (1)  :   1
6       5,      7       (1)  :   1
6       6,      7       (1)  :   1
6       7,      7       (1)  :   1
7       2,      5       (1)  :   1
7       3,      6       (1)  :   1
```

See Figures 4.5 and 4.6.

4.6 SIMULATION AND RESULT

4.6.1 Software: MATLAB® Version R2019a

The analysis of proposed model is carried out using MATLAB® simulation software, where numerical calculations and analysis of the integrated system are performed. MATLAB is a high-skilled language for technical computation, visualization, and programming in a simple way, where problems and solutions are expressed in familiar mathematical model.

- Data collection and data exploration along with analysing
- Data representation and engineering drawing along with graphical visualization
- Algorithmic analysing of functions
- Mathematical model and computational tools
- Prototyping of realistic models
- Graphical user interface (GUI) building environment.

4.6.2 Result Analysis and Simulation

See Figures 4.7–4.13.

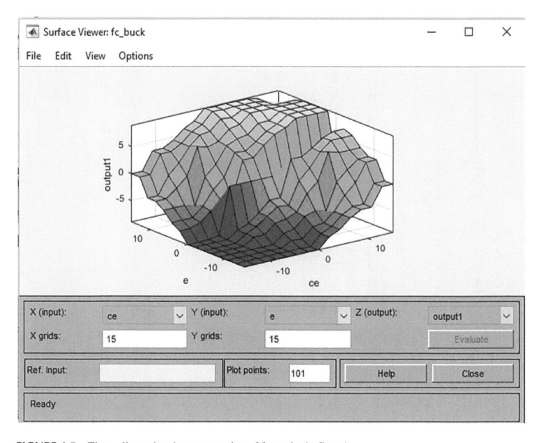

FIGURE 4.5   Three-dimensional representation of fuzzy logic Case 1.

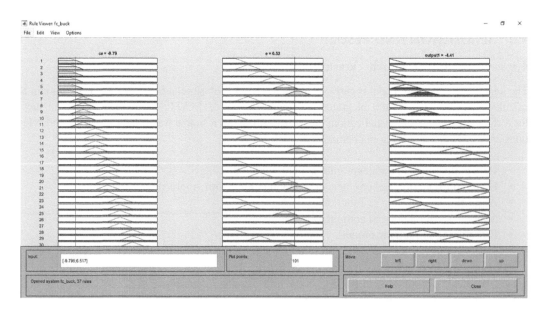

FIGURE 4.6   Fuzzy output 1 for value of $ce = -9.79$ and $e = 6.52$, output $= -4.41$.

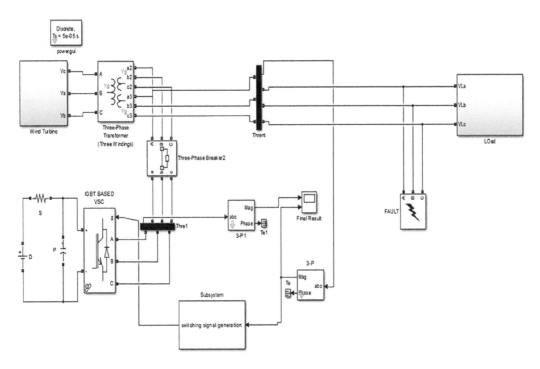

FIGURE 4.7 System block diagram of proposed model.

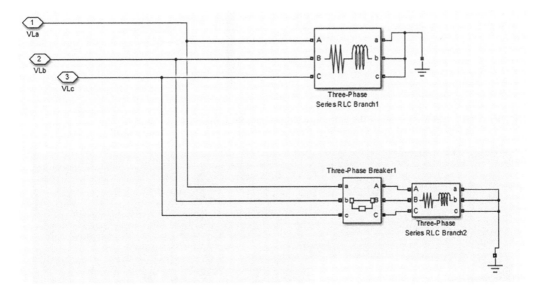

FIGURE 4.8 Block representation of subsystem load of system.

4.7 CONCLUSION

With steep advancement in technology, researchers are constantly striving to develop a high-end technology that is cost-effective as well as beneficial for the society. In our proposed work, we have integrated computational technology, namely fuzzy logic, with PI controller in STATCOM to elevate the stability rate of the wind system incorporated with grid system. The system is designed

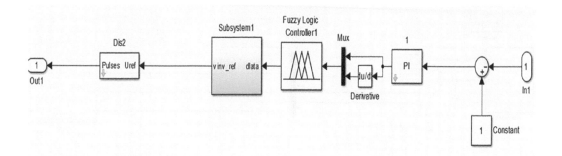

FIGURE 4.9   Integration of fuzzy logic with PI.

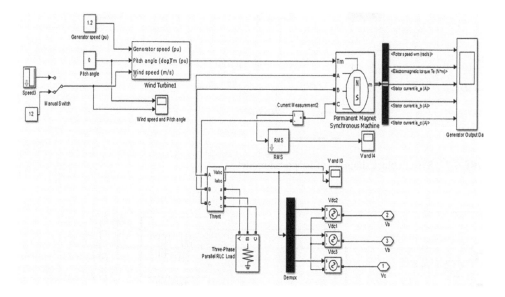

FIGURE 4.10   Representation of Wind Energy Conversion System (WECS).

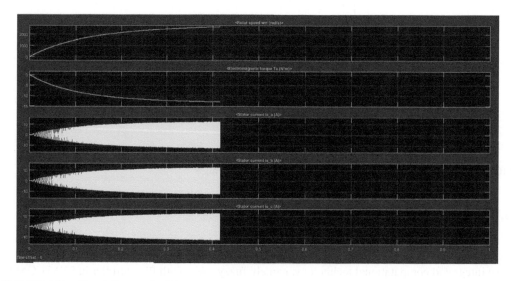

FIGURE 4.11   Simulation result of the generator output.

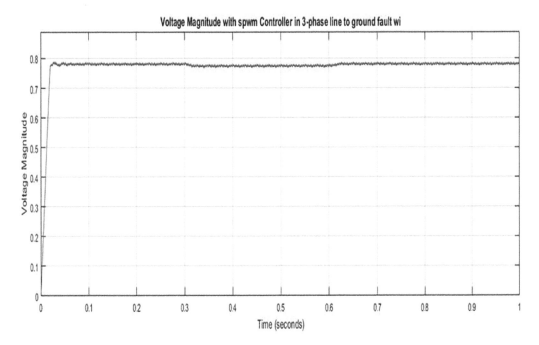

FIGURE 4.12 Voltage magnitude with SPWM controller in 3-phase line to ground fault.

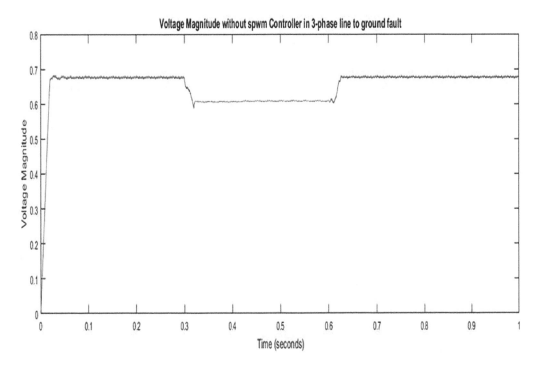

FIGURE 4.13 Voltage magnitude without SPWM controller in 3-phase line to ground fault.

and simulated in MATLAB along with various case study aspects such as involving controller, Switched Pulse Width Modulation (SPWM), and various fault cases.

The simulation results show that with the involvement of our proposed work, some sections have shown great improvement. Firstly, from the results of the analysis, it is observed that power quality gets improved, which is major factor in any power system. Secondly, and most importantly, it is observed that with the change in the speed of wind, the voltage remains constant in the grid side, as well as the reactive power compensation (snag and swelling) executed very efficiently and thus maintained the voltage profile.

BIBLIOGRAPHY

1. T. N. Mauboy, "Integrating Wind Power: Transmission and Operational Impacts," *Refocus*, Vol. 5, 36–37, 2015.
2. I. Manirule, "Transient Stability Enhancement of the Power System with Wind Generation," *TELKOMNIKA*, Vol. 9, No. 2, 267–278, 2015.
3. Pushbha. "Improving Stability of Multi-Machine Wind Turbine Generators Connected to the Grid," *Journal of Engineering for Thermal Energy and Power*, Vol. 26, No. 2, 241–245, 2015.
4. T. Yufi, *Power System Stability and Control*. McGraw Hill, 2015.
5. K. S. Nayna, *Thyristor-Based FACTS Controllers for Electrical Transmission Systems*. Wiley-IEEE Press, New York, 2014.
6. Q. Saleem, "Improvement of Voltage Stability in Wind Farm Connection to Distribution Network Using FACTs Devices," in *Proc. 32nd Annual Conference on IEEE Industrial Electronics*, Nov. 6–10, 2006 (pp. 4242–4247), Paris, 2014.
7. R. Jeevajothi, "Transient Stability Augmentation of Power System Including Wind Farms by Using ECS," *IEEE Transactions on Power Systems*, pp. 1179–1187, 2012.
8. ABB Report, "Power Quality Assessment of Wind Turbines by Matlab/Simulink," in Proc. *Asia-Pacific Power and Energy Engineering Conference (APPEEC)*, Chengdu, Mar., 2010.
9. Kerane, "A Comprehensive Comparative Analysis between STATCOM and SVC," in *Proc. International Conference on Applied Robotics for the Power Industry*, pp. 208–210, 2011.
10. H. M. Abdelhalim, "Representing Wind Turbine Electrical Generating Systems in Fundamental Frequency Simulations," *IEEE Transactions on Energy Conversion*, Vol. 18, No. 4, 516–524, 2010.
11. "20% Wind Energy by 2030 –Increasing Wind Energy's Contribution to U.S. Electricity Supply," U.S. Department of Energy May 2008. DOE/GO-102008-2567,
12. Haiya K. Anil, Wind Power: Modeling and Impact on Power S system Dynamics, Ph.D. Thesis, Technical University Delft, Delft, the Netherlands, 2010.
13. Peter W. Sauer and M. A. Pai , *Power System Dynamics and Stability*. Stipes Publishing L.L.C, Champaign, IL, 2006.
14. A. Aamczyk, "Impact of Wind Power Plants on Voltage and Transient Stability of Power Systems," in *IEEE Energy 2010*, Atlanta, Georgia, 2008.
15. Sharad and Mohan 2010, "Effects of Large Scale Wind Generation on Transient Stability of the New Zealand Power System," in *IEEE Power and Energy Society, General Meeting*, 2008.
16. Oskoui, MATLAB-based Power System Analysis Toolbox, 2010, available at http://thunderbox.uwaterloo.ca/_fmilano
17. M. Anju and R. Rajasekaran, 2013, "Co-Ordination of SMES with STATCOM for Mitigating SSR and Damping Power System Oscillations in a Series Compensated Wind Power System," in *IEEE 2013 International Conference on Computer Communication and Informatics (ICCCI – 2013)*.
18. W. J. Park, B. C. Sung, K. B. Song and J. W. Park, "Parameter optimization of SFCL with wind-turbine generation system based on its protective coordination," *IEEE Transactions on Applied Superconductivity*, Vol. 21, No. 3, pp.2153–2156, 2010.19. A. B. Arsoy, Y. Liu, P. F. Rebeiro, and F. Wang, "STATCOM-SMES," in *Industry Applications Magazine* (IX, Issue 2, pp. 21–28). IEEE, 2003.
20. E. Vittal, M. O'Malley, and A. Keane, "Rotor Angle Stability with High Penetrations of Wind Generation," *IEEE Transactions on Power Systems*, Vol. 27, No. 1, pp. 353–362, 2012.
21. G. D. Marzio, O. B. Fosso, K. Uhlen, and M. Þ. Pàlsson, "Large-Scale Wind Power Integration, Voltage Stability Limits and Modal Analysis," in *15th PSCC*, Liege, 2005.
22. A. Grauers, "Efficiency of Three Wind Energy Generator System," *IEEE Transactions on Energy Conversion*, Vol. 11, No. 3, 650–657, 1996.

23. Q. Salem, "Overall Control Strategy of Grid Connected to Wind Farm Using FACTS," *Bonfring International Journal of Power Systems and Integrated Circuits*, Vol. 4, No. 1, 5–13, February 2014.

24. T. Sun, Z. Chen, and F. Blaabjerg, "Voltage Recovery of Grid-Connected Wind Turbines with DFIG after a Short Circuit Fault," in *2004 IEEE 35th Annual Power Electronics Specialists Conference*, Vol. 3, pp. 1991–97, 20–25 June 2004.

25. M. Molinas, S. Vazquez, T. Takaku, J.M. Carrasco, R. Shimada, and T. Undeland, "Improvement of Transient Stability Margin in Power Systems with Integrated Wind Generation Using a STATCOM: An Experimental Verification," in *International Conference on Future Power Systems*, 16–18 Nov. 2005.

26. E. Muljadi and C.P. Butterfield, "Wind Farm Power System Model Development," in *World Renewable Energy Congress VIII*, Colorado, Aug–Sept 2004.

27. S.M. Muyeen, M.A. Mannan, M.H. Ali, R. Takahashi, T. Murata, and J. Tamura, "Stabilization of Grid Connected Wind Generator by STATCOM," *IEEE Power Electronics and Drives Systems*, Vol. 2, pp. 1584–1589, 2005.

28. M. Amiri and M. Sheikholeslami, "Transient Stability Improvement of Grid-Connected Wind Generator Using SVC and STATCOM," in *International Conference on Innovative Engineering Technologies (ICIET 2014)* Dec. 28–29, 2014 Bangkok (Thailand).

29. M. Shah, P. Khampariya, and B. Shah, "Application of STATCOM and SVC for Stability Enhancement of FSIG based Grid Connected Wind Farm," *International Journal of Emerging Technology and Advanced Engineering*, Vol. 5, No. 7, 296–304, 2015.

30. S. Teleke, T. Abdulahovic, T. Thiringer, and J. Svensson, "Dynamic Performance Comparison of Synchronous Condenser and SVC," *IEEE Transactions on Power Delivery*, Vol. 23, No. 3, 1606–1612, 2008.

31. P. Hippe, *Windup in Control*. Springer, Berlin, Germany, 2006.

32. W. Qiao, G. K. Venayagamoorthy, and R. G. Harley, Real-time implementation of a STATCOM on a wind farm equipped with doubly fed induction generators, IEEE Trans. Ind. Appl., Jan/Feb 2009.

33. K. Morrow, D. Karner, and J. Francfort, Plug-in hybrid electric vehicle charging infrastructure review, Final Rep., Battelle Energy Alliance, Contract 58517, Nov. 2008.

34. S. N. Deepa and J. Rizwana, "Multi-Machine Stability of a Wind Farm Embedded Power System Using FACTS Controllers," *India International Journal of Engineering and Technology (IJET)*, Vol. 5, No. 5, 3914–3921, 2013.

5 Solar-Powered Smart Irrigation System

Swapnil Jain and O.V. Gnana Swathika
Vellore Institute of Technology

CONTENTS

ABSTRACT

The aim of this paper was to build a solar tracker-powered smart irrigation system. Using a single-axis solar tracker circuit, which moves such that direction of sunlight is always perpendicular to the solar panel, an irrigation module has been powered, which is further connected to a mobile app. The irrigation module consists of an ESP8266 board as the brain of the system. It is connected via the Wi-Fi to the mobile app. The gardener/farmer can view real-time soil and surrounding information, such as temperature, humidity, and soil moisture. If he wants, he can manually command the system to water the plants. A 5 V single-channel relay is used for this purpose, which turns on the motor, draws water from the well, and waters the plant. The farmer/gardener can then turn off the watering when soil conditions reach the optimal level. This device would be very helpful to the farmers because this device is capable of monitoring the condition of the soil and then watering the plants according to the level of the soil moisture content. This device is cheap and highly efficient for farmers, and is a step towards renewable energy economy.

5.1 INTRODUCTION

The old method of watering plants is completely time-consuming and a tiring process. In order to save the farmer's/gardener's time, the paper proposes a device that can help farmers to water their plants regularly at certain intervals of time. With the use of this device, it can be ensured that the water given to the plant is sufficient as well as efficient. All of these are possible via a solar tracker, so it is eco-friendly and energy-efficient at the same time.

DOI: 10.1201/9781003331117-5

53

5.1.1 Literature and Background Survey

The authors of Reference [1] propose a Global System for Mobile Communication (GSM)-based smart irrigation system that sends the information of the surroundings including temperature and humidity of the soil. A fuzzy logic controller is used to compute these parameters. The authors then compare the proposed system against drip irrigation and manual flooding, and find that the proposed system conserves more water and power compared to other two. Reference [2] emphasizes how Internet of Things (IoT)-based systems are the necessity for industry 4.0. The authors further develop a smart irrigation system which uses a neural-network to optimize the sensor inputs and scheduling of tasks. The authors in Reference [3] also make a smart irrigation system, but use Arduino and ThingSpeak instead, to develop a system to help the farmer control and monitor the water flow. In Reference [4], the authors use the ESP8266 NodeMCU to perform the same task as done in [3]. Reference [5] indicates the importance of IoT in Agriculture 4.0, with the authors further demonstrating the technique of radiofrequency energy harvesting to deliver power to the IoT system. In one of the preceding papers, parameters like humidity, soil moisture, temperature, and light intensity were fed into a fuzzy logic to determine whether to water the plant [6]. The same logic has also been applied in healthcare for parameters like blood pressure and oxygen [7]. Applications of this analogy has also been found in fire-alarming applications [8,9]. Europe's SWAMP project [10] has developed an IoT-based smart irrigation system, based on four pilots in Brazil and Europe. The authors of Reference [10] also provide an analysis of FIWARE components used in the programme. The authors of References [11–17] did an on-field survey to test the water savings of a smart irrigation system and found 25% water saving in Mediterranean regions. Establishing food security is one of the United Nations Developmental Goals for 2030, and smart irrigation can potentially become a key enabler for this goal. The authors of References [18–23] explore this scenario with respect to Edge computing. Machine learning techniques like K-Nearest Neighbors (KNN) [24–26] also have the potential to predict water level requirement for soil types where farmers are not confident of the optimal water level.

5.1.2 Objectives

- The objective of the paper was to design an irrigation system that would:
  - Use water in a more well-organized way to prevent loss of water.
  - Minimize the labour expenses involved.
  - Make use of renewable energy.
  - Prevent over-watering/under-watering of plants.
- The issue of watering the plants, at the correct humidity level, when the owner goes out of station can be solved using this system.
- This product was determined to be the best choice through its installation cost, water saving, human intervention, reliability, power consumption, maintenance, and expandability.
- The most important consideration kept in mind is the overall cost, with the aim of powering the entire system using sun tracking solar panel to minimize power loss in a step towards renewable energy ecosystem.

5.1.3 Functioning of the Prototype

The prototype shown is able to deliver surrounding information like moisture content of the soil, temperature, and humidity. This information can be viewed by the concerned parties on their smartphone app (Figure 5.12). They can remotely water the plants if the conditions are suitable according to them. A proposed addition to the existing prototype (Figure 5.11) would be the addition of the feature of self-irrigation, which waters plant using pre-decided threshold values for soil moisture. It is estimated that the proposed system will be highly beneficial to all concerned parties as it is proven to be more efficient than traditional solutions [1], and at the same time help in water conservation. The whole system will be powered using a solar tracker, which will optimize itself for the best possible energy production as per direction of the sun.

5.2   DESCRIPTION

Modern-day irrigation systems are a combination of technologies, like water supply controllers that allows watering only when needed, and sensors that monitor moisture and temperature-related conditions in the farm/garden and automatically adjust watering to optimal levels. In the paper instead of using a constant dc supply source we have used a solar tracker system (Figure 5.1) to power the smart irrigation system (Figure 5.2). Also, the irrigation system gathers all the data like the moisture content, humidity, and temperature and sends it to the farmer's app (Blynk app) (Figure 5.12), which shows the farmers about the current state of their soil and accordingly they can water the plants (Figures 5.3 and 5.4).

5.3   DESIGN ASPECT

ESP-8266:
- ESP8266 is Wi-Fi microcontroller, and is a very inexpensive way to add internet connectivity to a small-scale system.
- It can fetch data using Application Programming Interface (API) and hence make a part of IoT.
- It can also be programmed very easily using Arduino IDE.
- Software like "Blynk" are often used alongside it.

DHT11:
The DHT11 sensor is used to measure the temperature and humidity around the surroundings.

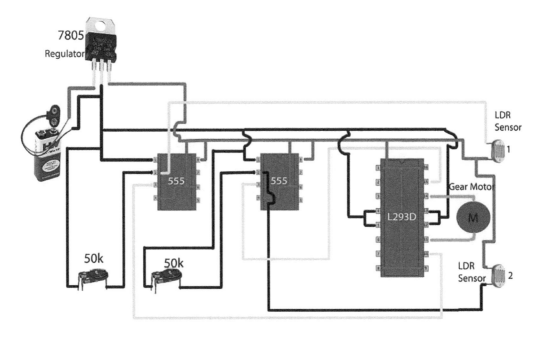

FIGURE 5.1   Solar tracker circuit diagram.

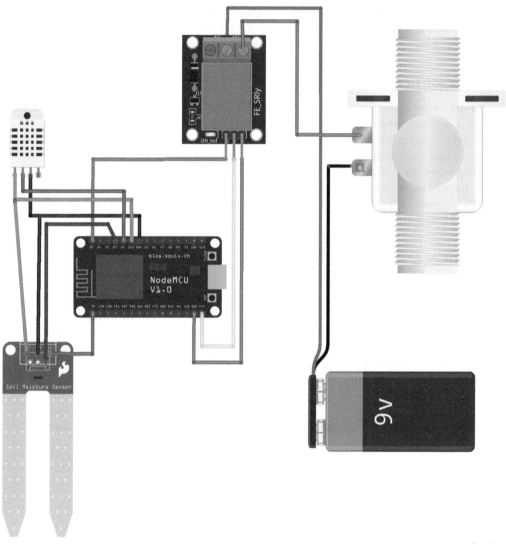

FIGURE 5.2   Irrigation module circuit diagram.

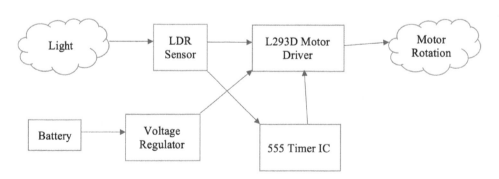

FIGURE 5.3   Solar tracker block diagram.

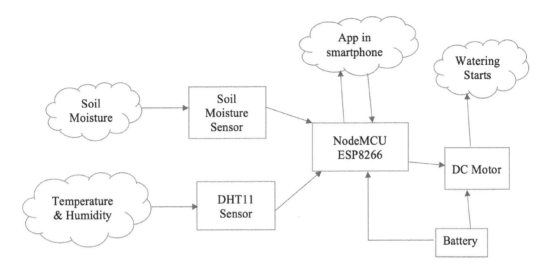

FIGURE 5.4 Irrigation module.

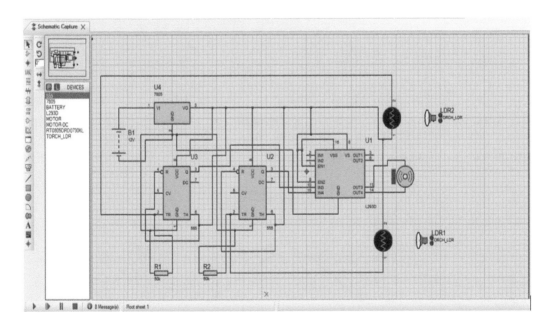

FIGURE 5.5 Simulation of solar tracker using Proteus.

Soil Moisture Sensor:

Soil moisture sensors are used to measure the moisture content level in the soil, often by using properties of resistance or dielectric constant.

5.4 DEMONSTRATION

5.4.1 SIMULATION

A simulation of the solar tracker was designed and run on Proteus (Figure 5.5) to demonstrate the proposed mechanism works.

5.4.2 GRAPHS OF IRRIGATION MODULE

As seen in Figure 5.6, the moisture in the soil is slowly getting reduced, with the effect being more prominent around 3 PM, where the temperature was higher than usual.

It can be seen from Figure 5.7 that the temperature peaked around 3 PM, and then started decreasing slowly thereafter.

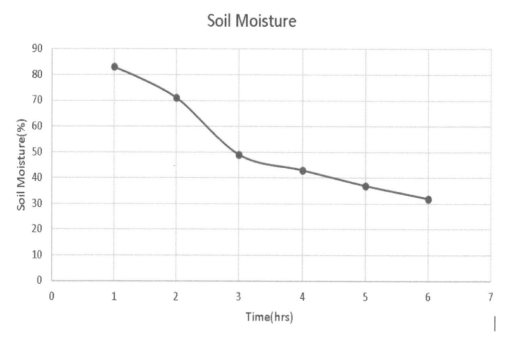

FIGURE 5.6 Soil moisture vs time graph.

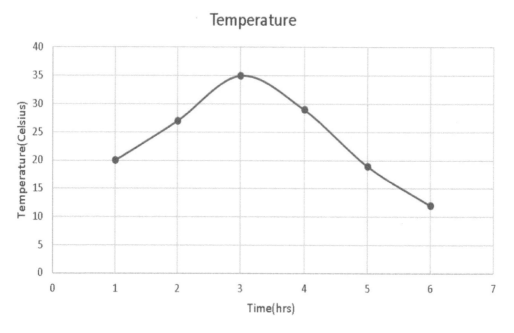

FIGURE 5.7 Temperature vs time graph.

In Figure 5.8, the humidity can be seen to have peaked around 4 PM, and decreasing thereafter, but increasing again at around 5:30 PM.

5.4.3 SOLAR TRACKER GRAPHS

Figure 5.9 shows a bar chart showing output energy comparison between conventional stationary solar panel and single-axis solar tracker.

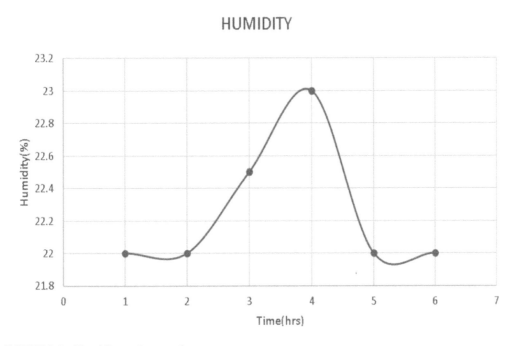

FIGURE 5.8 Humidity vs time graph.

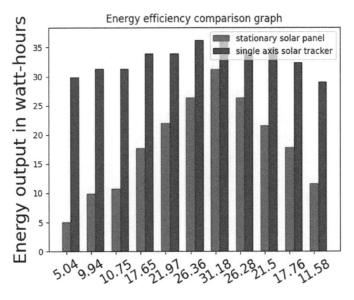

FIGURE 5.9 Bar chart showing output energy comparison between conventional stationary solar panel and single-axis solar tracker.

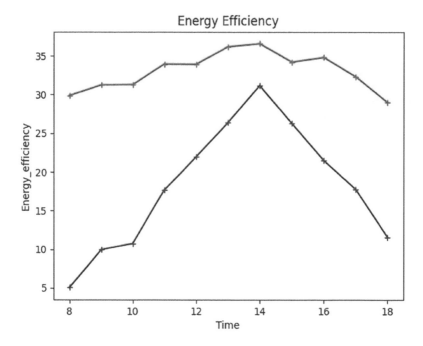

FIGURE 5.10 Graph showing energy efficiency comparison between conventional stationary solar panel and single-axis solar tracker.

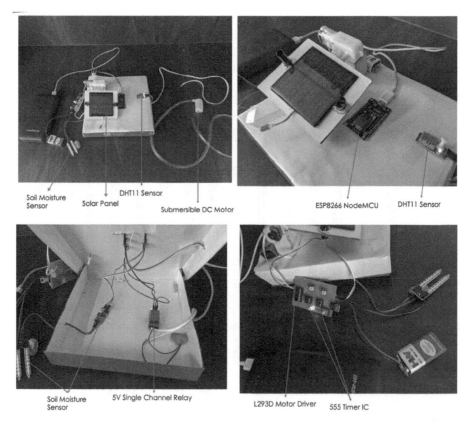

FIGURE 5.11 The entire hardware setup for the solar tracker-powered smart irrigation system.

Figure 5.10 shows a graph showing energy efficiency comparison between conventional stationary solar panel and single-axis solar tracker.

5.4.4 Hardware Setup

Figure 5.11 shows the entire hardware setup for the solar tracker-powered smart irrigation system.

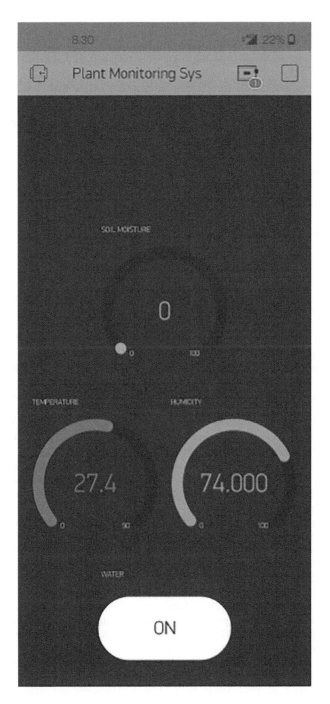

FIGURE 5.12 Sample readings taken in the Blynk App.

5.4.5 MOBILE APP

Using the android app, parameters like soil moisture, temperature, and humidity can be viewed as shown in Figure 5.12. The app updates in a real-time environment via the internet. The "ON" switch is used to further control the actions of the pump. Upon switching it on, the soil humidity starts increasing. The farmer can then switch off the pumping action when a satisfactory level of soil moisture level is reached.

5.5 CONCLUSION

5.5.1 FUTURE SCOPE

In India, agriculture is the primary source of livelihood for about 58% of India's population as of 2020 data, so our idea can potentially help a lot of people. Machine learning may be incorporated in the proposed prototype to remove the need of the farmer's manual intervention. The following are the benefits of proposed prototype:

- Easy to implement
- Environment-friendly
- Minimal maintenance and manual interaction
- Cost-friendly
- In the long run, the system is economical.

REFERENCES

1. Krishnan, R. S., Golden Julie, E., Harold Robinson, Y., Raja, S., Kumar, R., and Thong, P. H. 2020. Fuzzy logic based smart irrigation system using internet of things. *Journal of Cleaner Production*, 252, p. 119902.
2. Nawandar, N. K. and Satpute, V. R. 2019. IoT based low cost and intelligent module for smart irrigation system. *Computers and Electronics in Agriculture*, 162, pp. 979–990.
3. Benyezza, H., Bouhedda, M., Djellout, K., and Saidi, A. 2018. Smart irrigation system based ThingSpeak and Arduino. In *2018 International Conference on Applied Smart Systems (ICASS)* (pp. 1–4). IEEE.
4. Pernapati, K. 2018. IoT based low cost smart irrigation system. In *2018 Second International Conference on Inventive Communication and Computational Technologies (ICICCT)* (pp. 1312–1315). IEEE.
5. Boursianis, A. D., Papadopoulou, M. S., Gotsis, A., Wan, S., Sarigiannidis, P., Nikolaidis, S., and Goudos, S. K. 2020. Smart irrigation system for precision agriculture—The AREThOU5A IoT platform. *IEEE Sensors Journal*, 21(16), pp.17539–17547.
6. Munir, M. S., Bajwa, I. S., Naeem, M. A., and Ramzan, B. 2018. Design and implementation of an IoT system for smart energy consumption and smart irrigation in tunnel farming. *Energies*, 11(12), p. 3427.
7. Sattar, H., Bajwa, I. S., Amin, R. U. et al. 2019. An IoT-based intelligent wound monitoring system. *IEEE Access*, 7, pp. 144500–144515.
8. Sarwar, B., Bajwa, I., Ramzan, S., Ramzan, B., and Kausar, M. 2018. Design and application of fuzzy logic based fire monitoring and warning systems for smart buildings. *Symmetry*, 10(11), p. 615.
9. Sarwar, B., Bajwa, I. S., Jamil, N., Ramzan, S., and Sarwar, N. 2019. An intelligent fire warning application using IoT and an adaptive neuro-fuzzy inference system. *Sensors*, 19(14), p. 3150.
10. Kamienski, C., Soininen, J.-P., Taumberger, M. et al. 2019. Smart water management platform: IoT-based precision irrigation for agriculture. *Sensors*, 19(2), p. 276.
11. Saab, A., Therese, M., Jomaa, I., Skaf, S., Fahed, S., and Todorovic, M. 2019. Assessment of a smart-phone application for real-time irrigation scheduling in Mediterranean environments. *Water*, 11, p. 252.
12. Saqib, M., Almohamad, T. A., and Mehmood, R. M. 2020. A low-cost information monitoring system for smart farming applications. *Sensors*, 20(8), p. 2367.
13. OGrady, M. J., Langton, D., and O'Hare, G. M. P. 2019. Edge computing: A tractable model for smart agriculture? *Artificial Intelligence in Agriculture*, 3, pp. 42–51.
14. Munir, M. S., Bajwa, I. S., and Cheema, S. M. 2019. An intelligent and secure smart watering system using fuzzy logic and blockchain. *Computers & Electrical Engineering*, 77, pp. 109–119.

15. Bzdok, D., Krzywinski, M., and Altman, N. 2018. Machine learning: Supervised methods. *Nature Methods*, *15*, pp. 5–6.
16. Manu, D., Shorabh, S.G., Swathika, O.G., Umashankar, S., and Tejaswi, P. 2022, May. Design and realization of smart energy management system for standalone PV system. In *IOP Conference Series: Earth and Environmental Science* (Vol. 1026, No. 1, p. 012027). IOP Publishing.
17. Swathika, O. G., Karthikeyan, K., Subramaniam, U., Hemapala, K. U., and Bhaskar, S. M. 2022, May. Energy efficient outdoor lighting system design: Case study of IT campus. *In IOP Conference Series: Earth and Environmental Science* (Vol. 1026, No. 1, p. 012029). IOP Publishing.
18. Sujeeth, S. and Swathika, O. G. 2018, January. IoT based automated protection and control of DC microgrids. In *2018 2nd International Conference on Inventive Systems and Control (ICISC)* (pp. 1422–1426). IEEE.
19. Patel, A., Swathika, O. V., Subramaniam, U., Babu, T. S., Tripathi, A., Nag, S., Karthick, A., and Muhibbullah, M. 2022. A practical approach for predicting power in a small-scale off-grid photovoltaic system using machine learning algorithms. *International Journal of Photoenergy*, pp. 1–21.
20. Odiyur Vathanam, G. S., Kalyanasundaram, K., Elavarasan, R. M., Hussain Khahro, S., Subramaniam, U., Pugazhendhi, R., Ramesh, M., and Gopalakrishnan, R. M. 2021. A review on effective use of daylight harvesting using intelligent lighting control systems for sustainable office buildings in India. *Sustainability*, *13*(9), p. 4973.
21. Swathika, O. V. and Hemapala, K. T. M. U. 2019. IOT based energy management system for standalone PV systems. *Journal of Electrical Engineering & Technology*, *14*(5), pp. 1811–1821.
22. Swathika, O. V. and Hemapala, K. T. M. U. 2019, January. IOT-based adaptive protection of microgrid. In *International Conference on Artificial Intelligence, Smart Grid and Smart City Applications* (pp. 123–130). Springer, Cham.
23. Kumar, G. N. and Swathika, O. G. 2022. 19 AI Applications to. *Smart Buildings Digitalization: IoT and Energy Efficient Smart Buildings Architecture and Applications*, p. 283.
24. Swathika, O. G. 2022. 5 IoT-Based Smart. *Smart Buildings Digitalization: IoT and Energy Efficient Smart Buildings Architecture and Applications*, p. 57.
25. Lal, P., Ananthakrishnan, V., Swathika, O. G., Gutha, N. K., and Hency, V. B. 14 IoT-Based Smart Health. *Smart Buildings Digitalization: Case Studies on Data Centers and Automation*. 2022 Feb 23:149.
26. Chowdhury, S., Saha, K. D., Sarkar, C. M., and Swathika, O. G., 2022. IoT-Based Data Collection Platform for Smart Buildings. In *Smart Buildings Digitalization* (pp. 71–79). CRC Press.

6 Future Transportation
Battery Electric Vehicles and Hybrid Fuel Cell Vehicles

K. Jamuna, Harsh Jethwai, and Devasish Khajuria
Vellore Institute of Technology

CONTENTS

ABSTRACT

Since the number of vehicles is increasing on the road day by day, contributing to environmental pollution, and the climate change is getting worse, we need to find better solutions, which can make a difference, for a better society. Cars are responsible for approximately 10% of world's CO_2 emissions and one way to cut down this percentage is through electrifying them. Hence, all vehicle manufacturers should move towards battery electric vehicle (BEV), hybrid electric vehicle (HEV), plug-in HEV, and fuel cell vehicles (FCV). Among these, BEV and FCV are pure electrified vehicles. Although there are many types of fuel cell available, hydrogen fuel cells are more advantageous than other types. Hence, both BEV and hydrogen FCV (HFCV) have become the focus of this study. Both these offer many advantages and some disadvantages, and this paper studies them based on a few key parameters, namely emission and efficiency, materials availability for the battery, infrastructure requirements and its development, sustainability and vehicle weight, and compares them with those of Internal Combustion Engine vehicles. It is found that BEVs are a lot more efficient with their infrastructure being more than twice as efficient as HFCVs, though HFCVs are still more efficient than traditional vehicles and have higher range, and overall, BEVs are a better choice for adoption of clean EV technology in near future. HFCVs also have good future with the newly developed technologies for the production of hydrogen and its charging stations, since they have some key advantages such as fast charging, higher energy density and longer range.

DOI: 10.1201/9781003331117-6

6.1   INTRODUCTION

There are more than a billion vehicles on world's roads, and most of them are powered by combustion engines. The main cause of the climate change is the global CO_2 emission. The 22% of CO_2 emission is released from the transportation sector, as reported by International Energy Agency (IEA) [8]. In order to tackle these issues, the transportation industry should revolutionize, i.e., electrifying the transportation sector.

Electric vehicles offer a range of benefits over Internal Combustion Engine (ICE) cars like increased efficiency, lower maintenance because of simpler motors and a lot less moving parts, more space and ease to use. There are two main options of going electric: battery electric or hydrogen fuel cell. Both types of cars are electric with the difference being the way they store their energy – one uses permanently stored chemicals to store electric charge (battery) and the other just like ICE cars needs external fuel which is then converted into electricity (HFC).

Countries around the world should provide incentives to customers and set targets for manufacturers to help move towards the electric future. In 2003, Zero-Emissions Vehicle programme was introduced by California government for the auto manufacturers for claiming ZEV credits as small percentage incentives from the total vehicle sales. Due to this plan, the auto manufacturers gained million-dollar incentives, which encouraged them to develop fuel cell vehicle (FCV) technology. Recently, Delhi Government announced a 3-year policy [1], which plans to install charging stations for every 3 km in the city, and aims to make EVs account for 25% of the new vehicles registered in Delhi by 2024, i.e. around 500,000 vehicles. The benefits include incentives of up to 1.5 lakhs on electric cars, scrapping incentives and reduced interest rates.

Different organizations and countries are looking into these two directions of electrification of their automobile industry as well as other aspects of the society. Japan has adopted a Basic Hydrogen Strategy and plans to become a Hydrogen Society. This question is an important one: whether to invest in battery technology or go hydrogen. So, it's important to analyse which technology is better overall and can be sustainable over time.

In this paper, a small analysis has been done towards this approach. The major parameters taken for comparison are vehicle efficiency, materials availability, infrastructure, cost, vehicles' weight and its sustainability towards creating green environment. Based on this study, suggestions have been provided. The next section discusses about the basics of electric vehicle and its types.

6.2   ELECTRIC VEHICLE

Frenchman Gustave Trouve is the first inventor of EV in 1881 [2] to power a tricycle with lead acid batteries. The first commercial electric vehicle Electrobat was invented by Morris and Salom, which was used to drive taxis in the USA. Later, there were new inventions in this field as well, but due to its high cost compared to ICE vehicles, the latter dominates the market. However, the depletion of convention fuels at a faster rate, increase of global warming and to make the environment friendly all these factors pushed the inventors to think again of electric vehicles.

These electric vehicles are powered by batteries. Batteries can obtain power from any electrical sources. The electric vehicle stores its electric energy in on-board battery pack just like an ICE vehicle. This energy then powers the electric motors which propel the vehicle. The conceptual block diagram of an electric vehicle is shown in Figure 6.1.

6.2.1   BATTERY ELECTRIC VEHICLES

It is powered purely by electric energy [3, 4]. It stores its electric energy in on-board battery pack just like an ICE vehicle. The drive train mainly consists of three subsystems: energy source, propulsion and auxiliary system, as shown in Figure 6.1. The electric propulsion system is comprised of an electric motor, a power electronic converter, a mechanical transmission system through shafts

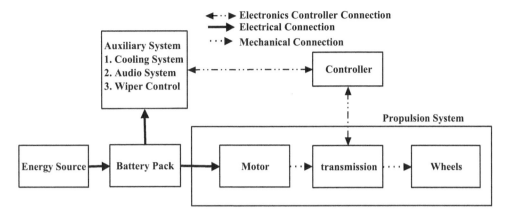

FIGURE 6.1 Conceptual diagram of electric vehicle.

and gears, and driving wheels. Battery packs are the major source for energy supply system. The auxiliary system consists of cooling system, audio system, steering unit, etc. These systems are monitored with suitable sensors that are connected to communicate to the controller which provides necessary control for the individual unit. The energy source for the vehicle is varied based on the type of vehicles. Battery-operated vehicles may have renewable sources, and they have regenerating braking system to recover the energy in order to improve the efficiency. The FCVs have energy source as fuel cells. The next section described about the battery-operated electric vehicles.

6.2.2 Hydrogen Fuel Cell Vehicles (HFCVs)

An electric drive system, Li-ion batteries, fuel cell system, and high-pressure hydrogen supplies are the parts of an HFCV powertrain [5]. The generated power is dispersed throughout the vehicle in order to drive the vehicle. The basic chemical reaction took place in the hydrogen fuel cell is expressed by equations (6.1) and (6.2).

$$\text{Anode Reaction}: H_2 \rightarrow 2H^+ + 2e^- \tag{6.1}$$

$$\text{Cathode Reaction}: 2e^- + 2H^+ + \frac{1}{2}O_2 \rightarrow 2H_2O \tag{6.2}$$

The produced electrons in the anode electrode travels to the cathode electrode through the electrical load circuit which indicates that electricity is generated by the fuel cell.

6.3 COMPARISON BETWEEN BATTERY ELECTRIC VEHICLE (BEV) AND HFCV

Both these vehicles have their own merits and demerits. However, it is necessary to study the effect of various parameters on their performance, which will be described in detail in the following subsections.

6.3.1 Efficiency and Emission

In order to address the environmental aspects of a vehicle from a global perspective, well-to-wheel (WTW) analysis is utilized by the researchers [6]. This analysis shows a clear understanding of the utilization of energy resources and their emissions. The energy has taken path from the major energy source extraction point denoted as well to the final utilization point of energy denoted as wheels. This analysis helps to calculate the total amount of energy required to move a vehicle. In

TABLE 6.1

Well-to-Wheel and Tank-to-Wheel (TTW) Emission and Fuel Consumptions for Different Fuel-Vehicle Combinations [7]

Fuel-Vehicle Combination	Fuel Consumption (l-gas/100 km)		Emissions (g CO_2/km)	
	WTW	TTW	WTW	TTW
Battery electric vehicles				
Germany E mix	*3.85*	1.38	69.73	0.00
Austria E mix	*2.48*	1.38	21.94	0.00
Sweden E mix	*3.48*	1.38	3.48	0.00
France E mix	*4.88*	1.38	9.72	0.00
EU mix E mix	*3.87*	1.38	50.43	0.00
USA E mix	*3.74*	1.38	68.24	0.00
Hydrogen FC				
NG 4,000 km OS reforming	4.53	2.21	83.66	0.00
NG 4,000 km central reforming	4.00	2.21	74.21	0.00

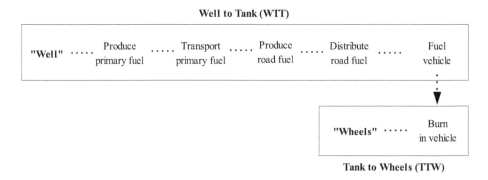

FIGURE 6.2 Well-to-wheel analysis.

this WTW pathway, the total energy required and the total Green House Gas (GHG) emitted were calculated. Based on these ratios, the best environmentally friendly vehicle is found.

The BEV's emissions are generally much lower than the ICE vehicles [7]. The global GHG emission is nearly 20% and the value of the carbon footprint is 3.5–70 g CO_2/km based on the energy mix to produce the electricity for driving the vehicle. Germany's electric power production is majorly dependent on fossil fuels (nearly 60%). Hence, the WTW emissions are as high as 61% of the existing diesel-based ICE engines and the value of the carbon footprint is 69.7 g CO_2/km. On the contrary, Sweden meets their electricity demand with carbon-free renewable energy sources and less than 3% fossil fuel sources as shown in Table 6.1. The emissions of HFC vehicles are higher than those of BEVs, but less than conventional vehicles. Like BEVs, HFCVs produce zero local emissions, but major source of these vehicles is hydrogen which is obtained from natural gas. The hydrogen production itself has higher emissions (104.4–117.7 g CO_2/MJ of fuel). The GHG emissions are 47%–66% higher than BEVs driven by electricity from EU mix (Figure 6.2).

With the available solar/wind energy, the BEV range could be achieved nearly three times that of a Fuel Cell Electric Vehicle (FCEV), because of its low efficiency in electrolysis process and the FC compared to batteries.

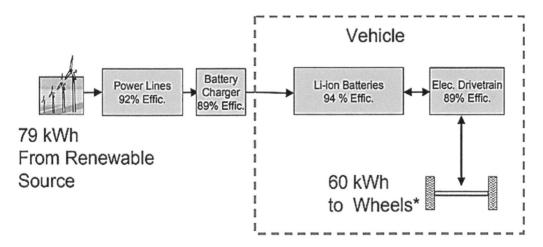

FIGURE 6.3 Power transmission process of battery-operated electric vehicle.

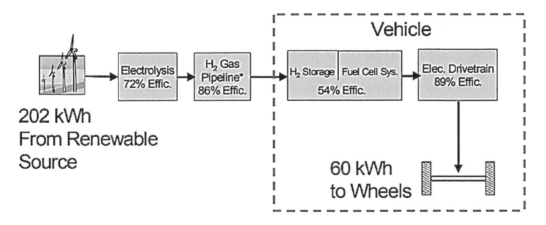

FIGURE 6.4 Power transmission process of hydrogen fuel cell-based electric vehicle.

Around 45% of energy is lost during electrolysis to produce hydrogen, and another 55% of the remaining is lost in converting that hydrogen to electricity in vehicle, giving a total efficiency of around 25%–35%, whereas, in BEVs only 8% energy is lost during transmission of electricity and to store it in batteries, and the remaining 18% loss is due to driving the electric motor which produces a total efficiency of 70%–80% [8]. Hydrogen is obtained by the process of electrolysis, and in BEVs, the batteries' chargers obtain the power from the power lines. The power conversion stages are 2 and 4 for BEV and HFCV, respectively.

In order to obtain 60 kWh vehicle output, the input energy required for BEV and HFCV is 79 and 202 kWh, respectively. It shows that BEVs are 2.56 times more efficient than HFCV. If the hydrogen pipeline stage is eliminated in HFCV, the overall efficiency increases and only 188 kWh is required as input. Its energy requirement is 2.37 times the energy required by the BEV. The power transmission stages of these vehicles are shown in Figures 6.3 and 6.4, respectively.

6.3.2 MATERIALS AVAILABILITY

The main drawback often discussed of EV battery is the availability of materials used in it. A typical lithium-ion battery has a cathode, anode and an electrolyte. The cathode can be made of different

TABLE 6.2

Metals Required for Batteries as Cathode and Anode (in kg/kWh) [9]

Materials	Li	Co	Ni	Mn	C
LCO	0.113	0.959	0	0	1.2
NCA	0.112	0.143	0.759	0	–
NMC-111	0.139	0.394	0.392	0.367	–
NMC-622	0.126	0.214	0.641	0.200	–
NMC-811	0.111	0.094	0.750	0.088	–

LCO, lithium cobalt oxide; NCA, lithium nickel cobalt aluminium oxide; NMC, lithium nickel manganese cobalt oxide; C, carbon; number denotes component's mole fraction basis.

combinations of materials apart from lithium like Co, Al, Mn and Ni. Table 6.2 shows required metals (in kg per kilowatt-hour): five cathode materials and graphitic carbon used as anode.

The various forms of these metals are used to make the cathode, like carbonates, hydroxides and sulphates, and these metals cannot be used in their metallic forms in the battery. The major source of lithium batteries is LCO, which has high energy density and is mainly used for portable electronics, and NCA and NMC are used in automobiles for their high-power capability.

Though the availability of lithium is an issue, limited production is the important concern, i.e., lithium is not being extracted to meet the current demand that impact the battery cost directly. In near future, the cost may be reduced by the increase in mass production.

The production of the materials in various countries is mentioned in Reference [9]. It is understood that the supply of cobalt and natural graphite is concentrated in the top countries, and 65% production is from Democratic Republic of the Congo (DRC) and China, respectively. There could be supply disruptions due to government policies and socio-political instability leading to supply gaps which causes price volatility. Both Ni and Mn supply show that well distributed and used in LIBs. Similarly, natural graphite also sees diversity in end-use but its production is concentrated in China which could be a concern. Because of high crustal abundance, the potential for increased production in other countries and ease of mining, its production is likely to increase as demand increases and synthetic graphite can also be used, further alleviating pressure.

Co is mainly produced as a by-product of the metals Ni and Cu which leads to supply risk. If the demand for Ni reduces, the co-production reduces, which affects the supply chain. But Cu has a relatively high concentration of Co (0.3% Co and 3% Cu). Hence, the producers valued both metals. But as more than 50% Co comes from DRC, its availability is greatly dependent on geopolitical stability of the region. This leads to significant price fluctuations. Co seems to be the main material risk for LIBs but these could be somewhat mitigated by future improvement in extraction technologies which will enable Co to be extracted primarily. Furthermore, variations in cathode material will be seen with more research in this area which can reduce or even eliminate Co from the cathode chemistry.

Li availability is often cited as a concern for BEVs, as Li cannot fulfil the future demand of batteries. But there are enough reserves to meet the demand. The estimation of the Li resources is around 33–64 million tonnes, and the expected reserves are nearly between 13 and 40 million tonnes [9, 10]. Based on a conservative forecast, lithium consumption is 5.11 million tons by considering only the growth of EVs from 2015 to 2050 [11]. A Deutsche Bank report predicted that over the next 10 years, the lithium supply market will triple, and even then, the lithium global reserves will sustain for another 185 years [12]. While land reserves may only contain 14–40 million tonnes of lithium, the ocean is nearly an infinite global lithium resource with 230 billion tons which will be able to extract in the future with further developments in the suitable technology [11].

6.3.3 INFRASTRUCTURE

Infrastructure is the major challenge which holds the EVs behind traditional ICE vehicles to be widely adopted. This is especially true for HFCVs even move than BEVs. BEVs have already taken a big lead with most major companies moving towards BEVs and only Toyota, which pioneered the hybrid electric vehicle with Prius seems to be the major player focused mainly on FC mobility. Honda and Hyundai in collaboration with others have also released HFC vehicles. US have more than 20,000 charging stations with 68,800 units in May 2019 [13, 14]. Whereas a meagre 43 hydrogen stations only available in California [15].

BEVs can either be charged with AC sources with converter for home charging (3–6 kW) or DC sources (fast chargers) that could able to deliver more than 150 kW [16]. For intermittent renewable sources, a buffer energy storage system (ESS) could be introduced to meet demand at all times. For households, a residential ESS like Tesla Powerwall can be installed to store solar power, and buffer batteries can be used for station operators. For a station serving 200 BEVs/day, the minimum energy required is 2.2 MWh battery sources needed.

Hydrogen infrastructure would consist of a hydrogen production unit that utilize water electrolysis as the renewable resource usage. Hydrogen can either be produced on site or will have to be transported in gaseous form which requires many electrolysis plants that are supported by intra-national pipelines or tube trailers. The hydrogen refuelling stations comprise hydrogen supply storage, compression system, hydrogen fuelling storage and dispenser. The dispenser consists of a fuelling nozzle, which connects to the car and is used for cooling where precool hydrogen is used.

6.3.4 COST

Hydrogen refuelling stations are complex and expensive to set up with the high cost of around $3 million for a station dispensing 450 kg of hydrogen in a day [17]. It can also be seen from a capacity basis that in turn leads the capital cost of $5,150 per kg/day. This is reduced from the initial station's cost of $16,570 per kg/day, which also dispensed less hydrogen per day – around 90 kg/day. By the increase of operating stations with larger capacity, the cost will reduce, which directly brings down the capital cost per capacity to nearly 69%. That is significant reduction and is estimated to be 80% if the station dispenses 1,500 kg of hydrogen a day. But still the capital for the station will be around $5 million, which is huge when compared to the costs of setting up a BEV charging station. A 50 kW DC fast charger could be as cheap as $10,000 for a unit and can go up around $35,000 with the average installation cost of $21,000 [18, 19]. Similarly, a 150 kW charging station could cost around $75,000–$100,000 per unit. Since the capital cost of BEV's charging station is less compared with hydrogen station, more stations can be opened up. Also, the cost of recharging a BEV is also very less than that of FCEV [20]. The hydrogen price in California is $16.51 per kg of hydrogen with the cost per mile of around $0.33. By assuming that the vehicle travels 100 miles, which consumes 34 kWh and the electricity costs $0.11 per kWh, then the cost per mile is about $0.04 for home charging and that of DCFC is nearly $0.145 per mile.

The major constraint to adoption of more EVs is the limited infrastructure whose development is in turn hindered by limited EVs on the road. But both EVs are not plagued with it equally. BEVs can be charged at home and offices, and around 82% of charging takes place, which is a great convenience and will aid in the adoption of BEVs, which can still be driven around town even with no dedicated infrastructure nearby, unlike HFCVs which can be refuelled only at dedicated stations and so are completely dependent on the infrastructure.

The autonomy flow rate is the range regained per minute of charge or refill that varies greatly across different types of vehicles. The autonomy flow rate of a BEV is 3–5 km/min (50 kW) or 9–15 km/min for fast charging (150 kW), but home charging gives only 0.2–0.6 km/min [16]. Compare that to FCEV of 160–220 km/min, it is clear that FCEVs have a great advantage in refuelling and

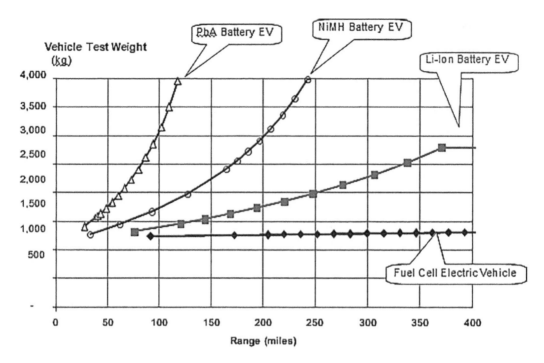

FIGURE 6.5   A plot of vehicle weight versus vehicle range.

can be greatly useful for people and vehicles that have to travel longer distances like commercial trucks.

6.3.5   VEHICLE WEIGHT AND SUSTAINABILITY

Specific energy (energy per unit weight) of various materials used in battery vehicles has been reported in the literature, like nickel metal hydride (NiMH), lead acid (PbA) batteries, the US ABC (Advanced Battery Consortium), etc. Among the two hydrogen pressure tanks of 10,000 and 5,000 psi, the higher psi tanks have heavier weight due to the addition of the extra fibre wrap to provide the required strength. Compared to lithium-ion batteries, fuel cell batteries produce more specific energy.

Vehicle range is the average kilometre covered by the vehicle. These battery vehicles are compared for various parameters, namely vehicle weight versus its range, which is shown in Figure 6.5. The weights of these vehicles are the function of the vehicle range. While the BEV weight escalates intensely for ranges higher than 100–150 miles due to its weight, whereas the extra weight for increasing the range of the FC electric vehicle is negligible. Extra weights of the battery increase the range which requires extra structural weight, a larger traction motor, heavier brakes and extra mass for additional batteries.

6.3.6   BENEFITS OF FCV

Nowadays, hydrogen (95% in the USA) is produced during the methane reformation. Hence, the hydrogen-powered vehicle provides a better solution for fighting against the climate change which reduces the usage of the FCV, since it produces carbon monoxide and carbon dioxide during its process. Current technology available for the extraction and transportation (via pipelines)of natural gas (a fossil fuel) is not promising. Even if the methane cracking process is improved, that is not a long-term solution. Hydrogen cars need twice the energy required for a BEV. As technology develops, the water electrolysis process of getting hydrogen improves the efficiency, but energy loss in this process is nearly 45% which includes compressing the hydrogen into a liquid and storing it. While electric cars utilize the energy from the grid directly, it further supports the usage of electric cars.

The development of new and innovative production technology of hydrogen might increase the proton exchange membrane efficiency up to 86% for FCs in future according to the scientists. Use of hybrid hydrogen-lithium-ion cars could be a promising solution at present, since the extra energy supply is needed for hydrogen production. In addition, a study reported that surplus water from the dams can be utilized to produce hydrogen . Electric cars could become more accessible vehicles if various models of cars are introduced and more viable charging points are installed. BEVs are the more efficient ones compared to hydrogen-powered cars, provided that the lithium batteries are re-used effectively for different applications. BEVs are still a more sustainable solution until for the next promising technology to be developed.

6.3.7 COMPARISON WITH ICE

There are many factors to be considered, when comparing ICE and electric vehicles. First, considering the capital cost, in 2010 BEVs and HFCVs were costlier than conventional ICE power engines. By 2030, HFCVs will have lower capital cost due to the technology advancement, followed by BEV. In 2030, the ICE engines are still cheaper and their lifetime fuel costs are varied based on the situation.

Second, both BEVs and HFCVs are more invariable to fuel cost changes, whereas ICEs cost changes based on the market of the hydrogen and gasoline costs. Third, HFCVs appear to be inexpensive compared with BEVs by considering the total lifecycle cost over 1 lakh miles but it exhibits a wider sensitive by combining both the capital and running costs. ICEs have very effective lifecycle costs compared to HFCVs and BEVs, i.e. nearly higher than 1.75 times.

When the BEV is considered, a possible study on battery size is conducted for this vehicle as a function of range and vehicle efficiency. The study concludes that cost of the BEVs lifecycle is very much sensitive to the size of the battery, and these vehicles will be cheaper if the battery size could be minimized mainly for the city vehicles that run around 50 miles every day.

6.4 CONCLUSION

The departure of ICE vehicles had started from a few years ago but the market was small and many new technologies come and go in short spans of time. EVs, though, are here to stay and dominate the market with major companies and governments planning cleaner and greener future. Most of the new technologies fail in scaling up EVs, which increases the production cost, whereas, for the battery technology, the cost is reduced to 19% with every doubling of the production. The market for BEVs is larger than hydrogen vehicles, nearly 1,000 times larger as on date.

In most areas, BEVs seem far better than HFCVs, and it makes sense why their market share is growing faster than that of HFCVs. BEVs already have a lead with new companies like Tesla, and their technology has been advanced farther than FCs in the commercial car market, and Tesla also has a massive supercharger network to support charging which the HFCVs lack significantly.

BEVs are far more efficient than their counterpart with their efficiency being 70%–80% while that of HFCVs come at only 25%–35%, which is still more than that of ICE vehicles. That makes it compelling along with obvious environmental reasons to replace some areas where BEVs still lack like the long charging time and less range compared to hydrogen-powered vehicles. So, it makes sense to use FC vehicles for long haul commercial vehicles like trucks. Also, due to its poor energy density, batteries are not a viable option for airplanes, ships and boats which can fare better with hydrogen power. Also because these vehicles need a lot of energy, it is possible to fit big sized batteries inside for longer range.

HFCVs are miles behind BEVs in terms of infrastructure development with a handful of stations around the world, and in whose respect the charging network seems massive. The other big drawback is that hydrogen infrastructure is less than half as efficient as charging infrastructure in converting renewable energy to mobility service.

From all this discussion, it is quite clear that BEVs are the way to go essentially, but HFCVs also have a place to fill. Most consumers will find BEVs fulfilling their needs, with HFCVs appealing to those who need to travel long distances and with some companies committed to HFCVs. Soon, both types of vehicles will run on the road in near future.

The following conclusions are made, based on this study:

1. Even though the hydrogen fuel cell electric vehicles have a good future in the road transport, it is better to integrate BEV with the fuel cell technology which has more benefits like building on a technology that initiates the hybrid plug-in ICE in the near future.
2. The main objective of using these electric vehicles is the reduction in capital cost for ongoing technology development. A better technology is needed for the minimization and recycling of lithium, platinum and other precious raw materials that are used in these vehicles manufacture.
3. The efficient range expanding technologies are ready to compete in an electrified transport network which is possible for ICE in near future. But the fuel cells may take a long period of time comparatively.

Therefore, both the BEVs and HFCVs should be pursued and supported in order to provide better policy making.

REFERENCES

1. Government of National Capital Territory of Delhi (Transport Department), "Delhi Electric Vehicles Policy," 2020. https://transport.delhi.gov.in/sites/default/files/All-PDF/Delhi_Electric_Vehicles_Policy_2020.pdf
2. Mehrdad Ehsani, Yimim Gao, Stefano Gongo and Kambiz Ebrahimi, *Modern Electric, Hybrid Electric and Fuel Cell Vehicles*, 3rd Edition, CRC Press, 2018.
3. MIT Electric Vehicle Team, "Summary of Electric Powertrains," April 2008 http://web.mit.edu/evt/summary_powertrains.pdf
4. Mehrdad Ehsani, Yimin Gao, Sebastien E. Gay and Ali Emadi, *Modern Electric, Hybrid Electric, and Fuel Cell Vehicles*, 1st Edition, CRC Press, 2005.
5. Minoru Matsunaga, Tatsuya Fukushima and Kuniaki Ojima, "Powertrain System of Honda FCX Clarity Fuel Cell Vehicle," *World Electric Vehicle Journal*, vol. 3, May 2009, pp. 0820–0829.
6. The European Commission's Science and Knowledge Service, "Well-to-Wheels Analyses," 2016. https://ec.europa.eu/jrc/en/jec/activities/wtw
7. Srikkanth Ramachandran and Ulrich Stimming, "Well to Wheel Analysis of Low Carbon Alternatives for Road Traffic," *Energy and Environmental Science*, vol.8, 2015, pp. 3313–3324.
8. A. G. Volkswagen, "Hydrogen or Battery? A Clear Case, until Further Notice," 2020. www.volkswagenag.com/en/news/stories/2019/08/hydrogen-or-battery--that-is-the-question.html
9. Elsa A. Olivetti, Gerbrand Ceder, Gabrielle G. Gaustad, and Xinkai Fu, "Lithium-Ion Battery Supply Chain Considerations: Analysis of Potential Bottlenecks in Critical Metals," *Joule*, vol.1, no.2, October 2017, pp. 229–243.
10. Paul W. Gruber, Pablo A. Medina, Gregory A. Keoleian, Stephen E. Kesler, Mark P. Everson and Timothy J. Wallington, "Global Lithium Availability: A Constraint for Electric Vehicles?" *Journal of Industrial Ecology*, vol.15, no.5, October 2011, pp. 760–775.
11. Sixie Yang, Fan Zhang, Huaiping Ding, Ping He and Haoshen Zhou, "Lithium Metal Extraction from Seawater," *Joule*, vol.2, no. 9, September 2018, pp. 1648–1651.

12. Mathew Hocking, James Kan, Paul Young and Chris Terry, "Lithium 101," *Deutsche Bank Markets Research*, 2016. http://www.metalstech.net/wp-content/uploads/2016/07/17052016-Lithium-research-Deutsche-Bank.compressed.pdf

13. Office of Energy Efficiency and Renewable Energy, FOTW #1089, July 8, 2019: There Are More Than 68,800 Electric Vehicle Charging Units in the United States, 2019. www.energy.gov/eere/vehicles/articles/fotw-1089-july-8-2019-there- are-more-68800-electric-vehicle-charging-units

14. Fred Lambert, "US now Has Over 20,000 Electric Car Charging Stations with more Than 68,800 Connectors," 2019. https://electrek.co/2019/07/09/us-electric-car-charging-station-connectors/

15. Energy Efficiency & Renewable Energy "Hydrogen Fueling Stations," *US Department of Energy*, 2019. https://afdc.energy.gov/fuels/hydrogen_stations.html

16. Yorick Ligen, Heron Vrubel and Hubert H. Girault, "Mobility from Renewable Electricity: Infrastructure Comparison for Battery and Hydrogen Fuel Cell Vehicles," *World Electric Vehicle Journal*, vol.9, no.3, 2018, pp. 1–12.

17. M. Melaina and M. Penev, "Hydrogen Station Cost Estimates," *National Renewable Energy Laboratory*, 2013. NREL/TP-5400-56412. www.nrel.gov/docs/fy13osti/56412.pdf

18. Chris Nelder and Emily Rogers, "Reducing EV Charging Infrastructure Costs," *Rocky Mountain Institute*, 2019. https://rmi.org/ev-charging-costs

19. Margaret Smith and Jonathan Castellano, "Costs Associated with Non-Residential Electric Vehicle Supply Equipment," *New West Technologies, LLC for the U.S. Department of Energy Vehicle Technologies Office*, 2015. https://afdc.energy.gov/files/u/publication/evse_cost_report_2015.pdf

20. California Energy Commission, California Air Resources Board, Joint Agency Staff Report on Assembly Bill 8: 2019 Annual Assessment of Time and Cost Needed to Attain 100 Hydrogen.

7 Application of AI to Power Electronics and Drive Systems
Mini Review

R. Atul Thiyagarajan, Adhvaidh Maharaajan,
Aditya Basawaraj Shiggavi, V. Muralidharan,
C. Sankar Ram, and Berlin V. Hency
Vellore Institute of Technology

Aayush Karthikeyan
University of Calgary

K.T.M.U. Hemapala
University of Moratuwa

CONTENTS

ABSTRACT

This chapter presents a review on application of artificial intelligence (AI) in power electronics (PE) and drive systems. With the advent of power electronics, efficiency and reliability of power systems have improved. Since power grids are getting smarter, a large number of PE devices are put to use, which degrades the power quality. AI gives machines human intelligence by allowing them to learn from experience. By this, the machines adapt to new changes in their atmosphere. The incorporation of AI with modern controlling techniques allows complex and dynamic control to enhance the system dynamics, power quality and efficiency. Design, control and maintenance are made easy with AI. Many research papers have been examined and studied in detail to comprehend the practical implementation and research opportunities in the application of AI in PE.

7.1 INTRODUCTION

In Reference [1], a brief review of the artificial intelligence (AI) applications for PE systems is presented. From the application point of view, the AI used in PE systems can be divided into plan, control and maintenance. Usage proportion, trend, features and necessities of AI in every life-cycle stage are seen thoroughly. From the method point of view, the AI used in PE systems could possibly be divided into expert systems, metaheuristic methods, fuzzy logic (FL) and machine learning

DOI: 10.1201/9781003331117-7

(ML). From the function point of view, the AI-related applications are primarily handling optimization, categorization, regression and data structure exploration. The achievements of pertinent algorithm variants and applications are found and arranged in the timeline chart. For all life-cycle stages, explanatory examples are discussed and also the disputes and future analysis opportunities are pointed out. Reference [2] presents a comprehensive overview and introduction of different AI techniques and applications of these techniques in harmonic detection and eradication, current supply inverters, power electronics (PE) expert systems, photovoltaic (PV) systems, etc. A revolutionary change has occurred in the power electronics system by integration of AI tools such as FL, artificial neural networks (ANNs) and genetic algorithms (GAs), etc. To spawn intelligence in a system that automatically adapts to the variations or changes is the goal of AI. The fuzzy controller is colossally wielded in converters and controllers, AC, DC and reluctance motor drives, pulse width modulation (PWM) inverters, and many other alternative implementations. The NNs are employed in rectifier regulators. Expert systems are used for algorithms training employed in automation systems. The growth and progress of digital signal processors (DSPs) and field programmable gate arrays (FPGAs) make the execution of fuzzy and neural systems practical and economical with improvement in quality performance. Reference [3] proposes a novel design layout approach based on multi-objective genetic algorithm (MOGA) for non-isolated interleaved multiport converters (MPCs). It minimizes the heaviness, losses and ripple in current that decrease the lifespan of energy sources. The Average Ranking technique is projected to establish a beneficial remedy among numerous Pareto-front solutions. The proposed design achieves superior execution than the standard one in inductor weight, current ripple and converter losses. The Finite Element Method like COMSOL software is employed to authenticate inductor design, which may be a pivotal step for future work. The optimization operation gives room for fresh layout in the optimization of MPC. The current ripples and losses distribution of the proposed design are compared to that of the basic design. It substantiates the view of the given optimization method, while taking into account rising technologies like wide bandgap semiconductors. Reference [4] aims to develop an AI research instrument for planning and analysis called power electronics expert system (PEES) as a problem eradicator. Its purpose, main operation blocks, duration results and performance contrast are reviewed. For initial evaluation and power supply loss assessment, PEES models and specialists' rules are precise. Loss of power in semiconductor devices and magnetic elements, control loop small signal analysis, virtual standard hardware circuit, closed-loop circuit electrical performance and rapid thermal simulation are consolidated. The study demonstrates expert system feasibility. Purpose, main operation blocks and run period performance are evaluated. The system is confirmed to be stable, flexible and extensible. PEES not only gives a remedy for technical aspects like design and analysis but also takes care of non-technical problems. Reference [5] emphasizes the effectiveness of monitoring power electronic circuits using a static NN. Input–output mapping is done by training the NN. From given input voltage and current and output current values, NN estimates the final output voltage. The NN is designed in such a way that when the performance properties of the elements within the rectifier circuit are modified, it gives an indication. This should prevent failure or performance deterioration of the circuit. A micro-level monitoring of the circuit is done to identify the type and source of the malfunction. A complicated circuit can also be mapped by having a large training data set or a vast number of neurons within the hidden layer. But the NN can be trained to model non-linear systems with high accuracy by proper and systematic design parameters. Reference [6] proposes a new hybrid control algorithm for a DC/DC boost converter to regulate the final voltage. The aim is setting the system to a stable limit cycle and also guaranteeing essential regulation in voltage. In hybrid-automaton representation, the problem in control is streamlined to a guard-selection problem. MATLAB® (simulink) using the state flow chart feature is used to simulate the system. It's a right selection for real-time control since involved computations in the algorithm are minimal. During continuous conduction mode, the system has a variable switching frequency with most permissible ripple in output voltage. During discontinuous conduction mode, the operation is smooth even under disturbances. With an output filter-capacitor value, the control scheme is demonstrated. Toggle switches are used to vary load, hence playing a part in transient response of the system. The algorithm provides fine regulation characteristics over a vast amount of line and load disturbance. Reference [7] discusses protection of high-voltage direct current (HVDC) converters by using neural networks (NNs). The potential of this network distinguishes between contrasting types of faults that may arise in the converter. Integration of ANN-based controllers improves the dynamic response of an AC/DC system. This paper discusses about how NN distinguishes the difference faults in three novel NN-based systems proposed for a HVDC converter in different system faults and disturbances. The identifier is tested with a powerful and frail AC side. This procedure does not rely on operating modes and the fault identifier can be utilized to design an integrated ANN-based controller with fault identifier. Reference [8] discusses principles of NN and applications of vector-controlled induction motor drive. Emerging AI networks like ANNs have influenced scientific and engineering applications. Power electronics field has advanced rapidly by the advent of this technology. Application-specific integrated circuit (ASIC) or DSP chips can be used for real-time implementation. ASIC chip is preferred over DSP because of fast parallel computation ability but non-availability of inexpensive ASIC chips has hindered the ANN applications. ANN is a type of network that requires large datasets and their computation time is time-consuming. ANN weights need to be adjustable by fast online training. Various methods like Extended Kalman filter (EKF) algorithms and random weight change have been attempted for online training. Hybrid AI techniques in which ANN is combined with FL or expert systems (ES) or GA to explore new areas of applications with a promise for the future. Reference [9] projects a curtailed review on three divisions of AI, i.e., ANN, expert systems and FL. AI networks like ANN, expert systems and FL have led to a new era in the discipline of power electronics and motion control. The paper primarily focuses on description of theoretical principles and applications of all three networks. Applications and principles have been presented in a comprehensible manner for readers from power electronics background. Applications have been described elaborately in each topic to supplement the concept. Integration of AI techniques will make the field more complex and interdisciplinary, thus providing a great challenge to power electronic engineers. Reference [10] proposes the design and implementation of full-load current (FLC) in PE converter circuits. Mathematical model for a system is necessary in traditional methods but in contrast to a traditional method, mathematical model is not necessary for an FLC and its implementation is very simple. Recently, there is a sudden rise in FL applications in the control of PE. The inherent characteristics of power electronic converter circuits like non-linearities, inconvenience of a definite model or its exorbitant complication, make FL control best suited. This paper focuses on designing a FLC for PE converter circuits. Matlab (simulink) is used to simulate the proposed converter and the performance is evaluated. Reference [11] discusses methods of application of NNs in power electronic systems. Comparator characteristics are decomposed into network structure by NNs and develop new fascinating characteristics by learning. The process is performed in simulation and results are validated. In comparison with the hysteresis comparator approach, this approach seems more effective due to smaller switching loss. The first stage indicates the application of NNs to inverter control. One of them is Hopfield's network process for optimization of PWM patterns. Issues which need very long processing time with basic sequential approaches are solved effectively by using Hopfield's networks. In the upcoming years, the application of these approaches will be for more complex control of the power converter system.

7.2   NEURAL NETWORK

In Reference [12], a new methodology aided with artificial techniques for developing an automatized design in power electronics is studied. This method develops an optimal design between design parameters and reliability metrics by establishing a functional relationship. In the initial stage, operant condition variables are characterized, and by characterizing the thermal stress of a converter, design parameters are converted into variables by creating a non-parametric surrogate model. To perform specific functions a dedicated NN named ANN is used. ANN_1 is deployed as the

Accurate Surrogate Converter Model (ACSM). The yearly mission profiles of any chosen device for a vast set of values of the design specification are mapped into thermal stress profiles by ANN_1. Design parameters are mapped explicitly into a yearly lifetime consumption by ANN_2. The parameters include diverse restriction in the design, the difference between system's size of filter to the output and optimal balance between the reliability. Reference [13] suggests a color adjust method on the basis of NN for attaining high color exhibiting signals for LED systems having multi-color. Even under ambient light this is achieved. Data points with high color rendering indexes are provided by the specific power consumption (SPC) method. After the updation of data points, NNs are trained on a microcontroller using these data points. The NNs are trained for the realization of controlling the color of multi-color LED systems. Red, green, blue, alpha (RGBA) LED system is used as the application specimen for the testing of proposed methods. Series connection of the RGBA LED channels is driven by a buck-type single-inductor multiple-output (SIMO) LED. Control of color with high color exhibiting is attained without requiring a precise Electrical Optical System Model (EOSM). This allows application in all many-colored LED systems with improvement in performance without adding any hardware. Results from the experiment illustrate that the efficacy along with validity of the suggested preventive method is evaluated. Reference [14] demonstrates a systematic and compact performance indicator for PE converters. The constant lifetime curves are used to represent performance indicators, which, by employing an ANN, are modeled and estimated. For an idea along with an assessment relevant plan for its solidness, functioning along with management action in PE systems the proposed performance index is provided as technical data. Performance index and reliability is predicted fastly and precisely by using ANN. The computation time has acceptable accuracy and the rapidity with five times more than the conventional approach has been depicted. The sharing of power method extends the lifespan by 30% by contemplating the lifespan of the converter while under functioning. From a reliability standpoint, the converter behavior in the conversion and reformation mode is illustrated, which should be contemplated during the plan and functioning of converters. The operating conditions of active and reactive power in power electronic converters have been considered for the long-term indicator in this paper. Reference [15] uses a NN algorithm to control the power electronic circuit, and the boost converter is used as Power conversion stage (PCS). First, the NN is trained using an administered assessment structure. The NN learns about the circuit itself. This paper uses ANN because of its potential for function estimation when trained appropriately. The inputs for the controller are capacitor voltage and load current, and the output will be the duty ratio which will be produced to the power electronic switches. Based on the user-specified input values, the NN will get trained by tuning its parameters based on the capacitor voltage and load current. As back-propagation requires a huge volume of historical dataset, so Evolutionary Computation is used for training the model. Reference [16] uses Bayesian Regularization-based ANN as well as Random Forest Model (RFM) for introduction of a current molding method for DC-DC converters. Unlike methods such as pulse width modulation switch (PWMS), state-space averaging (SSA), etc., simulated data or hardware assessment is used for evolving the model of the system in this approach. For the boost converter collection of data, filtering, priming, evaluation and verification are included in developmental steps of the model and discussed. Training of large data in the proposed methods helps in modeling of V across and I through the switches. Application of machine learning techniques in modeling of power converters is presented as an overview. Collection of different sets of system parameters such as R – resistance, V – voltages, C – currents, T – time, L – inductance, C – capacitance, etc., is done by using simulation prototype or hardware prototype of an existing converter The parameters are categorized into I/P-inputs and O/P-outputs of the design, and transient performance and state response are modeled by using Bayesian Regularization-Artificial Neural Network (BR-ANN) model and Random Forest (RF) with bootstrap aggregation, respectively. Training and testing data sets are performed for validation of the model, and new values of I/P-input specification are finally used to test the model. Creation of identical twins of power converters, escalation of system performance and forecasting the conditions of fault is done by this technique [17]. Accurate knowledge of system dynamics is essential to understand and emulate any system operation. To create mathematical models which represent the real system, usage of differential algebraic and equations helps to delineate the actions and system functional properties. This proposed paper emphasizes on using NNs to improve accuracy and get better results. An ANN is one such type that achieves a neurological associated performance, such as researching from incidence, creating similar situation inferences and determining states where outcomes were poor based on previous performance implementing algorithms. This paper enlists and discusses the most popular topologies of ANN and methods of training. Executional details, priming and validation of performance of such algorithms and their issues are discussed respectively. Reference [18] explores estimation of power electronic waveforms by applying NNs. ANN methods are intelligent techniques with a promise of future approach in PE systems. This paper takes two systems into consideration, one being deformed line current waves in a1-phase thyristor of an AC-controller and the other being a 3-phase diode rectifier that caters for a load of an inverter machine. For the estimation of the total root mean square (RMS) current, fundamental RMS current, displacement factor and power factor, NNs are trained. In comparison between NN estimators and actual values, they indicate excellent performance. The performance is rapid and has synchronized reaction of all the outputs with an extensibility of implementing them in dedicated analog or digital hardware. Reference [19] delineates the plan and methodology for implementing NN for Digital Current Regulation (DCR) of inverter drives and comparison of these methods with more conventional digital processor methods. This paper primarily focuses on designing one, two or three-layer network for DCR of inverter drives. Two different learning techniques have been developed for evaluation of learning requirements of various designs of such inverter current regulators. The model performance and learning techniques are analyzed through simulation. The observed results along with design consideration aids to determine the network that is best suited for the application. The implementation of NNs in both analog and digital circuitry depends on the network structure. Reference [20] demonstrates the achievement of vigorous execution of DC-DC converters by using a new optimal control. The suggested algorithm forecasts the transient response at optimal level for the converter using the theoretical idea of constant current (CC) balance under a large signal change in load current. Prediction of the minimal number of cycles in switching and their respective duty cycles (D) results in driving the O/P-output voltage to its nominal value in a limited time duration by minimizing the O/P-output voltage, undershoot or overshoot. The output capacitor size is reduced by using the algorithm while still matching the specifications. This reduces the cost without any change in dynamic performance. In steady-state conditions, current mode-based conventional proportional integral derivative (PID) is operated. The new control algorithm is put to use under transient conditions of large signal. Results prove that the latest derived method produces better execution than the current mode of a conventional PID controller. In Reference [21], a ML-based modeling of Gallium-nitride (GaN) power devices is presented. For increased accuracy, flexibility and scalability, a deep feed forward neural networks (FFNN) model is utilized. The usage of deep feedforward NNs accelerates the planning technique of GaN circuits. It predicts the device's switching behavior without the device's physics and geometry. Under different circuit conditions the changing voltage and current between the channel and origin of GaN devices are predicted and trained by using NN. The aim is to use a Stochastic Gradient Algorithm (SGA) by an algorithm to derive quickly to construct a GaN-based regression prototype. A mean absolute error (MAE) loss role is utilized for verifying the accuracy of the predictions. The NN with five hidden layers and 30 neurons was proved to be the finest for optimization and prediction. Reference [22] throws light on a method developed by the utilization of a NN based on controlling DC-DC converters digitally. The timing preventive technique is used for the improvement of the transient response of the DC-DC converters. The neural network-based committee (NNC) can be a travelling salesman problem (TSP) to reimburse for the transient response and hence, is an acceptable instruction-based method. To improve transient response, the NN is realized with the reference modification in PID control. In a PID control, the control of dynamic input–output conditions is determined by its gain parameters. The standard three-layer NN is embraced in the presented paper. Peak points are detected and are discussed and realized by taking into consideration the time taken and beginning – ceasing timing of the NNC term. The presented method suppresses the convergence and overshoot time together constructively and provides a steady transient reaction. Reference [23] delineates a new innovative approach in power electronics application by implementing hardware for NNs using FPGA. DSP or computers that proffer serial filtering offer hardware implementations by realization of NN. The puissant hardware choices, high execution speed, re-configurability, low cost and parallelism have nowadays made parallel programmable logic devices (PPLDs), such as the FPGA with embedded microprocessors, the best suited for hardware implementation of NN. The framework and innovative results are observed and used to depict the authenticity of utilizing the parallelism and transposability of a cheap-rate FPGA to execute a controller on the basis of NN for DC-DC converter [24]. The highly non-linear behavior of switching converters makes the conventional controllers not capable of obtaining good dynamical performance. Thus, system performance is improved by the proposed new NN control system. This paper addresses incorporating NNC for adaptive feedforward controller (AFC) and neural network-based ensemble (NNE) for converter recognition of switching converters and NN to observe overall dynamic response of the system. This technique comes into play when the attributes of the plant do not match the defined ones. The system is simulated under high-frequency pulsed voltage supply and reference signal and observed. The flexibility of this system makes it applicable to power conversion systems where behavior of the system is not clear cut. Reference [25] shows recognition and control of power converters using NNs. The dimensions of the converter in instances of unreliability in the variables of load are identified by means of a NN emulator by implementing a non-parametric prototype of DC-DC switching converters. By using a neural controller converter, control and closed-loop linear variable actions are also executed by a pseudo-linearization control technique. PWM boost converter is simulated under the working of large signals to show both applications. Additional research contemplates the extension of the identification plan and the planning technique to fourth-order converters under the working of large signals. Reference [26] demonstrates a digitally controlled method for PWM DC-DC converter by utilizing an NN. This paper primarily focuses on an overcompensation phenomenon for the transient response improvement caused by neural control term. Transient response can be improved by non-linear prediction but it also causes the overcompensation. Converter systems' time delay and its digital control computation cause the non-linear behavior. The study focuses on the use of NN control for suppressing the overcompensation by evaluating current info from a steady-state equation of the output voltage. Mode change in NN control items in transient condition is simply realized in the suppression method. The proposed method is simulated and results confirm that it simultaneously improves the transient response and suppresses the overcompensation phenomena. Reference [27] discusses the identification of power converters by using NNs. Implementation of a non-parametric model of the AC-DC inverter is done by means of a NN and used it for identification of the inverter dynamics during uncertainty in the input voltage supply and/or load parameter. The circuit comprises a main circuit, control system, along with the source inductance and the output filter. The final prototype is used in the analysis of the power system. The proposed converter is simulated in a PWM inverter under large-signal operation and the results are used for the illustration of applications and to detect the characteristics of the circuit. Reference [28] focuses on the efficacy of a constant N for I/P-input–O/P-output mapping of PE circuits. Modeling between the inputs and outputs of a Power Electronics (PE) circuit is formed by training a multilayer perceptron (MLP) NN. Elements like full-bridge diode rectifiers, along with the source inductance and the output filter, constitute the circuit. Dynamic models are used for the rectifier diodes, when one or more components' performance properties change the designed NN provides an indication. The proposed method is simulated and the results depict that the NN is able to map the circuit's I/P-input–O/P-output and identify functioning conditions that differ from the basic one. Reference [29] presents an overview of NN implementation in smart management and assessment of PE and motor drives. Artificial NNs and GAs opened a fresh gateway in PE and motor drives. The paper further reviews detailed descriptions of principal topologies of NNs that are appropriate for the power electronic applications. This paper also focuses on architectures like feedforward and feedback or recurrent architectures. Many application examples like non-linear function generation, delay-less filtering, space vector PWM of 2-level and multilevel inverters have been included in the paper discussion. According to the current trend in modern technology, it seems that NNs will spread its wings on application of PE and motor drives in future. Reference [30] depicts a solution for a DCR used in a 3-phase PWM inverter connected to an AC motor by using a 1-step OC controller. A space vector (SV) approach is used to make deadbeat management whenever feasible and good precision when it is not. A 4-neuron per phase NN is used to control. To enhance the mean squared current error, the NN is cascaded with a proportional integral (PI) controller regulator. The performance of the standard PI regulator is enhanced by the NN usage. Simple analog circuits are used to implement the NN. The developing graphical relationships are enhanced by comprehending classical PI and optimal space vector control strategies.

7.3 FUZZY

In Reference [31], the research work examines FL applications in analyzing PE waveforms. This paper takes two devices/systems into consideration, one being contort waves of line current in a triac light dimmer and another being an inverter machine load feed by the three-phase diode rectifier. Application of FL aids to monitor the RMS and fundamental RMS current, displacement factor and power factor for each system. For estimation, two methods have been put to use, i.e., a fuzzy rule-based methodology and a fuzzy relational or Sugano's methodology. Then the result readings are compared with actual readings to determine the correctness and they indicate a better accuracy. The fuzzy assessment has an upper hand of quick reaction, numerous results from rule of single premises. However, the two approaches has no effect of noise in the sensor as well as drift in the sensors, the comparative approach gives better precision, and development of algorithm is less strenuous. In Reference [32], a forward fuzzy logic control (FFLC) algorithm for light-emitting diode drivers is presented. Then main motive is to get a fine dynamic response and reduce the low-frequency ripple. It addresses concern that arises in LED voltage along with LED current. The paper introduces an FFLC algorithm for a Takagi–Sugeno–Kang fuzzy controller. Fuzzy control is familiar for its better performance; within the system it can appropriately handle variableness that is to be controlled. The FFLC uses a digital signal controller's high-speed mathematical engine that is capable of achieving quick multiplications during one common cycle of instruction. This eventually tweaks the precision of the multiplications and divisions in the digital signal controller mathematical engine. Consequently, the program implementation is short-lasting when the FFLC is carried out in DSC. The DSC was demonstrated physically with a relatively low-cost DSC. The LED current is controlled and it is possible to keep it constant. Reference [33] delineates the design of the power converter using a fuzzy expert system. FL technique aids in selection of a nearly flawless topology of power converter and aids in the determination of optimum component values using optimization programs. This paper primarily focuses on features of a fuzzy expert system, such as a CAD tool along with expert knowledge. The expert system aids the author to achieve required specifications by determining the optimum set of components to minimize the time as well as cost of the design. The demonstration of application of an expert system is simulated in various software like PSpice, MATLAB and Magnetics Designer. Reference [34] demonstrates a DC-DC boost converter's voltage tracking control using an adaptive fuzzy neural network control (AFNNC) strategy. To enhance system robustness, total sliding-mode control strategy is developed. A four-layer fuzzy NN mimics the total sliding mode control (TSMC) law to reduce the control chattering phenomena. The AFNNC scheme acquires online learning algorithms in the sense of Lyapunov stability and projection algorithm. This eliminates the need for controllers and ensures the stability of the regulated system. The duty cycle of the power switch in the boost converter receives the output from the AFFNC scheme. Manual tuning of parameters is avoided, since the online parameters learning mechanism is developed. The AFNNC scheme has a voltage tracking improvement of over 20.3% than the total sliding-mode control framework. The proposed Adaptive Fuzzy Neural Network Control scheme is more acceptable than the TSMC strategy for DC-DC boost converter's voltage tracking control [35]. On a low-cost 8-b microcontroller, a fuzzy controller for DC-DC converters is implemented. For real-time feedback control, modifications are done to the fuzzy control algorithm. The output voltage of the boost or buck converter is regulated to a required value without steady-state oscillations even when there is a change in input voltage or load. The fuzzy controller is built on the basis of linguistic details and hence, the algorithm is employed to control both the buck and boost converter. The DC-DC converter's duty cycle is refurbished every eight switching cycles due to the time needed for analog to digital conversion along with the control calculations. A narrow memory region of 2 kB and unsigned integer arithmetic are two key difficulties in programming the microcontroller. It is feasible to implement a FL-based controller in an economic manner. A non-linear controller like FL is inexpensively implemented with microcontroller technology. Reference [36] uses a fast forward neural network (FNN) in which a NN topology mimics fuzzy reasoning. Fuzzy rules are automatically identified and membership functions are tuned in such NNs. In a three-phase diode rectifier, distorted line current waves are a problem. The FNN is used to estimate RMS current including fundamental RMS current. The input signals that are given for evaluation are peak value and wave pulse. Good accuracy is achieved when estimated results are compared to actual values. The FNN is powerful because it integrates the numerical processing features of a NN with linguistic descriptions of FL. The estimation algorithm can be used for other waveforms and drives as well. In Reference [37], an alternative method using the fuzzy algorithm for tuning the PID controller for both linear and non-linear systems is proposed. A simple PID controller can be used as a controller for any power electronic circuit. If the circuit has more energy storing components, the order of the system will be higher, and finding the appropriate PID values will be arduous. By employing this technique, based on the current performance of the system, the weights are calculated instantaneously with the help of fuzzy loops, and the overall performance gets enhanced. Simulations are used to evaluate the working of the model and it is found to be very effective. In Reference [38], a novel adaptive fuzzy control is designed to control a DC-DC boost converter. To improve the sturdiness of the system in a transient state, TSMC is employed. For DC-DC boost converter, the adaptive fuzzy control is used to imitate the TSMC technique. The projection algorithms and Lyapunov method of stability are used to test the system's stability without using other auxiliary controllers. To mimic the TSMC law, a four-layer fuzzy network was used in this paper. One of the main advantages of using this method is that the output of the controller can directly be fed to the switch by varying the duty ratio of the switch which leads to a regulated output voltage.

7.4 FAULT

In Reference [39], a novel system is discussed, whose ultimate aim is to recognize defects in power electronic converters, which means finding the damaged insulated-gate bipolar transistors (IGBTs) in the rectifier by registering current periods in each phase circuit. Topology of NN was selected using GA, and Levenberg–Marquardt method was used for calculating weights. When multiple tests were performed, the circuit didn't generate any false alarms and the system performed correctly under power flow direction change. The system efficiency is analyzed and resulted two rectifier states. The proposed method is for low-power applications, and for high-power application, current shape and location of fault are important. Reference [40] focuses on designing ANN for testing different open-switch characteristics of three-phase pulse width modulators. Totally there are 21 different faults for 1 and 2-side open mistakes in converters. Diagnosis of open faults is generally needed, since abnormal currents can propagate to peripherals, causing secondary faults. In this paper, multiple faults are diagnosed using a two-step technique based on ANN. The first stage implies the process of receiving direct current and series current components using Adaptive Linear Neuron. According to the starting stage, DC components in three-phase planes are divided into six fault modes. The second step implies similar use of DC elements to determine total harmonic distortion (THD) for every sector to localize fault modes by using ANN. The given two-step process has a short execution time of 22 s and can be easily designed for the testing process [41]. Logic-based methods are proposed for the intelligent restoration and treatment of faults in PE systems at the base level. It successfully diagnoses various faults using FL and combinations. Combinatorial logic was shown to detect and use massive or alternate parts for system recovery in a variety of faults. The combination of national logic and FL achieved maximum system diagnostic ability. The signal's average value and RMS value are the two quantities observed in both methods. With the objective of reducing the number of measurements while maintaining effective diagnosis, a systematic methodology is introduced. For testing the platform for the proposed methods, we used a standalone solar PV inverter as an example. The proposed methods can be applied at any stage of the converter development process, ranging from concept design to testing and evaluating. In Reference [42], a robust system is developed to identify multiple fault types in an e-drive system with a 3-phase induction motor. For multiclass fault detection, the system network is carried out by a ML algorithm. Simulink is being used here to implement the model of the e-drive. A NN is created and trained on the generated signals at the operating conditions for fault response of inverters. SIM_DRIVE is used to generate representative training data. It has minimum sensors and control mechanisms to generate voltage and current signals. CP-Select is the ML technique used to choose the operating conditions so that the signals released by SIM_DRIVE can be used to train the system. The paper presents 2 NN techniques, a designed multilevel NN system and a basic single NN system. The former showed greater test results, as the latter is simpler to use and perform. The system prediction was 98% accurate and predicts the three post short-circuit fault classes. In Reference [43], to find the exact location of fault in a PV system, noise patterns with very high frequency are used. The PV cells are subjected to different faults like L-L, L-G, and arc faults on both the sides of the system. A multilayer (3) feedforward ANN was developed to identify fault location accurately. These faults are visualized using the real-time digital simulator (RTDS) simulator, and wavelets are used to analyze the data. An extra high-frequency generator is not required. The usage of voltage sensors is reduced in this method. Using various system parameters, the practicality of this method is tested. The results show good accuracy and robust performance in spite of noisy measurements. In Reference [44], a multi-layered ANN is developed to diagnose the faults in HVDC systems. Fault diagnosis is performed by training the ANN using the historical data of faults. Sixteen different faults are studied in this paper. This method proves to be an effective way for fault diagnosis. Reference [45] delineates the application of a radial basis function (RBF) NN for diagnosing the faults in a high-voltage DC power system. The pre-processed data are fed to the RBF network. An adaptive filter is present in pre-classifier to identify the system variables' average values and to monitor the proportions of the fundamental frequencies, and to aid the pre-classification of the signal, we use a signal conditioner which uses an expert knowledge base. The extraction of the feature of the local signals can give better symptoms for the fault diagnosis when observed. The proposed method for fault diagnosis is simulated using EMTP Package and evaluated. Reference [46] proposes a compact NN for a defect diagnosis module in a multilevel inverter. The difficulty in using mathematical models for diagnosis of multilevel inverter drive is because it consists of various modules and their system toughness has factors which are non-linear. MLP identifies the different types and specific places of common mistakes in inverter output voltage measurement. The importance is utilized to decrease the NN input size, and the time needed to train is reduced by using a lower dimensional input space and the reduced noise may improve the mapping performance. The performance of MLP, NN and PC-NN are compared. Two given systems are simulated, and the results are used to clearly get the faults, testing and results of a 4-level drive system. Reference [47] discusses the systematic investigation of the voltage-fed PWM inverter's various fault types. The knowledge about faults and the characteristics of a system is remarkably vital to the view-point of enhanced architectural layout or architecture, protection and fault-tolerant control. Once the fault methods are identified, an initial testing of the important fault types, such as single-phase input problems, rectifier diode short circuits and transistor short circuits, has been performed. In this study, fault performance predictions are simulated, the stresses in power circuit components are calculated, and the steady-state post-fault performance is evaluated. Outcomes are used to design protection systems and easy fault diagnosis.'

7.5   OTHER PREDICTION ALGORITHMS

In Reference [48], a Maximum Power Point Tracking (MPPT) algorithm for variable-speed Wind Energy Conversion System (WECS) using reinforcement learning method is proposed. A value-based learning algorithm is planned on an online basis for tuning of the control parameters of a WECS by changing the control values, and the Maximum Power Points (MPPs) are updated based on the learning. For controlling MPPT for wind turbines, based on the MPPs learned, speed power curve is obtained. By employing a reinforcement learning algorithm, even without the knowledge of parameters of the turbine, the WECS learns on its own in a certain environment. The model algorithm has been tested using simulations for a permanent-magnet synchronous generator (PMSG) wind turbine. Reference [49] proposes an ANN-based-Reinforcement Learning (RL)-MPPT rule for PMSG-based WECSs. ANNs and the Q-learning algorithms are used to learn the relationship between the speed of the rotor and output power of the generator. The WECS is managed primarily on the basis of relationships realized. The WECS learns on its own by interacting with the environment. The digital RL model can be reinstated whenever a change is observed. The effectiveness of the proposed MPPT preventive model is depicted via simulation.

Reference [50] focuses on different approaches to trace the MPP of photovoltaic systems. The individuals owning a PV system can provide excess energy back to the grid. These individuals are called 'prosumers' (producer+consumer). MPPT methods are conducted to maximize power output from the PV systems. AI-based approaches prove to be a feasible and effective solution. The disadvantages of traditional MPP tracking methods and various AI-based MPPT schemes are reviewed. Evaluation is done to assess the complexity, capacity to track Global Max Power Point. A comparison study on various systems under different conditions has been carried out in this paper. The proposed paper ensures that the basics of each method are understood and selected according to the requirements. Reference [51] emphasizes the absolute power loss adaptation and systematic design of the converter related to AI technology. The resonant converters are used judiciously as a DC transformer in the hybrid AC-DC microgrid to work within the AC/DC buses. An AI model-based 2-stage function method is allocated to reduce the overall power loss. In Stage 1, the absolute power loss is managed by the identified AI process, and the allocations like the leakage and managing inductances and resonating capacitances are derived. In the second stage, the optimal leakage inductances and magnetizing inductance are observed. The equations of two types of inductances are derived. In Reference [52], a control method is proposed to optimally control buck and boost converters. By utilizing the idea of Convertible Circuit Breaker (CCB) the change in load current under a large signal, the optimal transient response for the converter is predicted by the proposed algorithm. To obtain a regulated output voltage quickly, the duty cycle is varied. The output capacitor size is reduced by using the algorithm without altering the basic requirements. This reduces the cost without any change in dynamic performance. In steady-state conditions, a classical PID controller is used, whereas, in transient state, the proposed algorithm is used. The proposed algorithm outperforms the classical PID method by producing a regulated output voltage waveforms [53]. Generally, for obtaining approximate models, Markov parameters are used as a mathematical criterion in the model reduction field. In this paper, a reference model is approximated using a new control methodology. Here, the transfer function of the specified model and pre-specified parameters of Markov parameters and Pade coefficients are mapped. The proposed algorithm is known for its flexibility. For example, a higher-order system can be matched with a lower-order reference. In this way, for higher-order systems, complex calculations can be avoided. The algorithm also supports lower-order controller design. If the Pade and Markov coefficients are known, the transfer function and reference model can either be continuous or discrete. Reference [54] depicts usage of dynamic decoupling method. Induction motors are designed to operate at high speeds and the characteristics of the motor's torque response will be impaired due to the coupling of d–q current dynamics. A dynamic decoupling is a method of integrating two PI controllers. By employing this method, the estimation errors of the motor will be reduced. The proposed model consists of both classical and cross-coupling PI controllers. When compared with the feedforward controller, the proposed model will be more robust to the variations of estimation errors. This paper derives a sensitivity function to showcase the robustness of the decoupling action of the proposed model. Even if there is a mismatch in parameters, no steady-state errors shall be observed. In Reference [55], a novel method for controlling the voltage source inverter (VSI) is presented. The main aim is to obtain a modulated output waveform with a fixed switching frequency for the VSI. In classical control methods, ripples and noise are generated, so the overall power quality of the converter reduces. The expressions of the cost functions in other controlling techniques are bound to be complex. So, no new objectives can be introduced in the cost function. This paper uses space vector modulation with a PI controller. The common-mode voltage can be reduced naturally by modifying the applied vectors. In References [56–60], different predictive controls for power electronics-based applications are described. Due to the advent of the microprocessor, predictive algorithms are widely used because of their advantages when used as a control scheme in power electronics and drives applications. By employing predictive control, non-linear systems can be controlled, and complex calculations can be avoided. The predictive control is divided into four broad classifications: deadbeat control, hysteresis-based, trajectory-based and MPC. The prediction algorithms can be applied in current control, torque and flux control, power control, control of NPC converter, matrix control converter, etc. In References [61–64], a novel active damping method is introduced that uses fewer sensors compared to the active damping method. A Lathing Current Limiter (LCL) filter or an inductor is widely used as an interface between the grid and the power electronic device. LCL filters are used to filter the output voltage ripples from the rectifiers. To overcome the stability problem due to the current control loop, resistors can be used. The disadvantages of employing the active damping method include reduction in efficiency as the resistor dissipates energy in the form of heat, and more sensors are to be used. With the help of a GA, based on a wide range of sampling frequencies, the filter is tuned. In Reference [65], for tuning the PID controller embedded in programmable logic controllers (PLCs), a recursive least square (RLS) and GA-based tool is presented. The P, I and D parameters are found based on the plant model after identifying the plant. The least squares is responsible for identifying the plant, whereas for tuning the PID parameters, the GA tool is used. Once the connection is established with the PLC and the data are acquired, the GA sets in the new PID parameters, and the simulated results are evaluated. In Reference [66], PI controller parameters are tuned using optimal fuzzy gain scheduling that controls the induction motor drive. Due to the advent of power electronics and complex advanced controlling techniques, there is a significant development in AC and DC machines. The fuzzy control-based control technique is a non-linear control approach in which the PI parameters are determined based on the instantaneous values fed from the sensors. If the operating condition keeps changing, the controller needs more information; hence, the complexity of the rule increases. In order to control the negatives of PI controllers and FL, the paper proposes a method where the control parameters can be made online for controlling the induction motor using an adaptive FL mechanism. Finally, in order to optimize the fuzzy controller, the GA is employed as an optimization technique. References [67–71] demonstrate the use of GAs to tune fuzzy control rules. This method suits the membership function of the fuzzy rules made by the experts. By minimizing the error function, high-performance membership functions can be obtained. FLC on the basis of control systems with fuzzy control rules (FCRs) associated by means of fuzzy inference. An appropriate FLC is defined, which modulates the definitions of fuzzy sets to predict the membership functions. These, in turn, generate maximal FLC performances as per the inference system and defuzzification strategies. The behavior of the FLC systems is improved by the obtained FCRs. The quadratic and line errors are decreased.

7.6 CONCLUSION

The focus of this chapter is to study the works of literature on implementation of AI in power electronics, and also the recent trend in this field is thoroughly reviewed. The chapter is divided into five sections: introduction, NN-based control, fuzzy control, fault detection and other prediction algorithms. Due to the advent of power electronics and AI, different control techniques using AI are used in power electronic circuits. By employing AI-based controlling techniques, non-linear systems can be controlled, and complex calculations for finding the optimal control parameters can be avoided. In a classical approach, the control parameters are determined based on the instantaneous values fed from the sensors and if the order of the system is high, then finding the appropriate PID values will be arduous, whereas, in any AI-based control technique, the control parameters can be made online for controlling.

REFERENCES

1. Zhao, S., Blaabjerg, F., and Wang, H. An overview of artificial intelligence applications for power electronics. *IEEE Transactions on Power Electronics*, vol. 36, no. 4, pp. 4633–4658, April 2021. doi:10.1109/TPEL.2020.3024914
2. Khandelwal, A. and Kumar, J. "Applications of AI for Power Electronics and Drives Systems: A Review." *2019 Innovations in Power and Advanced Computing Technologies (i-PACT)*, 2019, 1, 1–6.
3. Tran, D., Chakraborty, S., Lan, Y., Van Mierlo, J., and Hegazy, O. Optimized multiport DC/DC converter for vehicle drivetrains: Topology and design optimization. *Applied Sciences*, 2018, 8. doi:10.3390/app8081351
4. Li, W. and Ying, J. "Design and Analysis Artificial Intelligence (AI) Research for Power Supply—Power Electronics Expert System (PEES)," *2008 Twenty-Third Annual IEEE Applied Power Electronics Conference and Exposition*, 2008, pp. 2009–2015. doi:10.1109/APEC.2008.4523003
5. Mohagheghi, S., Harley, R. G., Habetler, T. G., and Divan, D. Condition monitoring of power electronic circuits using artificial neural networks. *IEEE Transactions on Power Electronics*, vol. 24, no. 10, Oct. 2009, pp. 2363–2367.
6. Sreekumar, C. and Agarwal, V. A hybrid control algorithm for voltage regulation in dc-dc boost converter. *IEEE Transactions on Industrial Electronics*, vol. 55, no. 6, pp. 2530–2538, Jun. 2008.
7. Bawane, N. and Kothari, A. G. "Artificial Neural Network Based Fault Identification of HVDC Converter." *4th IEEE International Symposium on Diagnostics for Electric Machines, Power Electronics and Drives, 2003. SDEMPED 2003*, 2003, pp. 152–157. doi:10.1109/DEMPED.2003.1234564
8. Bose, B. K. "Artificial Neural Network Applications in Power Electronics." *IECON'01. 27th Annual Conference of the IEEE Industrial Electronics Society (Cat. No.37243)*, 2001, 1631–1638, vol.3.
9. Bose, B. K. Expert system, fuzzy logic, and neural network applications in power electronics and motion control. *Proceedings of the IEEE*, vol. 82, no. 8, pp. 1303–1323, Aug. 1994. doi:10.1109/5.301690
10. Rathi, K.J. and Ali, M.S. "Design and Simulation of Fuzzy Logic Controller for Power Electronics Converter Circuits." *International Journal of Science Technology & Engineering*, vol. 2, no. 8, pp.353–356, 2016.
11. Harashima, F. et al. "Application of Neutral Networks to Power Converter Control." *Conference Record of the IEEE Industry Applications Society Annual Meeting*, 1989, 1086–1091, vol.1.
12. Dragičević, T., Wheeler, P., and Blaabjerg, F. Artificial intelligence aided automated design for reliability of power electronic systems. *IEEE Transactions on Power Electronics*, vol. 34, no. 8, pp. 7161–7171, Aug. 2019. doi:10.1109/TPEL.2018.2883947
13. Zhan, X., Wang, W., and Chung, H. A neural-network-based color control method for multi-color LED systems. *IEEE Transactions on Power Electronics*, vol. 34, no. 8, pp. 7900–7913, Aug. 2019.
14. Peyghami, S., Dragicevic, T., and Blaabjerg, F. Intelligent Long-Term Performance Analysis in Power Electronics Systems. Scientific Reports, 2021, 11. doi:10.1038/s41598-021-87165-3

15. Wang, W. et al. "Training Neural-Network-Based Controller on Distributed Machine Learning Platform for Power Electronics Systems." *2017 IEEE Energy Conversion Congress and Exposition (ECCE)*, 2017, pp. 3083–3089. doi:10.1109/ECCE.2017.8096563

16. Krishnamoorthy, H. S. and Narayanan Aayer, T. "Machine Learning based Modeling of Power Electronic Converters." *2019 IEEE Energy Conversion Congress and Exposition (ECCE)*, 2019, pp. 666–672. doi:10.1109/ECCE.2019.8912608

17. Meireles, M. R. G., Almeida, P. E. M., and Simoes, M. G. A comprehensive review for industrial applicability of artificial neural networks. *IEEE Transactions on Industrial Electronics*, vol. 50, no. 3, pp. 585–601, June 2003. doi:10.1109/TIE.2003.812470

18. Kim, Min-huei et al. "Neural Network Based Estimation of Power Electronic Waves." In *Proceedings of IECON '95-21st Annual Conference on IEEE Industrial Electronics* (vol. 1, pp. 353–358). IEEE, 1995.

19. Buhl, M. R. and Lorenz, R. D. "Design and Implementation of Neural Networks for Digital Current Regulation of Inverter Drives." *Conference Record of the 1991 IEEE Industry Applications Society Annual Meeting (1991)*, pp. 415–421, vol.1.

20. Lin, W. M. and Hong, C. M. A new Elman neural network-based control algorithm for adjustable-pitch variable-speed wind-energy conversion systems. *IEEE Transactions on Power Electronics*, vol. 26, no. 2, pp. 473–481, Feb. 2011.

21. Hari, N., Chatterjee, S., and Iyer, A. "Gallium Nitride Power Device Modeling Using Deep Feed Forward Neural Networks," *2018 1st Workshop on Wide Bandgap Power Devices and Applications in Asia (WiPDA Asia)*, pp. 164–168, 2018.

22. Maruta, H., Motomura, M. and Kurokawa, F. "A Novel Timing Control Method for Neural Network Based Digitally Controlled DC-DC Converter." In *Proc. IEEE ECCE Europe European Power Electronics Conference on Applications (EPE)*, CD-ROM 8 pages, Nov. 2013.

23. Bastos, J. L., Figueroa, H. P., and Monti, A. "FPGA Implementation of Neural Network-Based Controllers for Power Electronics Applications." *Twenty-First Annual IEEE Applied Power Electronics Conference and Exposition, 2006. APEC '06*, 2006, pp. 6. doi:10.1109/APEC.2006.1620729

24. Chan, H.-C., Chau, K. T., and Chan, C. C. "A Neural Network Controller for Switching Power Converters." In *Proceedings of IEEE Power Electronics Specialist Conference - PESC '93* (pp. 887–892), 1993. doi:10.1109/PESC.1993.472026

25. Leyva, R., Martinez-Salamero, L., Jammes, B., Marpinard, J. C., and Guinjoan, F. Identification and control of power converters by means of neural networks. *IEEE Transactions on Circuits and Systems I: Fundamental Theory and Applications*, vol. 44, no. 8, pp. 735–742, Aug. 1997, doi:10.1109/81.611270

26. Maruta, H. and Hoshino, D. "Transient Response Improvement of Repetitive-trained Neural Network Controlled DC-DC Converter with Overcompensation Suppression." *IECON 2019–45th Annual Conference of the IEEE Industrial Electronics Society*, 2019, pp. 2088–2093. doi:10.1109/IECON.2019.8927501

27. Yan, Cheng. "A Neural Network Identification Method for the Inverter." *2008 International Conference on Electrical Machines and Systems*, 2008, pp. 1810–1813.

28. Mohagheghi, S., Harley, R. G., Habetler, T. G., and Divan, D. "A Static Neural Network for Input-Output Mapping of Power Electronic Circuits." *2007 IEEE International Symposium on Diagnostics for Electric Machines, Power Electronics and Drives*, 2007, pp. 329–334. doi:10.1109/DEMPED.2007.4393116

29. Bose, B. K. Neural network applications in power electronics and motor drives—An introduction and perspective. *IEEE Transactions on Industrial Electronics*, vol. 54, no. 1, pp. 14–33, Feb. 2007, doi:10.1109/TIE.2006.888683

30. Manu, D., Shorabh, S. G., Swathika, O. G., Umashankar, S., and Tejaswi, P. Design and realization of smart energy management system for standalone PV system. In *IOP Conference Series: Earth and Environmental Science* (Vol. 1026, No. 1, p. 012027). IOP Publishing, 2022.

31. Swathika, O. G., Karthikeyan, K., Subramaniam, U., Hemapala, K. U., and Bhaskar, S. M. Energy efficient outdoor lighting system design: case study of IT campus. In *IOP Conference Series: Earth and Environmental Science* (Vol. 1026, No. 1, p. 012029). IOP Publishing, 2022.

32. Sujeeth, S. and Swathika, O. G. "IoT Based Automated Protection and Control of DC Microgrids." In *2018 2nd International Conference on Inventive Systems and Control (ICISC)* (pp. 1422–1426). IEEE, 2018.

33. Patel, A., Swathika, O.V., Subramaniam, U., Babu, T.S., Tripathi, A., Nag, S., Karthick, A., and Muhibbullah, M. A practical approach for predicting power in a small-scale off-grid photovoltaic system using machine learning algorithms. *International Journal of Photoenergy*, 2022, pp. 1–21.

34. Odiyur Vathanam, G. S., Kalyanasundaram, K., Elavarasan, R. M., Hussain Khahro, S., Subramaniam, U., Pugazhendhi, R., Ramesh, M., and Gopalakrishnan, R. M. A review on effective use of daylight harvesting using intelligent lighting control systems for sustainable office buildings in India. *Sustainability*, vol. 13, no. 9, p.4973, 2021.

35. Swathika, O. V. and Hemapala, K. T. M. U. IOT based energy management system for standalone PV systems. *Journal of Electrical Engineering & Technology*, vol. 14, no. 5, pp. 1811–1821, 2019.

36. Swathika, O. V. and Hemapala, K. T. M. U. "IOT-Based Adaptive Protection of Microgrid." In *International Conference on Artificial Intelligence, Smart Grid and Smart City Applications* (pp. 123–130). Springer, 2019.

37. Kumar, G. N. and Swathika, O. G. 19 AI Applications to. *Smart Buildings Digitalization: IoT and Energy Efficient Smart Buildings Architecture and Applications*, p.283, 2022.

38. Swathika, O. G. 5 IoT-Based Smart. *Smart Buildings Digitalization: IoT and Energy Efficient Smart Buildings Architecture and Applications*, p.57, 2022.

39. Lal, P., Ananthakrishnan, V., Swathika, O. G., Gutha, N. K., and Hency, V. B. 14 IoT-Based Smart Health. Smart Buildings Digitalization: Case Studies on Data Centers and Automation. 2022 Feb 23:149.

40. Chowdhury, S., Saha, K. D., Sarkar, C. M., and Swathika, O. G. "IoT-Based Data Collection Platform for Smart Buildings." In *Smart Buildings Digitalization* (pp. 71–79). CRC Press, 2022.

41. Seidl, D. R., Kaiser, D. A., and Lorenz, R. D. "One-Step Optimal Space Vector PWM Current Regulation Using a Neural Network." In *Proceedings of 1994 IEEE Industry Applications Society Annual Meeting* (vol. 2, pp. 867–874), 1994. doi:10.1109/IAS.1994.377520

42. Smioes, M. G. and Bose, B. K. "Applications of Fuzzy Logic in the Estimation of Power Electronic Waveforms." In *Conference Record of the 1993 IEEE Industry Applications Conference Twenty-Eighth IAS Annual Meeting* (1993) (vol. 2, pp. 853–861). IEEE.

43. Osorio, R. et al. Fuzzy logic control with an improved algorithm for integrated LED drivers. *IEEE Transactions on Industrial Electronics*, vol. 65, no. 9, pp. 6994–7003, Sept. 2018, doi:10.1109/TIE.2018.2795565

44. Wang, S. J. et al. "Applying Expert System and Fuzzy Logic to Power Converter Design." In *Proceedings of 1997 IEEE International Symposium on Circuits and Systems. Circuits and Systems in the Information Age ISCAS '97* (vol. 3, pp. 1732–1735). IEEE, 1997.

45. Wai, R. J. and Shih, L. C. Adaptive fuzzy-neural-network design for voltage tracking control of a DC-DC boost converter. *IEEE Transactions on Power Electronics*, vol. 27, pp. 2104–2115, 2012

46. Gupta, T., Boudreaux, R. R., Nelms, R. M., and Hung, J. Y. Implementation of a fuzzy controller for DC-DC converters using an inexpensive 8-b microcontroller. *IEEE Transactions on Industrial Electronics*, vol. 44, no. 5, pp. 661–668, Oct. 1997.

47. Simes, M. G. and Bose, B. K. Application of fuzzy neural networks in the estimation of distorted waveforms. *Rec. IEEE International Symposium on Industrial Electronics (ISIE)*, vol. 1, pp. 415–420, Jun. 1996.

48. Mitra, P., Dey, C., and Mudi, R.K. Fuzzy rule-based set point weighting for fuzzy PID controller. *SN Applied Sciences*, 3, 651, 2021. doi:10.1007/s42452-021-04626-0

49. Wai, R. -J. and Shih, L. -C. Adaptive fuzzy-neural-network design for voltage tracking control of a DC–DC boost converter. *IEEE Transactions on Power Electronics*, vol. 27, no. 4, pp. 2104–2115, April 2012. doi:10.1109/TPEL.2011.2169685

50. Sobanski, P. and Kaminski, M. Application of artificial neural networks for transistor open-circuit fault diagnosis in three-phase rectifiers. *IET Power Electronics*, 12, 2189–2200, 2019. doi:10.1049/iet-pel.2018.5330

51. Kim, W.-J. and Kim, S.-H. ANN design of multiple open-switch fault diagnosis for three-phase PWM converters. *IET Power Electronics*, 13, 4490–4497, 2020. doi:10.1049/iet-pel.2020.0795

52. Chen, W. and Bazzi, A. M. Logic-based methods for intelligent fault diagnosis and recovery in power electronics. *IEEE Transactions on Power Electronics*, vol. 32, no. 7, pp. 5573–5589, July 2017, doi:10.1109/TPEL.2016.2606435

53. Murphey, Y. L., Masrur, M. A., Chen, Z., and Zhang, B. Model-based fault diagnosis in electric drives using machine learning. *IEEE Transactions on Mechatronics*, vol. 11, no. 3, pp. 290–303, Jun. 2006.

54. Karmacharya, I. M. and Gokaraju, R. Fault location in ungrounded photovoltaic system using wavelets and ANN. *IEEE Transactions on Power Delivery*, vol. 33, no. 2, pp. 549–559, April 2018. doi:10.1109/TPWRD.2017.2721903

55. Lai, L. L., Ndeh-Che, F., Chari, T., Rajroop, P. J., and Chandrasekharaiah, H. S. "HVDC Systems Fault Diagnosis with Neural Networks." In *1993 Fifth European Conference on Power Electronics and Applications* (vol. 8, pp. 145–150). IEEE, 1993.

56. Narendra, K. G., Sood, V. K., Khorasani, K., and Patel, R. Application of a radial basis function (RBF) neural network for fault diagnosis in a HVDC system. *IEEE Transactions on Power Systems*, vol. 13, no. 1, pp. 177–183, Feb. 1998, doi:10.1109/59.651633

57. Khomfoi, S. and Tolbert, L. M. "Fault Diagnosis System for a Multilevel Inverter Using a Principal Component Neural Network." In *2006 37th IEEE Power Electronics Specialists Conference* (pp. 1–7), 2006. doi:10.1109/pesc.2006.1712246

58. Kastha, D. and Bose, B. K. Investigation of fault modes of voltage-fed inverter system for induction motor drive. *IEEE Transactions on Industry Applications*, vol. 30, no. 4, pp. 1028–1038, July–Aug. 1994. doi:10.1109/28.297920

59. Wei, C., Zhang, Z., Qiao, W., and Qu, L. Reinforcement-learning-based intelligent maximum power point tracking control for wind energy conversion systems. *IEEE Transactions on Industrial Electronics*, vol. 62, no. 10, pp. 6360–6370, Oct. 2015.

60. Wei, C., Zhang, Z., Qiao, W., and Qu, L. Y. An adaptive network-based reinforcement learning method for MPPT control of PMSG wind energy conversion systems. IEEE *Transactions on Power Electronics*, vol. 31, no. 11, pp. 7837–7848, Nov. 2016.

61. Seyedmahmoudian, M. et al., State of the art artificial intelligence-based MPPT techniques for mitigating partial shading effects on PV systems—A review. *Renewable and Sustainable Energy Reviews*, vol. 64, pp. 435–455, Oct. 2016.

62. Zhao, B., Zhang, X., and Huang, J. AI algorithm-based two-stage optimal design methodology of high-efficiency CLLC resonant converters for the hybrid AC–DC microgrid applications. *IEEE Transactions on Industrial Electronics*, vol. 66, no. 12, pp. 9756–9767, 2019. doi:10.1109/TIE.2019.2896235

63. Feng, G., Meyer, E., and Liu, Y. F. A new digital control algorithm to achieve optimal dynamic performance in dc-to-dc converters. *IEEE Transactions on Power Electronics*, vol. 22, no. 4, pp. 1489–1498, 2007.

64. Aguirre, L. New algorithm for closed-loop model matching. *Electronics Letters*, vol. 27, pp. 2260–2262, 1991. doi:10.1049/el:19911398

65. Jung, J. and Nam, K. A dynamic decoupling control scheme for high-speed operation of induction motors. *IEEE Transactions on Industrial Electronics*, vol. 46, no. 1, pp. 100–110, 1999. doi:10.1109/41.744397

66. Rivera, M., Morales, F., Tarisciotti, L., Zanchetta, P., Wheeler, P., Baier, C., and Munoz, J. "A Modulated Model Predictive Control Scheme for a Two-Level Voltage Source Inverter." In *Proceedings of the IEEE International Conference on Industrial Technology*. IEEE, 2015. doi:10.1109/ICIT.2015.7125425

67. Cortes, P., Kazmierkowski, M. P., Kennel, R. M., Quevedo, D. E. and Rodriguez, J. Predictive control in power electronics and drives. *IEEE Transactions on Industrial Electronics*, vol. 55, no. 12, pp. 4312–4324, 2008. doi:10.1109/TIE.2008.2007480

68. Liserre, M., Dell'Aquila, A., and Blaabjerg, F. Genetic algorithm-based design of the active damping for an LCL-filter three-phase active rectifier. *IEEE Transactions on Power Electronics*, vol. 19, no. 1, pp. 76–86, 2004. doi:10.1109/TPEL.2003.820540

69. Galotto, L., Pinto, J. O. P., Filho, J. A. B., and Lambert-Torres, G. "Recursive Least Square and Genetic Algorithm Based Tool for PID Controllers Tuning," *2007 International Conference on Intelligent Systems Applications to Power Systems*, 2007, pp. 1–6. doi:10.1109/ISAP.2007.4441623

70. Bousserhane, I. K., Hazzab, A., Rahli, M., Kamli, M., and Mazari, B. "Adaptive PI Controller Using Fuzzy System Optimized by Genetic Algorithm for Induction Motor Control," *2006 IEEE International Power Electronics Congress*, 2006, pp. 1–8. doi:10.1109/CIEP.2006.312162

71. Herrera, F., Lozano, M., and Verdegay, J. (1995). Tuning fuzzy logic controllers by genetic algorithms. *International Journal of Approximate Reasoning* 12, 299–315.

8 Analysis of Economic Growth Dependence on Energy Consumption

R. Rajapriya, Varun Gopalakrishnan,
and Milind Shrinivas Dangate
Vellore Institute of Technology

Nasrin I. Shaikh
Nowrosjee Wadia College

CONTENTS

ABSTRACT

The study divides economy of the state into categories based on which of the preceding hypotheses is expected to hold. This classification can be used to determine the expected growth consequences of energy policies. Stakeholders will benefit from greater knowledge of the growth implications of energy conservation. Policies targeted are more likely to be implemented if they aim to reduce energy usage when energy conservation is predicted to be growth neutral. The assessments are based on Granger's work. To group economies according to the energy output assumptions, much of the available work uses bi or trivariate analysis. The theoretical foundation of the work is an energy market state-specific partial equilibrium model, which adds to the current literature. Important transmission channels can be accounted for using a partial equilibrium technique. The role of supply and demand-side aspects on how energy use affects state economic growth is isolated. The second extension is in the area of the analytical level. Important sub-national distinctions may be overlooked in macroeconomic analyses. This could be one reason for the lack of consensus on causality directions in the literature. According to Metcalf, energy intensity is decreasing among states, but intensity variation is growing. At the state level, differences in intensity are becoming increasingly obvious. Diverse intensities mean that energy use will likely have a different impact on state growth. The study focuses on a group of states in the Western United States. Arizona was picked because of its low energy use per capita. In addition, from 1970 to 2007, the state saw remarkable economic growth. During this time, real output has roughly quadrupled. In 2007, it was ranked 12th in terms of energy intensity across all states. Total energy use is divided by the real gross state product to compute energy intensity.

8.1 INTRODUCTION

During 1970–2007, the economy of California in actual terms nearly doubled. California has the largest economy and the sixth highest energy intensity in 2007 (Energy Information Agency [EIA]). Simultaneously, energy-intensive businesses including chemical production, forest product development, and petroleum refining are critical to the state's economy (EIA). California is likewise a policy-driven state when it comes to energy management. Wyoming was picked because of its significant position in the energy economy. The Powder River Basin in Wyoming produces 40% a quarter of all coal mined in the United States (EIA). The state is also a significant natural gas exporter. Wyoming was ranked 48th in terms of energy intensity in 2007; as a result, it is one of the least energy-efficient states in the country (EIA). This position is most likely determined by industries that use a lot of energy, like fossil fuel extraction as well as processing. Due to the two factors, the study was not expanded to other states. For certain Western states, the preliminary study did not yield theoretically consistent supply and demand curves. Colorado, for example, was discovered to have a demand curve with a favourable slope. This is a common problem with energy-related state-specific estimations [1]. Although the immensity of spreading the use of the methodology to a growing number of states surrounded by single research would have resulted in the document being circulated more widely than the planned, if theoretically sound curves had been estimated, the paper would have become more widely distributed than the author intended. The author concentrated on a small number of unique states rather than a vast number of states.

8.2 LITERATURE REVIEW

To test for Granger causation, this research, like many others in the literature, uses bi- and trivariate estimations. When evaluating the demand side of a market, trivariate studies are frequently performed. In many cases, a demand curve is determined in a supply and demand system by data availability. If a market's demand side is the only thing considered, the model has a chance of being underspecified. Changes in quantity and price that occur as a result of supply-side shocks may be mistakenly recognized as adjustments in the demand curve. Shifts in the respective curves are regulated in a fully described supply and demand system. There are a few situations where estimating both curves isn't essential. Any short-run changes in the market or demand fluctuations lead to a fully elastic supply curve. Second, if the predicted demand curve takes into consideration the most important demand drivers and is completely described, there will be no omitted variable bias in the estimation findings. Market changes will be accurately attributed to the correct demand variables if a fully described curve is used. The geographic breadth of the research, the data's temporal span, and the energy sources, represented either by the data, will most likely determine if the projected demand curve is adequately described. Important transmission mechanisms, according to Peach [2] and Zachariadis [3], aren't taken into account in these kinds of standards. To remedy this problem, Zachariadis recommends using general equilibrium, partial equilibrium, or production function-based techniques. For many economies, Peach and others have approximated production functions. Peach's estimate is for the United States, and he confirms the growth idea. Finding the growth hypothesis is not guaranteed by estimating a production function. Research by Bartlett and Gounder [4] uses an approach for production functions to support the conservation hypothesis. It's worth noting that their demand-only specification backs up this conclusion. This form of analysis has only been attempted in a few publications in the literature. The question of whether a method that increases it is still debatable if the complexity of a production function is superior. Payne [5] showcases a study that has looked at the economics of a single state. Payne uses employment and energy consumption statistics to evaluate the direction of causality in Illinois, finding evidence for the growth hypothesis. Employment in the United States as a whole is used as a control variable as well. The author validated Payne's findings in preliminary research.

8.3 MATERIALS AND METHODS

A basic energy price index will also be employed. For Wyoming and California, the price of primary energy (Prim) is utilized, while for Arizona, the price of coal (Coal) is used. The data span the years 1970–2007. The data shown with notation is hidden, as it is with natural logs. The nominal figures (gross state product [GSP] and prices) were converted to 2,000 USD using the Bureau of Labor Statistics' Consumer Price Index (CPI). Gross State Product (GSP) is the equivalent of Gross Domestic Product (GDP) at the state level, and it is measured in millions of 2,000 dollars. The time series is affected by the switch from Standard Industrial Classification (SIC) to the North American Industrial Classification System (NAICS) system of industrial classification. A seamless time series is not produced as a result of the reclassification. From 1997 onwards, NAICS metrics of both GSP and Mfg are used. For reasons of secrecy, there is no data on manufacturing Wyoming employment in 2002. Producing is a modest part of Wyoming's economy. Manufacturing output accounted for an average of 5% of real GSP from 1997 to 2007, according to the author's calculations. As a surrogate for the missing observation, the average employment in the sector in 2001 and 2003 is utilized. EIA provides data about energy. Energy consumption is estimated as that of the annual use of primary energy inside a state and is quantified in billion British thermal units (BTUs). Primary energy consumption is defined by the EIA as the energy used directly from the source. Both a utility's and a household's usage of coal to generate electricity and natural gas for heating are examples. To avoid double-counting, this is done. Data on sales and distribution in each state are used to generate consumption measurements. Consumption figures include net electricity imports as well. Consumption should be equal to net production when using renewable energy sources like hydro or solar. The total quantity of electricity generated minus any power utilized in its production is characterized as renewable energy consumption. The average price of total energy is used to compute the price of energy. At a rate of 2,000 dollars per billion BTU, nominal data are converted into dollars per billion BTU in this study. Taxes are included in the pricing whenever possible. Municipal sales taxes are frequently not included, as are excise and per-gallon taxes. The prices used as inputs are estimates of the costs of energy generation. Prim is a metric for comparing fundamental energy sources like coal and natural gas. This could be owing to its prominence in the field of electricity; in 2008, coal generated 40% of the electricity consumed in the state. Prices are at state level; therefore, they do not reflect disparities between states, but rather overall patterns over time. The kinds and quantities of energy utilized within the state will affect the discrepancies between E_Price and Prim. A crucial distinction between the pricing variables will be the within the state, the price of motor fuel. The consumption of motor gasoline is a significant component of total energy consumption. Due to taxes, refining costs, and other factors, the cost will be different from the cost of petroleum. A second distinction between the variables is how buildings are heated. The discrepancy between the absolute values of the variables will grow as more heating demands are provided with electricity. If primary kinds of energy are used to meet heating needs, the price variables will be closer together. The price, as well as quantity generated by the interaction of energy supply and demand, can be utilized to deduce what is happening in the output energy connection. The market for energy, like the markets for capital and labor, can be regarded as a factor market. Consider the demand relationship in isolation for a moment. In all likelihood, this scenario will result in a decrease in output.

The following is a general description of a state's energy demand:

$$Q_D = \beta_0 + \beta_1 P_E + \beta'_X, \tag{8.1}$$

where income, climate urbanization of the state's population, and other factors are included in X, as well as the state's industrial structure, X indicates a vector of demand-side variables, P_E indicates the price of energy, and Q_D is the quantity of energy demanded.

$$Q_S = \alpha_0 + \alpha_1 P_E + \alpha' Y, \tag{8.2}$$

where Y is a vector of other influences, P_E indicates the price of energy, and Q_S is the quantity of energy supplied. Input costs, natural endowment, and technology would all be included in Y. Equations (8.1) and (8.2) are identical in equilibrium.

Between 1970 and 2007, both energy use and the real price of energy grew. This would be caused by changes in demand or supply and demand. State actual output has increased at the same time.

The theoretical link between energy and output is depicted by a conventional neoclassical production function. Take into account the following:

$$Y = f(A,K,L,E),$$ (8.3)

where E is the energy, L is the labor, K is the capital, and A is the technological progress. According to theory, energy has a declining marginal product, which means that as consumption rises, so does output, but at a slower rate. Particularly:

$$\frac{dY}{dE} > 0 \text{ and } \frac{d^2Y}{dE^2} < 0,$$ (8.4)

where E is the energy, and Y is the output. The goal of the research is to see if the first derivative is statistically important positive. This is accomplished in an understated rather than overt manner. Conventionally, the derivatives in (8.4) have been calculated using a production function. Granger causality tests are used similarly in the study. The first derivative being equal to zero could be an example Granger is not the source of energy that causes output.

8.4 METHODOLOGY

The application of co-integration techniques in the analysis is motivated by the calculation of supply and demand curves. For long periods, co-integrated variables do not diverge from each other, showing the existence of an equilibrating mechanism among them. This process could be triggered by a theoretical econometric relationship or the one proposed according to economic theory. Evidence of co-integration for non-stationary variables implies that the stochastic trends in the data are related. It states that "Equilibrium theories incorporating non-stationary variables necessitate the existence of a stationary combination of variables."

The preliminary co-integration test was created by Corcoran et.al. [6]. The Johansen approach, based on Johansen [7], eventually supplanted Engle and Granger's. The Johansen approach has been used in a variety of studies. After doing early data testing, Johansen's method requires estimating a vector auto-correction model (VECM). A vector auto-regression model (VAR) that contains any data co-integrating relationships is referred to as a VECM (also known as co-integrating equations). The long-run relationship between the variables is represented by the co-integrating equations. Sims [8] was the first to write about VAR. Sims argued that external and endogenous variables are separated in macroeconomic models were arbitrary at the time. All variables in a system can be viewed as endogenous using his VAR method. Many of the challenges linked with VAR's ambivalent acceptability are discussed by Keppler [9]. VARs, according to critics such as Cooley and LeRoy [10], have a place in analysis. Hypothesis-testing is the strength of this modelling approach. VECMs have the potential to be a useful statistical tool for detecting states based on energy increase theories because evaluating one requires estimating a VAR.

8.5 ESTIMATION

The VECM in general looks like this:

$$\Delta y_t = \alpha_i + \delta_j CE_{j.t-1} + \sum_{t=1}^{P} \Gamma_l A \Delta y_{t-1} + \in_{i,t},$$ (8.5)

where y_t is a vector of the variables of interest, Δ is the difference operator, α_i is a constant, CE_i is the co-integrating equation, δ_i is the speed of adjustment parameter, j will equal 1 or 2, and $\varepsilon_{i,t}$ is a vector of the independent disturbances. $\Gamma S \Delta y_{t-s}$ represents stationary variation in the variables. The speed of adjustment factors in the co-integrating equation defines how the dependent variable reacts to disequilibrium. The VECM is estimated in two different ways. $j=1$ is the only co-integrating equation in the first version. This illustration is based on the demand-only standard. This estimation results when the total supply and demand are compared (forecast). There are two co-integrating equations in the supply and demand specification, $j=2$. E Con, GSP, and Mfg are used to calculate the demand curve. E Con, the price of energy input, and Mfg were used to predict the supply curve. Prim (pricing of primary energy) is utilized in California and Wyoming. Mfg is expected in the demand curve function as a positive shifter, according to theory. Mfg isn't usually incorporated as a supply-side variable; its presence in the supply curve is just for the sake of solving the system of equations. There are several reasons why manufacturing could operate as a positive or negative supply curve shifter in this industry. A rise in manufacturing employment corresponds to a rise in employment in this industry. This sector's increase in employment comes on the backs of energy-related businesses; this could result in a decline in energy supply. An increase in manufacturing could boost the energy supply if it coincides with more productive energy-related businesses.

The demand curve based on demand-alone specification uses the general notation of equation (8.1).

$$E\_Con = \beta_0 - \beta_1 * E\_Price + \beta_2 * GSP \tag{8.6}$$

When the co-integrated equation (8.2) is estimated, it becomes:

$$CE : GSP_{t-1} - b_1 * E\_Con_{t-1} - b_2 * E\_Price_{t-1} + c = v_{t-1}, \tag{8.7}$$

where v is an independent disturbance and c is an intercept term. GSP is used to normalize the co-integrating vector. In the supply and demand framework, the same approach is applied.

The demand and supply curves are given by equations (8.1) and (8.2):

$$E\_Con = \beta_0 - \beta_1 * E\_Price + \beta_2 * GSP + \beta_3 * Mfg \tag{8.8}$$

$$E\_Con = \alpha_0 + \alpha_1 * E\_Price - \alpha_2 * IC + \alpha_3 * Mfg. \tag{8.9}$$

Because the quantity demanded, the quantity supplied, and prices for individual curves are not recorded. E Con and E Pricing are proxies for quantity and pricing in equilibrium. As a guide, use the VECM equations (8.4) and (8.5).

$$CE_1 : GSP_{t-1} - \beta_1 * E\_Price_{t-1} - \beta_2 * E\_Con_{t-1} - \beta_3 * Mfg_{t-1} + d = u_{t-1} \tag{8.10}$$

$$CE_2 : IC_{t-1} - \beta_4 * E\_Price_{t-1} - \beta_5 * E\_Con_{t-1} - \beta_6 * Mfg_{t-1} + h = v_{t-1}, \tag{8.11}$$

where h and d are the intercepts terms, and u and v are the independent disturbances. The co-integrating vector is made up of the beta coefficients. They are not the beta from the previous equations. Energy input costs are referred to as IC. The demand and supply curves are represented by co-integrating equations (8.1) and (8.2). The equations are standardized on GSP and IC in the estimation. For California and Wyoming, Prim will be used, but for Arizona, Coal will be used.

Many ways in which energy use affects growth are accounted for by the supply and demand system. The time series' energy output combinations reflect previous efficiency advancements. In this approach, the time series encapsulates capital efficiency benefits. As a result, the conservation theory overtakes the growth hypothesis as a more powerful economic classification. The impact of manufacturing on the state's energy supply and demand is reflected in employment figures. It also represents the industry's long-term employment trends. The manufacturing-to-output ratio is therefore taken into account. In contrast, manufacturing economies are predicted to utilize more energy per dollar of production than service economies. Granger causality tests determine inter-temporal causality directions. They're frequently utilized in predicting applications. As a forecasting technique, the tests take into consideration previous relationships to predict future events. The mechanism used creates an energy market at the state level explicitly. This hypothetical model is useful for recording changes at the state level, but it would be constrained by the geographic market it serves. Energy markets can be found on a wide range of geographical scales, from local to worldwide. Electricity prices, for example, are generally controlled at the state (or sub-state) level, whereas petroleum prices are determined on international markets. The theoretical model describes a regional market that is halfway between local as well as global markets. As a result, the model is incapable of clearly accounting for sub-state or global market shifts. This isn't to say that these modifications go unnoticed; rather, they have an unspoken impact on the data. The price variables in the model are the key mechanism through which the model is affected by changes in global energy markets. Changes in global energy demand affect energy's final price, as determined by E Price measures. The input price proxy captures (IC) changes in the price of primary fuels such as coal, natural gas, and petroleum. If the price of petroleum (or another significant energy) rose, state supply curves would shift to the left. For example, if China's demand for petroleum rises, the price of petroleum rises in response, resulting in a reduction in energy supply at the state level, ceteris paribus. Reduced natural gas prices, on the other hand, would result in an increase in energy supply due to higher extraction rates, especially in the United States.

8.6 RESULTS

The sequence of integration of the data must be specified before estimating. The augmented Dickey–Fuller (ADF) test is used to do this. The ADF takes the following form:

$$\Delta Y_t = \beta_1 + \beta_2 t + \delta Y_{t-1} + \alpha_i \sum_{i=2}^{m} \Delta Y_{t-i} A = \pi r^2 + \varepsilon_t, \tag{8.12}$$

where t denotes the lagged dependent variable's coefficient in a deterministic temporal trend, and Δ denotes a lagged variable; the coefficient of the lagged differenced dependent variable is denoted by I, while the random error term is denoted by t. The non-stationary of the series is the null hypothesis of the test. The author is unaware of any research that does not include economic and energy variables (1). Data at the state level should be no different.

The ADF was calculated using a drift and a temporal trend for GSP and energy usage. Price variables were estimated using a random walk and adrift. At the 5% level, the series is found to be I (1) in general, with both specifications confirming each other. The results are presented in Table 8.1. In Arizona and Wyoming, GSP is I (1) with drift, but the results are equivocal with drift and trend. Under both specifications, the pricing factors discovered are I (1). The Schwarz (SC) and Akaike information criterion (AIC) are used to calculate the appropriate lag length for the system of equations, respectively. Both assess a regression's overall fit while accounting for the number of observations. Each state has one or two delays, according to the SC and AIC. To provide sufficient interaction between the variables, two delays are required. The Johansen tests are performed once the data have been strong-minded to be I (1), as well as the total amount of system delays has been calculated. In the supply and demand framework, these tests are more conclusive than those conducted in response to the demand-only scaffold. Only one is in demand, according to the evidence. Arizona in co-integrating with California had performed a trace test at 10% threshold, whereas Wyoming had not yielded any test results. The trace and max tests reveal four co-integrating equations in the supply as well as demand framework for Arizona. Both tests reveal three California equations and two Wyoming equations. Every state will have two co-integrating equations due to the estimation of a

TABLE 8.1

Augmented Dickey-Fuller Tests

State	Variable	Intercept ADF Test Stat	Intercept P Value	Intercept & Trend ADF Test Stat	Intercept & Trend P Value	Random Walk ADF Test Stat	Random Walk P Value
AZ	GSP	0.2034	0.968	−5.1796	1.0008	−	−
	d(GDP)	−4.7383	0	−5.6533	1.0002	−	−
	E_Con	−3.0701	1.7174	−4.0445	1.1346	−	−
	d(E_Con)	−5.2958	1.0002	−6.227	1.0007	−	−
	E_Price	−2.4124	1.1457	−	−	1.0558	0.9207
	d(E_Price)	3.1784	1.0013	−	−	−4.019	0.0003
	Mfg	−3.3588	1.6103	−2.7012	1.7302	−	−
	d(Mfg)	−4.6805	1.0088	−6.4772	1.0003	−	−
	Coal	−3.1702	1.2202	−	−	0.2147	0.744
	d(Coal)	−5.4584	1.0013	−	−	−6.2387	0
WY	GSP	−2.5467	0.1135	−6.5114	0	−	−
	d(GDP)	−3.0712	0.0378	−	−	0.5216	0.8235
	d(E_Con)	−6.5686	0	−	−	−4.1118	0.0003
	E_Price	−1.6143	0.4647	−3.0525	0.1332	−	−
	d(E_Price)	−4.3594	1.0016	−5.9989	1.0002	−	−
	Mfg	−5.9757	0	−	−	−	−
	d(Mfg)	−5.9758	0	−	−	−	−
	Prim	−1.6189	0.462	−	−	0.5268	0.8248
	d(Prim)	−4.3225	0.0017	−	−	−4.0765	0.0003
CA	GSP	−1.4868	1.8822	−4.3833	1.0693	−	−
	d(GDP)	−5.3517	1.0016	−5.278	1.0093	−	−
	E_Con	−2.2296	1.6514	−3.96	1.1596	−	−
	d(E_Con)	−6.4259	1.0002	−6.3398	1.0007	−	−
	E_Price	−3.3175	1.1723	−	−	2.558	1.9685
	d(E_Price)	−5.2289	1.0022	−	−	−5.0674	1.0003
	Mfg	−2.9115	1.3236	−2.9594	1.603	−	−
	d(Mfg)	−3.0162	1.0036	−6.0959	1.0012	−	−
	Prim	−2.9595	1.3028	−	−	1.1242	1.9293
	d(Prim)	−5.5833	1.0009	−	−	−4.4715	0

Note: MacKinnon's *P*-values are reported in EViews (1996). The log of a variable is denoted by L_, while the first difference is denoted by d(L_). I failed to reject the null on levels and differenced data, indicating that I failed to reject the null on both (1).

supply and demand framework. Table 8.2 shows the GSP equations co-integrating vector estimated (estimated coefficients within the co-integrating equation). The supply and demand framework's results are the only ones offered. The co-integrating vectors describe supply and demand curves that are logically compatible. A few parameters in the co-integrating equations are not important. These are Mfg in the supply equation for California and Mfg in the demand equation for Wyoming.

In California, increases in manufacturing work as a positive demand shifter, but in Wyoming, they act as a positive supply shifter. Manufacturing harms demand and has a positive impact on supply in Arizona. This may be related to increasing manufacturing efficiency. (1) and (2) can be represented using the co-integrating vector (2). Table 8.3 shows these representations. The mean coefficients reflect elasticities, and the data are in natural logs. For Arizona, California, and Wyoming, –0.14, –0.16, and –0.23 are the price elasticities of demand, respectively. They are all inelastic; the price elasticity of supply varies greatly between the states, with 0.28, 2.3, and 0.54, respectively, being the most common (same order). The supply curve in California is the most elastic. Additional tests are required to guarantee that the estimates are accurate. The demand-side elasticities are calculated in using a panel data technique. The computed coefficients in Tables 8.4 and 8.5 describe whether disequilibrium within the co-integrating equation has a statistically significant impact on GSP. In response to positive disequilibrium, the signs in the co-integrating equation reflect whether GSP

TABLE 8.2
Estimated Co-integrating Vectors

GSP (– 1)	IC (– 1)	E_Price (– 1)	E_Con (– 1)		Mfg (–1)	
AZ	1	0	–0.196	***	–1.384	–0.151
			[–3.581]	[–21.462]	[–1.959]	
	0	1	0.702	***	2.543	3.152
			[–2.413]	[7.441]	[–7.697]	
CA	1	0	0.296	***	–1.799	0.531
			[–4.156]	[–13.702]	[3.719]	
	0	1	–1.338	***	0.582	0.078
			[–34.861]	[8.238]	[1.012]	
WY	1	0	–0.441	***	–1.92	0.133
			[–5.022]	[–8.329]	[0.477]	
	0	1	–1.434	***	2.639	1.621
			[–12.773]	[8.950]	[–4.552]	

Note: The approximated co-integrating vectors are presented in tables. The letters IC and C stand for input costs and constant terms, respectively. In California and Wyoming, IC represents the cost of primary energy, while in Arizona, it represents the cost of coal. The demand curve is denoted by CED, whereas the supply curve is denoted by CES. In brackets are T-statistics. The levels of significance are ***, **, and *, respectively, and are symbolized by the letters ***, **, and *.

TABLE 8.3
Estimated Supply and Demand Curves

AZ QD : E_Con = –0.143*E_Price + 0.724*GSP-0.108* Mfg
    QS : E_Con = 0.275*E_Price-0.394*IC + 1.23*Mfg
CA QD : E_Con = –0.275*E_Price + 0.555*GSP + 0.296*Mfg
    QS : E_Con = 2.298*E_Price-1.719*IC
WY QD : W_Con = –0.24*W_Price + 0.522*GSP
    QS : E_Con = 0.542*E_Price-0.378*IC + 0.613*Mfg

Note: Curves are constructed using the general form shown in equations 8.1 and 8.2. Coefficients are understood as elasticities because all data is in natural logs. When variables aren't significant, they're left out.

TABLE 8.4
Granger Causality Test, Demand Side Only

	Wald F-Tests & CE-T Test				
	d(E_Con)	d(E_Price)	CE	CE & d(E_Con)	CE & d(E_Price)
AZ					
GSP	2.404	2.235	−1.133 [−0.761]	2.853	2.248
CA					
GSP	0.142	1.025	0.172* [1.993]	5.543	5.603
WY					
GSP	0.226	1.357	−0.246 [−2.652]	8.19**	8,188**

Note: Granger causality to GSP is represented in the table. On the speed of adjustment parameter, T-tests are used.

The levels of significance are ***, **, and *, respectively, and are symbolized by the letters ***, **, and *.

TABLE 8.5
Granger Causality Tests Supply and Demand Specification

	Wald F-Tests & CE-T Test					
State	d(IC)	d(E_Price)	d(E_Con)	d(Mfg)	CED	CES
AZ						
GSP	2.473	1.654	1.796	6.796**	0.433* [−2.014]	−0.023 [−0.613]
CA						
GSP	3.973	5.923**	0.068	6.027*	0.188*** [−2.926]	−0.008 [−0.046]
WY						
GSP	1.506	1.716	0.496	0.236	0.292** [2.403]	0.277** [2.188]

Note: Granger causality to GSP is represented in tables. Input costs are referred to as IC. In California and Wyoming, it's the cost of primary energy, while in Arizona, it's the cost of coal. CED is the demand curve in the supply and demand specification, which is CE1 in the estimation. CE2 is the supply curve, and CES is the demand curve. On the speed of adjustment parameters, T-tests are used.

The levels of significance are ***, **, and *, respectively, and are symbolized by the letters ***, **, and *.

is increasing or decreasing. Consider positive disequilibrium in the demand curve to illustrate the co-integrating equation's disequilibrium process. GSP and perhaps Mfg could be excessively large in contrast to E Price and E Con, or E Price and E Con may be inadequate in comparison to GSP and/or Mfg, resulting in positive disequilibrium. The values that make up the stationary long-run series are either too large or too tiny. The effect of positive disequilibrium in the demand curve in Wyoming supply and demand specification is that GSP accelerates. Tables 8.5, 8.6, and 8.7 show the study's focus, Granger causality testing. The complete findings of the demand-only specification are shown in Table 8.5, while the short-run results are shown in Table 8.6. Table 8.7 depicts the

TABLE 8.6

Supply and Demand Interaction Terms, Granger Causality Tests

	Wald F-Tests & CE-T Test					
	CED & d (E_Price)	CED & d (E_Con)	CED & d (Mfg)	CES & d (IC)	CES & d (E_Price)	CES & d (E_Con)
AZ						
GSP	5.273	6.631	7.948*	3.762	2.162	3.07
CA						
GSP	9.363**	9.172**	8.504**	5.642	5.958*	2.323
WY						
GSP	7.123	7\8.158	6.966	9.179**	8.072*	9.578**

Note: The levels of significance are ***, **, and *, respectively, and are symbolized by the letters ***, **, and *.

TABLE 8.7

Diagnostic Test of GSP Solution

GSP - LHS

Demand Only

State	R2	adj.R2	F-Stat.	
AZ	1.385466	1.226157	3.4196	**
CA	1.476553	1.340845	4.51159	***
WY	1.506384	1.378407	4.9568	***

Supply & Demand

State	R2	adj.R2	F-Stat.	
AZ	1.566227	1.329607	3.39306	**
CA	1.590546	1.367207	3.64417	**
WY	1.560546	1.320843	3.3386	**

Note: The levels of significance are ***, **, and *, respectively, and are symbol-ized by the letters ***, **, and *.

supply and demand framework's interconnected terms. Short-run causation is represented by the different terms, but long-run causation is represented by the co-integrating equations. Oh and Lee [11, 12] dubbed tests on the interaction terms; Granger causality is strong because they capture. In his study of Asian developing economies, Asafu-Adjaye [13] used Granger's strong causality test. California and Wyoming are classified as such under the growth concept. In both conditions, the co-integrating equation is crucial. For Wyoming, the Granger tests of E Price and E Con, as well as the co-integrating equation, confirm this.

The supply and demand framework is used to evaluate the various supply and demand drivers. This enables the recording of specific curve impacts, as well as supply and demand components. Arizona is no longer neutral, in contrast to the demand-only definition. Granger causality from any of the factors to GSP is not observed, but it is discovered when Mfg interacts with the demand curve. This shows that the consumption of energy by manufacturing is driving growth. There is no indication of Granger or strong Granger causation from any of the supply curve's components to growth. The growth hypothesis is used to classify California; however, this is primarily due to the supply–demand connection. On the demand side, E Price, E Con, and Mfg all interact with the demand curve to lead GSP. GSP is caused by every single component of Granger's demand curve. In the short run, manufacturing outperforms GSP, but in the long run, to drive growth, it interacts with the demand curve. California's expansion has been aided by manufacturing's energy use.

Economic growth in the state will be aided by a decrease in E Price and increases in E Con. Within the GSP equation, on the supply curve for California, the speed of adjustment coefficient is irrelevant. Granger creates GSP when the supply curve interacts with the pricing. Furthermore, E Price outperforms GSP in short-run Granger tests. Thus, the cumulative impact on E Price, rather than the individual components of the supply curve, is driving the increase in California. The demand curve's findings back up this theory. Due to the demand relationship, supply shocks that raise E Price lead GSP to decline. Increases in supply, on the other hand, will cut prices and help the state's economy expand. In the GSP equation, Wyoming's supply and demand curves are critical. None of the variables resulted in growth in the short term. On the supply side, we can observe that when it comes to fuelling growth, consumption levels matter more than price. When price interacts with the demand curve, no growth occurs; however, when consumption interacts with demand, growth occurs. Wyoming's economy is classified as "very energy-intensive" by the EIA. In driving growth, energy use is more significant than energy pricing. In Arizona and California, production has a different impact on energy consumption than it does in other states. Mining contributed 20% of GSP on average from 1997 to 2007, while manufacturing contributed 5%. In Wyoming, supply-side factors such as supply interact with input costs, E Price, and E Con to drive growth. Important growth leaders that aren't covered in the demand-only specification are included in the supply-side connection. The growth hypothesis categorizes Wyoming. In terms of policy implications, a rising energy supply will result in increased state expansion. A decline in energy input pricing, as well as the supply shift that ensues, is one scenario that could occur. Diagnostic tests for the GSP equations are listed in Table 8.7 as two specifications. Two additional diagnostic tests are used in addition to the $R2$, corrected $R2$, and F-tests. The Jarque–Bera test is used to determine whether the residuals are normal. The normality of residuals is required for statistical tests to be correct. The multivariate version of this test will be displayed and described instead of the individual equation test because the solution to a system of equations has already been calculated. This test's null hypothesis is that mistakes are normally distributed. Table 8.8 displays the multivariate statistic. Small samples are compensated for utilizing Doornik and Hansen corrections in EViews (the tool used in the analysis) (1994). The test demonstrates a failure to reject the null hypothesis in all conditions.

8.7 POTENTIAL LIMITATIONS OF RESULTS

There is one exception to the estimations' results to be aware of. Market equilibrium, as well as input price proxies, has a 0.94 pair-wise connection in California. The pricing pair-wise correlations for each state are listed in Table 8.9. For Arizona and Wyoming, the association between pricing variables is not very strong. The presence of a high degree of the possibility of multi-co-linearity in estimations is increased by the correlation between right-hand side variables. Multi-co-linearity does not always imply a high pair-wise correlation between regresses. "Pair-wise correlations may be a sufficient but not required condition for the occurrence of multi-co-linearity," Gujarati (2003) writes. Parks [14] explains a few of the probable repercussions that can happen when multi-co-linearity is present. The impact on estimated coefficients is particularly significant to our study. Because variances and co-variances are greater, t ratios are statistically insignificant. This result is frequently accompanied by a high $R2$. Another possibility is that Ordinary Least Squares regression (OLS)-estimated coefficients are sensitive to modest changes in data (Gujarati, 2003, p. 350). The coal series was used to see if California's results held up when additional input price proxies were introduced. It's important emphasizing that coal only accounts for a minor portion of California's electricity. In 2008, coal accounted for about 0.7% of the state's overall energy consumption (author's calculation). In Arizona, the figure was at around 30%. The speed of adjustment parameter for both supply and demand is constrained when coal is used as an input price proxy in the GSP equation. Table 8.10 shows the results. Table 8.10 reveals that none of the energy factors have changed. Granger induces an increase in the state, according to Granger tests of short-run

TABLE 8.8
Jarque-Bera and ARCH Tests

Demand Only

Jarque-Bera	ACH(1)			
State	Stat.	P-Value	Stat.	P-Value
AZ	3.4118	1.8783	1.3547	1.9487
CA	3.7703	1.8372	5.5173	1.6873
WY	7.3578	1.3844	8.6696	1.3778

Supply & Demand Jarque-Bera

State	Stat.	P-Value	Stat.	P-Value
AZ	4.2238	1.8758	23.4662	1.6087
CA	6.9818	1.8169	23.657	1.5977
WY	12.0926	1.2186	30.3222	1.1787

Note: The multivariate Jarque-Bera test is 2(6) for the demand-only specification and 2(10) for the supply-plus demand specification, as shown in the table. ARCH(1) is 2(9) in the demand specification and 2(25) in the supply and demand specification. This statistic is for the third lag. A complete list of all estimated statistics can be found in the appendix.

TABLE 8.9
Pair-Wise Correlation between Price Variables AZ

AZ	NATGAS	PRIM	COAL	E_PRICE
NATGAS	1	0.0507	1.0659	1.0871
PRIM	1.0058	1	1.062	1.0730
COAL	1.0659	1.063	1	1.0593
E_PRICE	1.0871	1.0732	1.05932	1

CA	NATGAS	PRIM	COAL	E_PRICE
NATGAS	1	0.813	1.005	1.0914
PRIM	1.8119	1	1.0397	1.0943
COAL	1.1005	1.397	1	1.0275
E_PRICE	1.9148	1.945	1.0275	1

WY	NATGAS	PRIM	COAL	E_PRICE
NATGAS	1	1.092	−1.0117	1.0768
PRIM	91	1	1.0668	1.0661
COAL	−1.0117	1.0669	1	1.0402
E_PRICE	1.0768	1.0662	1.0402	1

causality. Furthermore, the adjusted R2 and F-statistic, 0.2284 and 1.8387, respectively, are relatively low. Table 8.11 shows the results. The model only explains a small portion of the state's economic growth. It effectively represents the relationship between energy use and economic growth in California, although Prim and E Price have a strong favour towards the variables in California's supply curve that have the predicted negative relationship. The calculations for California using Prim do not have a high R2, in the presence of multi-co-linearity; this is usual. In the short run, there is no Granger causality when coal is used.

TABLE 8.10

Alternative Input Proxy Price for California

	Granger Causality Tests Wald F-Tests & CE T-Test					
State	d(IC)	d(E_Price)	d(E_Con)	d(Mfg)	CED	CES CA
GSP	0.8358	2.218	0.6244	0.1647	0.0085	−0.0088
				[0.0645]	[−0.7381]	

Note: Coal is employed as the input price proxy. Granger causality to GSP is represented in the table. On the speed of adjustment parameter, T-tests are used.

TABLE 8.11

Diagnostic Tests Using the CA GSP Equation with an Alternative Input Price Proxy

GSP-LHS

Supply & Demand

State	R2	adj.R2	F-Stat.
CA	0.5008	0.2283	1.8398

Note: ***, **, and * represent the levels of significance of 1%, 5%, and 10%, respectively. The GSP equation is the subject of all of the tests offered.

The levels of significance are ***, **, and *, respectively, and are symbolized by the letters ***, **, and *.

8.8 CONCLUSION

This chapter has contributed to knowledge in two key ways. The two estimated specifications do not agree on how to classify economies based on alternative energy-growth theories. The demand-only specification, in particular, assigns the conservation hypothesis to Arizona's economy; consequently, energy conservation has no bearing on economic growth. The demand-only criteria were adopted to ensure that the results could be directly compared to the majority of current research. Arizona is characterized as a growth hypothesis in the supply and demand framework. This emphasizes the importance of using a more completely stated demand equation when studying the energy-growth link. Manufacturing is the primary consumer of energy in Arizona, and energy is the primary driver of growth. According to this analysis, the state's economic growth is unaffected by energy use in other sectors of the economy, such as transportation and the home. Consequently, efforts to cut energy use in these industries are unlikely to have a significant impact on growth. Arizona's economic classification also reveals that its economy is less vulnerable to energy shocks than that of California and Wyoming. This connection could be used to address a variety of Arizona's energy challenges. In contrast to Arizona, total energy use and manufacturing energy usage are both on the rise. The state's economy will be slowed as energy prices rise. Increases in energy supply could help California overcome growth limits caused by high energy demand. The state's economy has grown as its energy supply has increased. This is most likely due to an increase in the number of goods demanded as a result of the increased demand. Energy conservation efforts do not have to be limited to industry to encourage economic growth. Wyoming is unusual in that supply-side factors play a larger role in the growth process. This is likely due to the government's extensive involvement in the energy sector. This finding is most likely the result of market interaction and the importance of supply considerations, as demand-side factors drive growth via the demand curve. Increases in supply are expected to have an impact on the state's economic growth. Granger causes economic growth by lowering input costs, rising energy prices, or increasing the quantity supplied. Wyoming is well positioned to gain from rising non-renewable energy scarcity. Improvements in technology that boost supply or lower input prices will also lead. This research showed significant discrepancies between states. These distinctions point to crucial areas for further research. Energy consumption equations, for example, have poor $R2$ and F-stats, indicating that many crucial factors of energy usage have been overlooked. These topics will be the focus of future research. The policy implications of this study could become more targeted with a greater understanding of the sources of energy use. The study's findings also raise the question of whether different energy sources have varying effects on growth. The use of disaggregated energy metrics can reveal significant source disparities. If different fuels respond to expansion in different ways, policymakers will have more specialized instruments at their disposal.

REFERENCES

1. Maddala, G.S.; Trost, R.P.; Li, H.; Joutz, F. Estimation of Short-Run and Long-Run Elasticities of Energy Demand from Panel Data Using Shrinkage Estimators. *J Bus Econ. Stat.* **1997**, *15* (1), 90–100.
2. Peach, N.D. *Three essays on energy and economic growth* (Doctoral dissertation, Colorado State University), **2011**.
3. Zachariadis, T. Exploring the Relationship between Energy Use and Economic Growth with Bivariate Models: New Evidence from G-7 Countries. *Energy Econ.* **2007**, *29* (6), 1233–1253. https://doi.org/10.1016/j.eneco.2007.05.001
4. Bartleet, M.; Gounder, R. Energy Consumption and Economic Growth in New Zealand: Results of Trivariate and Multivariate Models. *Energy Policy* **2010**, *38* (7), 3508–3517. https://doi.org/10.1016/j.enpol.2010.02.025
5. Payne, J. E. On the Dynamics of Energy Consumption and Employment in Illinois. *J. Reg. Anal. Policy* **2009**, *39* (2), 126–130.
6. Corcoran, P. A.; Weiner, J.; Vitagliano, E. M.; Hoagland, E. M.; Holbrook, E. R.; Keller, J. R.; Moulton, H. E.; Diamond, L. K. Linkage Study of Gamma Globulin Groups. *Vox Sang.* **1966**, *11* (5), 620–622. https://doi.org/10.1111/j.1423-0410.1966.tb04259.x
7. Johansen, S. Statistical Analysis of Cointegration Vectors. *J. Econ. Dyn. Control* **1988**, *12* (2–3), 231–254. https://doi.org/10.1016/0165-1889(88)90041-3
8. Sims, C. A.; Kydland, F. Macro-Economics, and Reality. *Cent. Econ. Res. Discussion Pap. No, 77-91* **1977**, No. 77.
9. Keppler, J. H.; Bourbonnais, R.; Girod, J. The Econometrics of Energy Systems. *Econom. Energy Syst.* **2006**, 1–266. https://doi.org/10.1057/9780230626317
10. Cooley, T. F.; Leroy, S. F. Atheoretical Macroeconometrics: A Critique. *J. Monet. Econ.* **1985**, *16* (3), 283–308. https://doi.org/10.1016/0304-3932(85)90038-8
11. Oh, W.; Lee, K. Causal Relationship between Energy Consumption and GDP Revisited: The Case of Korea 1970–1999. *Energy Econ.* **2004**, *26* (1), 51–59. https://doi.org/10.1016/S0140-9883(03)00030-6
12. Oh, W.; Lee, K. Energy Consumption and Economic Growth in Korea: Testing the Causality Relation. *J. Policy Model.* **2004**, *26* (8–9), 973–981. https://doi.org/10.1016/j.jpolmod.2004.06.003
13. Asafu-Adjaye, J. The Relationship between Energy Consumption, Energy Prices, and Economic Growth: Time Series Evidence from Asian Developing Countries. *Energy Econ.* **2000**, *22* (6), 615–625. https://doi.org/10.1016/S0140-9883(00)00050-5
14. Parks, J. J.; Champagne, A. R.; Costi, T. A.; Shum, W. W.; Pasupathy, A. N.; Neuscamman, E.; Flores-Torres, S.; Cornaglia, P. S.; Aligia, A. A.; Balseiro, C. A.; Chan, G. K. L.; Abruña, H. D.; Ralph, D. C. Mechanical Control of Spin States in Spin-1 Molecules and the Underscreened Kondo Effect. *Science.* **2010**. https://doi.org/10.1126/science.1186874

9 Artificial Intelligence Techniques for Smart Power Systems

Sanjeevikumar Padmanaban
University of South-Eastern Norway

*Mostafa Azimi Nasab, Tina Samavat, Mohammad Zand,
and Morteza Azimi Nasab*
CTiF Global Capsule

CONTENTS

ABSTRACT

The soaring increase in the modern world's population has led to a growth in electricity demand. Also, due to the emergency of climate change and the ordinary power plant's destructive impact, replacing environmentally friendly sources with fossil fuel-based power plants is vital, leading to a decrease in power generation.

As a solution for overcoming mentioned shortage, significant changes must be applied to usual power systems and consumption patterns.

DOI: 10.1201/9781003331117-9

There are some practical choices for adding to the conventional power systems. Foremost, renewable sources must be added to the grid to raise the generated power. However, these sources are challenging to operate solely; for instance, photovoltaic systems are just available on sunny daytimes, or wind turbines are only working in the presence of an acceptable level of wind. Secondly, emergence of electric vehicles, charge stations, and their ability to connect the grid carry an opportunity to store electricity and take advantage of this enormous capacity during pick hours.

The solution does not conclude in the generator section. Advanced techniques of demand-side managing should also be involved. For instance, intelligent usage management can be regarded instead of generating more power to meet the pick time markets.

By adding mentioned parts to power systems, the conventional controlling sections must be adapted, as they can no longer manipulate the grid and guarantee its stability. In other words, a fundamental change in demand-side managing and dispatching is required to diminish the loss and supervise the consumption usage pattern during the pick hours, based on generated power and limitations of sources.

The smart power system is a satisfying solution for encountering mentioned concerns and has provided people with stable, secure, and clean power during the last decade. Internet of Things (IoT) is a fundamental part of a smart grid and provides an appropriate platform for all sections to store their records, which must be used for anticipating future conditions and following the change of consumption and generation patterns even in long periods like a decade. Forecasting can be labeled as the most critical phase of stabilizing smart grids. The reason is rooted in the management bases in the previous conditions. For instance, based on previous data, pick hours can be defined, so prevent it by cutting unnecessary consumption or using reserved capacities. Also, it is possible to use different monitoring equipment such as sensors and cameras to supervise the smart grid more accurately. Also, as mentioned earlier, IoT stores recorded data to predict future conditions; therefore, it can be considered a security matter in a grid.

Furthermore, conventional methods used to monitor, manage, and protect the power grid could not operate precisely, and an adaption is requested. Artificial intelligence techniques have expanded the system's ability to govern the network. Adding these practical tools, a stable and integrated network is formed that, by anticipating the future status of the network, can feed the consumer in an optimal condition. This chapter focuses on the critical position of smart grids and the impact of an integrated network on the energy industry. Further, how artificial intelligence and IoT increase smart grid efficiency and reliability (in mentioned zones such as stability analysis, demand-side managing, dispatching, and security) are discussed.

9.1 INTRODUCTION

In today's modern world, the old power system cannot meet the diverse needs of societies. The advent of high-speed computer systems, new high-precision sensors, artificial intelligence, and the Internet of Things (IoT) enabled experts in power systems to use their capabilities to reduce power generation costs in addition to reducing losses. In addition, an increasing understanding of the importance of renewable resources due to environmental consequences has led countries to look for ways to eliminate fossil fuels from their energy portfolios. For the above reasons, the issue of smart power grid has become a hot topic in the world.

It can be said that an intelligent network without artificial intelligence is useless because by examining each of its implementations, at least one of the artificial intelligence algorithms is responsible for a massive portion of work. In various studies, artificial intelligence is divided into four general categories [1, 2]:

- Artificial neural networks
- Reinforcement learning and deep reinforcement learning
- Metaheuristic algorithms
- Expert system.

For instance, reference [3] predicts smart grid stability and Support Vector Machines (SVMs), K-Nearest algorithms, Neighbor, Logistic Regression, Naive Bayes, Neural Networks, and Decision

Tree Classifiers are examined, and the best result is obtained with the Decision Tree classification algorithm. Each of the above is used in a specific sector, comprehensively referred to in reference [2]. This article also points to the evolution of energy management, which helps understand the impact of artificial intelligence. In reference [4], subjects such as Transmission Line, stability in Renewable Energy Systems such as wind turbines or photovoltaic systems, and fuzzy logic system and ANN algorithms are mentioned. Artificial intelligence is present in various parts of intelligent styles. A comprehensive article that addresses this issue is reference [5]. For example, the prediction of irradiation received in solar panels is indicated by an algorithm and a database based on the system's previous state. Also, many factors are involved in determining the size of the panel, which complicates this task, but artificial intelligence can calculate the scale of required panels for a defined project. Utility energy planning, control, and demand-side management are other topics addressed. The application of artificial intelligence in regulating the load of consumers has been considered by experts, and in reference [6], a model for this purpose to automate the scheduling of smart home appliances has been presented. This study shows that relying on a renewable source is not efficient enough, and it is better to use a hybrid manufacturer because it reduces the peak-to-average ratio (PAR) of overall energy demand.

This chapter discusses the role of artificial intelligence in some parts of an intelligent network. We will first look at the reasons for the importance of a smart grid in power systems. Then we examine one of the various categories of artificial intelligence to get a relative insight into the method of each operation. Finally, the role of artificial intelligence in solving the complexities and problems of these networks is discussed to clarify the importance of advancing algorithms and their undeniable role in providing power to a variety of consumers for the reader.

9.2 SMART POWER SYSTEM

The modern world has witnessed rapid changes in recent decades, in most of which the role of electricity is undeniable. These changes have caused traditional systems, both in the production and distribution sectors, to no longer be able to meet human needs, and we are witnessing a lot of losses and reduced efficiency. Smart grids have become a hot topic for developing power grids and their adaptation to new consumers and today's life. The smart grid gives the electricity industry a unique opportunity to enter a new level of reliability, accessibility, and efficiency that will contribute to protecting our environment and empower our economy. Telecommunication and digital technologies that enable two-way communication between power generators and customers through various lines make a network smart. A huge change in the power production sector has made expanding renewables resources in an intelligent system easier. Due to the depletion of fossil fuels and environmental damage, in order to survive, human beings must find a permanent replacement for it as soon as possible, which was not possible in traditional networks, but in smart networks, the use of green resources will be maximized. Also, the connection of all sections, close monitoring of all components, and the development of artificial intelligence algorithms have created a great change in the electricity industry to provide all types of consumers with secure and trustworthy power. Electric vehicles are also a new component of the network, which, in addition to evolving the transportation industry, has added energy storage capability, which until now was considered very expensive and unreasonable, to the conventional power supply systems. Figure 9.1 provides an overview of a conventional and smart power system. It clearly shows the differences in resources, power transmission, and the consumer [7, 8].

As mentioned earlier, although the basis of a smart grid is connections of all components and control section, this is known as one of the drawbacks that leads to weakness of networks and raises the issue of cyber-attacks and network hacking. So far, much research has been conducted, and some solutions have been provided to eliminate this weak point and meet the needs of today's global community with the least possible risk [9].

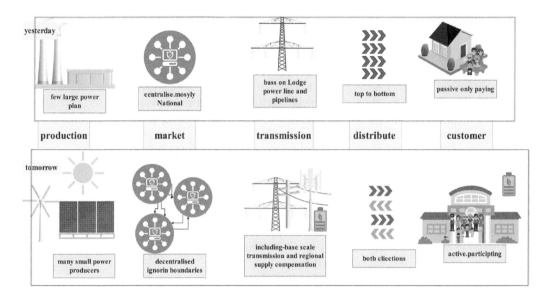

FIGURE 9.1 Conventional and smart power systems.

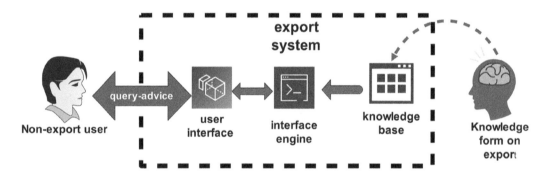

FIGURE 9.2 Expert systems.

9.3 ARTIFICIAL INTELLIGENCE

9.3.1 EXPERT SYSTEMS

One of the most successful forms of artificial intelligence is expert systems that simulate the behavior of an expert in the field used. These systems consist of two main parts, the database and the inference engine (Figure 9.2) [10].

9.3.2 DATABASE

The database is the part in which the various states of the system are defined, so that the system can have reasoning power and make decisions like an expert. This database is usually defined as a set of "if-then" rules designed in modern systems to experience and learn from decisions like a human. In this way, the system improves over time.

The most important thing to pay special attention to in an expert system is the database that is provided to the system. This importance is rooted in the strong dependence on the correctness of system decisions and provided information. In general, the database should be integrated, comprehensive, and accurate. If the database is not integrated and cannot cover all situations, the system has no reference in the face of undefined situations, which severely threatens the system's performance. The comprehensiveness of the database means that it is large enough for the system to be accurate enough and not too large to increase the complexity of the system and ultimately reduce the speed of decision-making. The last case is accuracy. The more precise the defined rules, the more accurate the system's decisions are.

9.3.3 INFERENCE ENGINE

The second part, the inference engine, is an automated reasoner. In this section, with the help of defined rules, the cognitive system derives from the current situation. This section can have the power of inference and reach a certain conclusion by examining several arguments. Neural networks are one of the most well-known categories that use this mechanism.

9.3.4 SUPERVISED LEARNING

The first step in a supervised learning algorithm is training, through which the algorithm looks for patterns in the data that relate to the desired outputs. After training, the algorithm will be given new inputs that have not been encountered and determine which new inputs will be classified based on previous training data (Figure 9.3) [11]. In this step, the accuracy of the training process is assessed. In its most basic form, a supervised learning algorithm can be written as follows:

$$Y = f(x).$$

Y is the output predicted by a function that assigns a class to an input value x. The function used to connect the input properties to the output predicted by the machine learning model is created during the training.

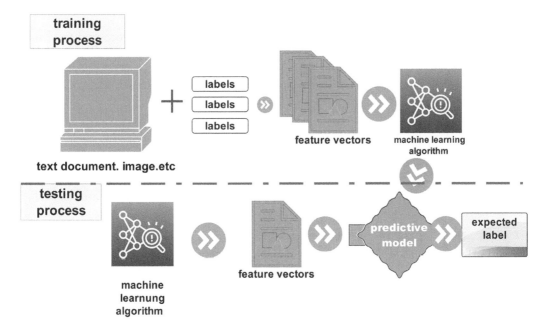

FIGURE 9.3 Supervised learning algorithms.

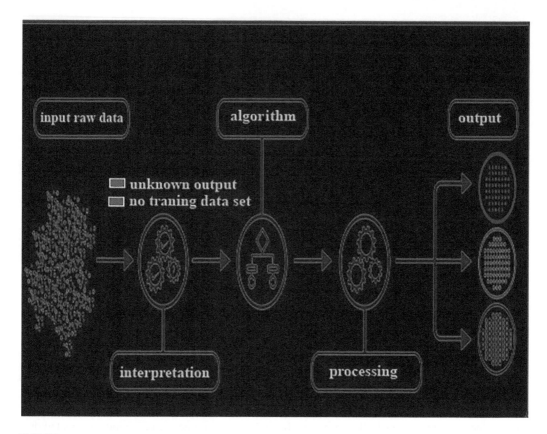

FIGURE 9.4 Unsupervised learning algorithm.

9.3.5 UNSUPERVISED LEARNING ALGORITHMS

Unsupervised learning algorithms allow users to perform more complex processing tasks than supervised learning. However, unsupervised learning can be unpredictable compared to other natural learning methods [12]. Unsupervised learning algorithms include clustering, continuity, dimensionality reduction, anomaly detection, and neural networks. Figure 9.4 shows the steps of this type. The reason for the popularity of this method can be considered the following:

- Unsupervised machine learning finds all sorts of unknown patterns in data.
- Unsupervised methods find features that can be useful for categorization.
- Unsupervised learning is done in real-time, so all input data must be analyzed and labeled in the presence of learners.
- It is easier to get unlabeled data from a computer than labeled data that requires manual intervention.

9.3.6 REINFORCEMENT LEARNING

Reinforcement Learning is currently one of the hot topics of research, and its popularity is increasing day by day. This method enables an "agent" to learn in an interactive environment using trial and error and using feedback from his or her actions and experiences. Although both supervised learning and reinforcement learning use mapping between input and output, reinforcement learning uses rewards and punishments as signals for positive and negative behavior. Reinforcement learning

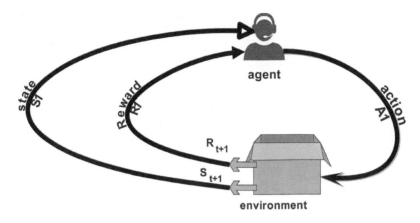

FIGURE 9.5   Reinforcement learning.

has different goals than unsupervised learning. While unsupervised learning aims to find similarities and differences between data points, in reinforcement learning, the goal is to find the right data model that maximizes the total cumulative reward for the agent (Figure 9.5) [13, 14].

9.4 ARTIFICIAL INTELLIGENCE IN SMART POWER SYSTEMS

9.4.1 SMART POWER SYSTEM

A smart grid is an automation system between production and consumption. This smart grid consists of an intelligent digital system, automation, computer, and control that establish two-way communication between power generation and consumption (Figure 9.6). An intelligent energy grid can be introduced as a new generation of power systems that have been introduced to eliminate the weaknesses of traditional systems. In these networks, energy sources have been changed mainly from fossil fuels to renewable energy, and to reduce the operation costs, extensive changes have been made in the consumption pattern. The most important feature of these systems is being smart. This means that the system with various artificial intelligence methods can collect the data needed to understand the system's status, and finally, orders are issued that the network continues operating in the most optimal condition. It should be noted that the optimal situation in power networks means the establishment of system stability or the balance of supply and demand in the electricity market at the lowest cost.

Intelligent energy grid innovation is a broad form of analog technology that paves the way to control the use of devices through two-way communication links. However, the prevalence of Internet access in most homes makes the smart grid more reliable for operations. Smart grid devices transmit information to enable ordinary users, operators, and automated devices to respond quickly to changes in the smart grid system [15].

9.4.2 FORECASTING

One of the most controversial issues in intelligent power systems is system prediction. System prediction is usually divided into long term, short term, and mid-term, leading to a piece of accurate and complete knowledge about the network. Acquiring such knowledge leads to proper management and making the right decisions in the shortest time. The reason for this is the importance of keeping these power networks stable. Smart grids are composed of various resources depending on their location, time, and consumers, each with maximum efficiency in specific situations. For example, in networks that use solar panels or wind turbines, the weather conditions must be considered

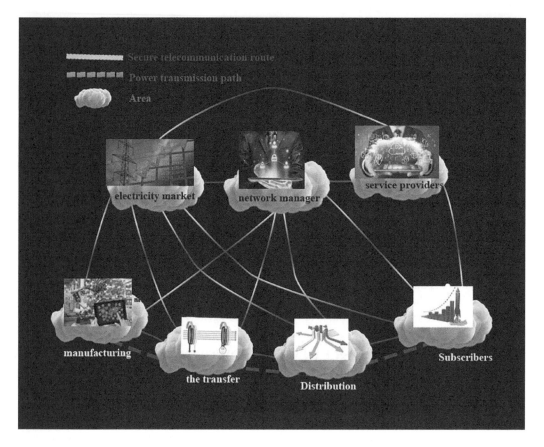

FIGURE 9.6   Smart power system layouts.

to balance the supply and demand of energy. This issue is also seen in the consumer sector. For example, power is devoted to industrial or domestic sectors, or which season is planned for because it is one of the important factors in consumption pattern. Also, electric vehicles and charging stations have become a hot topic in the prediction process and energy management because they have the potential to generate, consume, and store electricity.

Therefore, due to the high importance of prediction, many studies have conducted the best results using different artificial intelligence algorithms. So, the high ability of artificial intelligence is utilized to predict the future state of the network [16, 17].

9.4.3   NETWORK SECURITY

From the beginning of the emergence of power grids, the issue of grid security has been considered by experts because all the vital infrastructure of a country is dependent on sufficient electricity supply. Also, the energy storage presented in such centers can meet the needs of consumers only in the early hours of power outages. This issue in smart networks is more complex than in conventional power grids, because, firstly, the possibility of system hacking and cyber-attacks is higher due to the structure of smart networks. Secondly, due to the comprehensive connection of components, disruption in one part of the network can threaten the stability and security of the entire network. In this area, we see the presence of various artificial intelligence methods. To properly manage an intelligent network to achieve the highest efficiency and reduce losses, the connection between the power control network and the database is critical. The basis of an intelligent network is the communication of components with each other and with the control center, database, and telecommunication system to transfer data from sensors and transfer commands to operators. This connection is usually designed in layers and is connected to the Internet, which leads to the location of network hacking and other attacks. For example, smart grids use smart meters that connect production and consumption dynamically. Some smart meters are easily hacked and may control the power source of a building or neighbors. These attacks may occur in power plants and disrupt the power supply phase.

Regarding error detection by artificial intelligence algorithms, simple methods such as decision trees and simple feed-forward NNs or more complex methods train SVMs or semi-supervised networks to analyze data taken from voltage, current, sensors, power, and high-frequency filters that are used to detect abnormal conditions. However, in the case of cyber-attacks, it allowed the data falsified by the enemy, who could hack the network, to be provided to the control center and disrupt the system. It can also be solved with the help of artificial intelligence and data analysis. Furthermore, in case of any disturbance in the system, artificial intelligence decision algorithms can identify the exact location of the disturbance, determine the type of problem by examining the situation, and solve the problem by providing appropriate solutions and issuing appropriate commands [18].

9.4.4 Economic Dispatching

In a smart grid, the main goal is balancing the energy demanded and the amount produced. Like other markets, the price of goods, which is generated power here, is defined by the amount of supply and demand. Stability and equity of supply and demand are the determining parameters of a desirable network, and acceptable efficiency and economy are significant issues. So, experts try to manage the network with the help of artificial intelligence algorithms so that the price does not fluctuate much. This fluctuation, which occurs due to a lack of consumption management and is seen in nonstandard distribution networks, increases the pressure on the generator sector and secondary costs. In the subject of economic dispatch, manufacturers are carefully studied. The limitation is that a situation is defined for the generators in which the system is in optimal condition. This requires a meticulous telecommunications system to quickly provide information to the controller so that the best analysis and, therefore, the most appropriate commands for network management can be applied. For this purpose, various algorithms have been proposed to be presented with the help of economic network optimization to identify and achieve optimal conditions [19, 20].

9.4.5 Consumer and Resource

Smart grids are very flexible and are designed to operate optimally in any situation. Flexibility in smart grids is one of the things that makes the big difference between old and smart distribution systems. One of the characteristics that show this feature well is the definition of the role of producer and consumer. Some elements are defined so that by creating a bi-directional relationship, they help improve network performance to provide a trustworthy network in all respects to the consumer and reduce production costs.

In older systems, the consumer and the producer were usually fixed. However, in smart networks, we see the replacement of these two roles based on different situations prevailing in the network. In addition to the benefits of this new and practical feature, the system's increasing complexity is undeniable. Artificial intelligence algorithms helped solve this problem in such a situation, and amazing results have been achieved. The energy storage section is a good example of this feature. In this section, artificial intelligence, by examining the amount of demand and the status of resources, determines the status of the storage sector, which includes three general modes of consumer-producer and the status of the reservation. In reference [21], a method based on linear regression is presented to manage such components.

9.4.6 RESOURCES MANAGEMENT

Resources connected to power distribution networks and their management are critical. Figure 9.7 illustrates an example of the cooperation layout among the present resources, distribution network, and consumers. In smart grids, this layout is more important because the variety of available resources is significantly wide which increases the system's efficiency and complexity. Firstly, in such networks, power generation from renewable sources such as solar panels or wind turbines is maximized to reduce the need for conventional generators, usually based on fossil fuels. Secondly, the sources are connected to the network to keep the balance of the supply and demand and also dedicate the best possible periods to charge and discharge the storage units. Reserve resources are also important to the network due to emergency status to prevent any threat to the stability of the network. However, due to the unique characteristics of each resource, management is not an easy task and requires attention to various parameters. Also, the uncertainty in this part of the network is more than that in other parts [22, 23].

For example, the productivity of the most common renewable sources in today's systems is highly dependent on weather conditions. Therefore, artificial intelligence must have an accurate knowledge of the geographical location of the area under its control, which is usually done with classification algorithms. In addition, it must be able to predict the future status of each resource in which artificial intelligence prediction algorithms are used. Finally, after identifying the system, the need to decide on network-connected resources, connection duration, and reserved resources must be clarified. In this stage, artificial intelligence algorithms that perform practically in decision-making take all the vital factors into account and apply the best possible plan [24, 25].

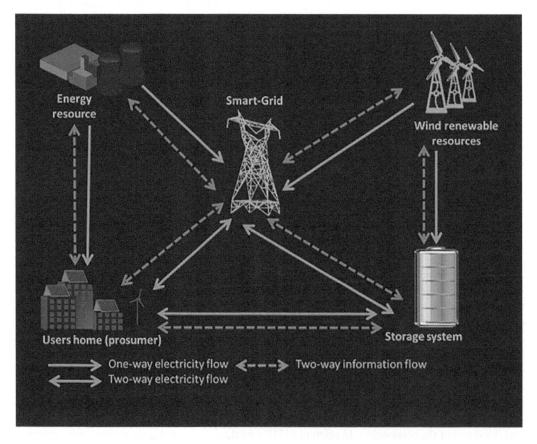

FIGURE 9.7   Resources, distribution network, and consumers' layout in smart power systems.

Therefore, artificial intelligence is very useful in resource management; for example, an artificial neural network is a method that researchers have studied. Different structures have been introduced and exploited, but the important issue is to have a reliable and accurate database that is highly dependent on parameters such as time and geographical location. The network must be trained with a reliable database for proper operation. Any defect in the management of this part of the system causes the management to jeopardize the electricity supply, upset the balance between supply and demand, and ultimately destroy the stability of the network [1, 26–28].

9.4.7 HOME ENERGY MANAGEMENT

In modern intelligent control networks, consumption management is not the responsibility of users alone, and the network is allowed to plan and change the schedule to increase efficiency and approach the ideal consumption pattern. Artificial intelligence algorithms make analyses and decisions related to this issue. In smart grids, the minor the difference between peak hours and PAR, the better the management of consumers. As a result, reducing the pressure on the power generating units can be achieved, and the cost of power generation significantly decreases. For example, in traditional networks, each user was responsible for managing consumption, and they were always asked to reduce consumption at certain periods during the day. Unfortunately, not everyone followed these instructions, which was an exterminating factor in conventional power systems. Nevertheless, in a smart grid, all appliances are known as a consumer in a network (smart appliances) are connected to the control unit, so during peak hours, the connection of unnecessary electrical appliances is disconnected, and their work schedules are moved to off-peak hours; therefore, the performance of the network enhance (Figure 9.8). To do so, accurate calculations consisting of considering various factors are vital, so these processes of planning, forecasting, and decision-making are completed by artificial intelligence algorithms such as Heuristic Algorithm, such as Earth Worm Algorithm, and Harmony Search Algorithms [6, 29, 30].

9.4.8 ENERGY STORAGE SYSTEM

One of the distinguishing features of smart grids is the energy storage sector, which is infrequently seen in time-honored networks. In the past, the use of energy storage devices was not economically justified due to the high cost of production and maintenance, and experts in the electricity industry tended to boost production capacity at particular intervals. Even this unit could not compete with traditional generators to provide energy for use in emergencies such as power outages. With the increasing development of manufacturing technology, smart grids, and artificial intelligence, the presence of these systems has become economically logical. In this section, the task of monitoring and planning the charging and discharging of batteries is the responsibility of artificial intelligence algorithms. By identifying and learning the complex relationship between consumer behavior, the status of the power generation sector, and the structure of the power storage system, this algorithm has been able to play a significant role in the eminence of the entire network, preventing severe peaks at certain intervals on daily bases and reducing network costs (Figure 9.9).

The general path of the artificial intelligence algorithms usage is the aim of optimizing the system in a way that electricity is purchased and stored by this section at intervals when consumption is low, and as a result, the price is lower, and stored power is injected into the network during peak hours (Figure 9.10). The network needs to have full control of supply and demand, changing trends during the year to have a short-term and long-term prediction to achieve the best performance. So far, artificial intelligence algorithms are the most reliable and accurate tools to forecast complex systems even with uncertainties. Therefore, in addition to the direct profit from the purchase and sale of energy, this activity reduces the damage to power generators and reduces energy production costs in long-term analyses [31–35].

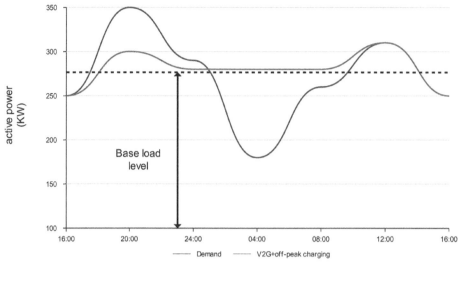

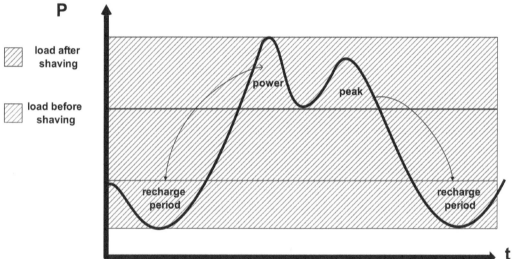

FIGURE 9.8 Pick shaving.

9.4.9 EV CHARGING STATION

This component of smart grids has attracted much attention in recent years. The reasons are, firstly, the expansion of the production of electric vehicles and, as a result, the need to create places to charge these vehicles, and secondly, the possibility of using these stations to balance the network. The key component is batteries, which can act as a consumer at lower demand intervals, such as midnight to morning, to charge; and inject power into the grid during peak hours. In many research studies, the behavior of stations is assumed to be identical to power generation in dams and plays an important role in peak hours. They are also referred to as a subset of the energy storage sector. However, a remarkable difference is the ability to generate energy, consequently reducing the costs of energy purchases related to the required amount for charging the batteries (Depending on the power generation capacity of the station and the pattern of vehicles present in the station). It should be noted that some stations do not have a production department, in which case the station

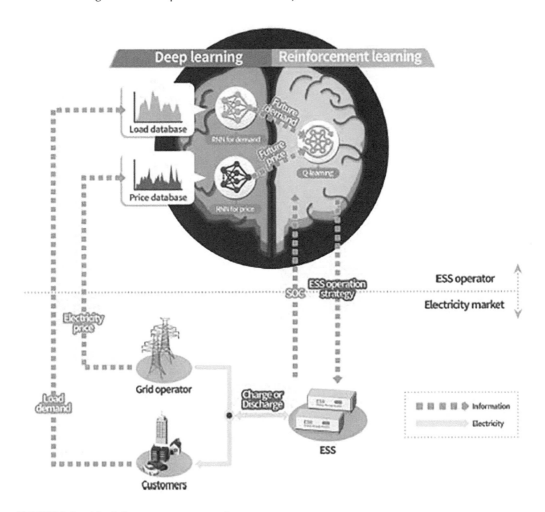

FIGURE 9.9 AI role in energy storage section.

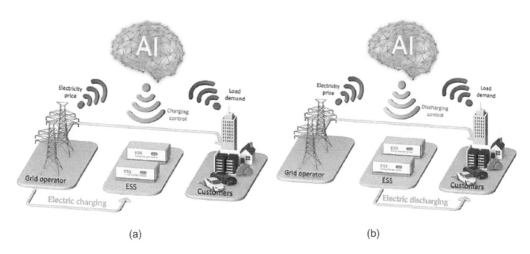

FIGURE 9.10 Energy storage connection during (a) charging mode and (b) discharging mode.

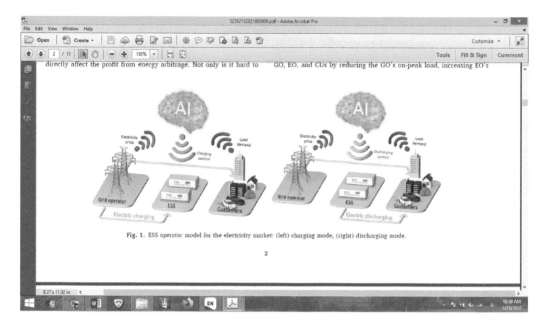

FIGURE 9.11   Smart power system with EV charging station.

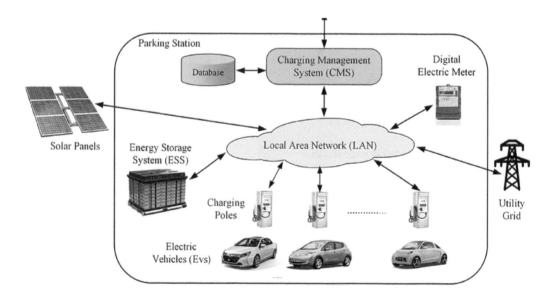

FIGURE 9.12   Smart power system with EV charging station.

can be considered an energy store where car batteries are assumed as storage devices. In addition to conventional consumers connected to the electricity distribution network, the owners of electric vehicles are known as Energy Buyers. Figures 9.11 and 9.12 show the network in which the EV charging station is located.

Managing the charging stations of electric vehicles confront many challenges, such as the time of entry and exit of vehicles, the charging capacity of the station, the type and life time of the battery. In such a complex system, various algorithms have been proposed to maximize stations' productivity and help the stability of the network. Algorithms are designed in such a way that by knowing the number of available vehicles and the duration of their presence in stations, the batteries are charged and discharged in such a way that, firstly, the cost of charging is minimized, and secondly, the station owner gains the highest profit from buying and selling energy. For example, the chicken swarm optimization algorithm has been used in Guwahati, India, and this two-way relationship has been managed in a new and practical way [36–42].

9.5 CONCLUSION

One of the problems that power systems face today is the issue of increasing demand, which has led to an increase in production capacity. Nevertheless, this method is no longer practical, because first, seeing the deteriorative impacts of fossil fuels and the risk of depletion of these reserves in the next few decades, we have to update the production sector with renewable resources. Also, with the advent of a new generation of technologies such as electric vehicles, smart sensors, and energy storage sections, it is possible to increase efficiency and reduce costs and losses. Therefore, smart networks must replace the old generation of power systems to use these facilities. It should be considered that these networks are much more complex, and traditional methods cannot monitor and manage the complex layout. Artificial intelligence algorithms have proven to be the best option for this. In this chapter, the main parts of smart grids are mentioned, and it is stated how artificial intelligence algorithms absolutely solve the specific problems of each part. Using artificial intelligence in smart power networks makes it possible to provide the required stable electricity with minimum resources, costs, and losses.

REFERENCES

1. Lin, Z. and Liu, X. Wind power forecasting of an offshore wind turbine based on high-frequency SCADA data and deep learning neural network. *Energy*, 2020, 201, 117693.
2. Zand, M., Nasab, M. A., Hatami, A., Kargar, M., and Chamorro, H. R. Using adaptive fuzzy logic for intelligent energy management in hybrid vehicles. In *2020 28th ICEE* (pp. 1–7). doi:10.1109/ICEE50131.2020.9260941.IEEE Index
3. Ahmadi-Nezamabad, H., Zand, M., Alizadeh, A., Vosoogh, M., and Nojavan, S. Multi-objective optimization based robust scheduling of electric vehicles aggregator. *Sustainable Cities and Society*, 2019, 47, 101494.
4. Zand, M., Nasab, M. A., Sanjeevikumar, P., Maroti, P. K., and Holm-Nielsen, J. B. Energy management strategy for solid-state transformer-based solar charging station for electric vehicles in smart grids. *IET Renewable Power Generation*, 2020, 14(18), 3843–3852.
5. B. W. Kennedy, "Integrating Wind Power: Transmission and Operational Impacts," *Refocus*, Vol. 5, 36–37, 2004.
6. Dashtaki, M. A., Nafisi, H., Pouresmaeil, E., and Khorsandi, A. Virtual inertia implementation in dual two-level voltage source inverters. In *2020 11th Power Electronics, Drive Systems, and Technologies Conference, PEDSTC 2020*. IEEE, 2020.
7. Bagdadee, A. H. and Zhang, L. A review of the smart grid concept for electrical power system. *Research Anthology on Smart Grid and Microgrid Development*, 2022, 1361–1385.
8. Ghosh, S., Das, J., and Chanda, C. K. Performance analysis of latency on wide area monitoring and control for a smart power grid. In *Advanced Energy and Control Systems* (pp. 113–122). Springer, 2022.
9. Ghasemi, M., Akbari, E., Zand, M., Hadipour, M., Ghavidel, S., and Li, L. An efficient modified HPSO-TVAC-based dynamic economic dispatch of generating units. *Electric Power Components and Systems*, 2019, 47(19–20), 1826–1840.
10. Zangeneh, M., Aghajari, E., and Forouzanfar, M. Implementation of smart fuzzy logic strategy to manage energy resources of a residential power system integrating solar energy and storage system using arduino boards. *Journal of Intelligent Procedures in Electrical Technology*, 2022, 13(49), 115–131.
11. Nasri, S., Nowdeh, S. A., Davoudkhani, I. F., Moghaddam, M. J. H., Kalam, A., Shahrokhi, S., and Zand, M. Maximum power point tracking of photovoltaic renewable energy system using a new method based on turbulent flow of water-based optimization (TFWO) under Partial shading conditions. In *Fundamentals and Innovations in Solar Energy* (pp. 285–310). Springer, 2021.

12. Liu, T., Yu, H., and Blair, R.H. Stability estimation for unsupervised clustering: A review. *Wiley Interdisciplinary Reviews. Computational Statistics*, 2022, e1575.

13. Ganesh, A.H. and Xu, B. A review of reinforcement learning based energy management systems for electrified powertrains: Progress, challenge, and potential solution. *Renewable and Sustainable Energy Reviews*, 2022, 154, 111833.

14. Rohani, A., Joorabian, M., Abasi, M., and Zand, M. Three-phase amplitude adaptive notch filter control design of DSTATCOM under unbalanced/distorted utility voltage conditions. *Journal of Intelligent & Fuzzy Systems*, 2019, 37(1), 847–865.

15. Zand, M., Nasab, M. A., Neghabi, O., Khalili, M., and Goli, A. Fault locating transmission lines with thyristor-controlled series capacitors by fuzzy logic method. In *2020 14th International Conference on Protection and Automation of Power Systems (IPAPS)*, Tehran, Iran (pp. 62–70). IEEE, 2019. doi: 10.1109/IPAPS49326.2019.9069389

16. Roy, S. and Sen, A. A self-updating K-contingency list for smart grid system. In *2021 IEEE 11th Annual Computing and Communication Workshop and Conference (CCWC)*. IEEE, 2021.

17. Dashtaki, M. A., Nafisi, H., Khorsandi, A., Hojabri, M., and Pouresmaeil, E. Dual two-level voltage source inverter virtual inertia emulation: A comparative study. *Energies*, 2021, 14(4), 1160.

18. Kovács, L.A.T., Cyberattack on the Smart Power Grid. Military National Security Service, p. 224.

19. Tightiz, L., Nasab, M. A., Yang, H., and Addeh, A. An intelligent system based on optimized ANFIS and association rules for power transformer fault diagnosis. *ISA Transactions*, 2020, 103, 63–74.

20. Zand, M., Neghabi, O., Nasab, M. A., Eskandari, M., and Abedini, M. A hybrid scheme for fault locating in transmission lines compensated by the TCSC. In *2020 15th International Conference on Protection and Automation of Power Systems (IPAPS)* (pp. 130–135). IEEE, 2020.

21. Ertuğrul, Ö.F., Tekin, H., and Tekin, R. A novel regression method in forecasting short-term grid electricity load in buildings that were connected to the smart grid. *Electrical Engineering*, 2021, 103(1), 717–728.

22. Zand, M., Azimi Nasab, M., Khoobani, M., Jahangiri, A., Hossein Hosseinian, S., and Hossein Kimiai, A. Robust speed control for induction motor drives using STSM control. In *2021 12th (PEDSTC)* (pp. 1–6). 2021. doi:10.1109/PEDSTC52094.2021.9405912

23. Sanjeevikumar, P., Zand, M., Nasab, M. A., Hanif, M. A., and Bhaskar, M. S. Spider community optimization algorithm to determine UPFC optimal size and location for improve dynamic stability. In *2021 IEEE 12th Energy Conversion Congress & Exposition - Asia (ECCE-Asia)* (pp. 2318–2323). IEEE, 2021. doi: 10.1109/ECCE-Asia49820.2021.9479149

24. Joos, M. and Staffell, I. Short-term integration costs of variable renewable energy: Wind curtailment and balancing in Britain and Germany. *Renewable and Sustainable Energy Reviews*, 2018, 86, 45–65.

25. Fast, M. and Palme, T. Application of artificial neural networks to the condition monitoring and diagnosis of a combined heat and power plant. *Energy*, 2010, 35(2), 1114–1120.

26. Jiang, D. et al. Importance of implementing smart renewable energy system using heuristic neural decision support system. *Sustainable Energy Technologies and Assessments*, 2021, 45, 101185.

27. Azimi Nasab, M., Zand, M., Eskandari, M., Sanjeevikumar, P., and Siano, P. Optimal planning of electrical appliance of residential units in a smart home network using cloud services. *Smart Cities*, 2021, 4, 1173–1195. doi:10.3390/smartcities4030063

28. Ramasamy, K. and Ravichandran, C. S. Optimal design of renewable sources of PV/wind/FC generation for power system reliability and cost using MA-RBFNN approach. *International Journal of Energy Research*, 2021, 45(7), 10946–10962.

29. Hassan, C.A.U. et al. Smart grid energy optimization and scheduling appliances priority for residential buildings through meta-heuristic hybrid approaches. *Energies*, 2022, 15(5), 1752.

30. Nasab, M. A., Zand, M., Padmanaban, S., Bhaskar, M. S., and Guerrero, J. M. An efficient, robust optimization model for the unit commitment considering renewable uncertainty and pumped-storage hydropower. *Computers and Electrical Engineering*, 2022, 100, 107846.

31. Azimi Nasab, M., Zand, M., Padmanaban, S., Dragicevic, T., and Khan, B. Simultaneous long-term planning of flexible electric vehicle photovoltaic charging stations in terms of load response and technical and economic indicators. *World Electric Vehicle Journal*, 2021, 190. doi:10.3390/wevj12040190

32. Zand, M., Nasab, M. A., Padmanaban, S., and Khoobani, M. Big data for SMART Sensor and intelligent electronic devices–building application. In *Smart Buildings Digitalization* (pp. 11–28). CRC Press, 2022.

33. Padmanaban, S., Khalili, M., Nasab, M. A., Zand, M., Shamim, A. G., and Khan, B. Determination of power transformers health index using parameters affecting the transformer's life. *IETE Journal of Research*, 2022, 1–22.

34. Ng, M.-F., et al., Predicting the state of charge and health of batteries using data-driven machine learning. *Nature Machine Intelligence*, 2020, 2(3), 161–170.

35. Sanjeevikumar, P., Samavat, T., Nasab, M. A., Zand, M., and Khoobani, M. Machine learning-based hybrid demand-side controller for renewable energy management. In Kumar, K., Shringar Rao, R., Kaiwartya, O., Kaiser, S., and Padmanaban, S. (eds) *Sustainable Developments by Artificial Intelligence and Machine Learning for Renewable Energies* (pp. 291–307). Elsevier, 2022.

36. Khalili, M., Ali Dashtaki, M., Nasab, M. A., Reza Hanif, H., Padmanaban, S., and Khan, B. Optimal instantaneous prediction of voltage instability due to transient faults in power networks taking into account the dynamic effect of generators. *Cogent Engineering*, 2022, 9(1), 2072568.

37. Gayathri, M. N. A smart bidirectional power interface between smart grid and electric vehicle. In Vinoth Kumar, B., Sivakumar, P., Rajan Singaravel, M. M., and Vijayakumar, K. (eds) *Intelligent Paradigms for Smart Grid and Renewable Energy Systems* (pp. 103–137). Springer, 2021.

38. Asadi, A. H. K., Jahangiri, A., Zand, M., Eskandari, M., Nasab, M. A., and Meyar-Naimi, H. Optimal design of high density HTS-SMES step-shaped cross-sectional solenoid to mechanical stress reduction. In *2022 International Conference on Protection and Automation of Power Systems (IPAPS)* (Vol. 16, pp. 1–6). IEEE, 2022.

39. Chen, A., Zhang, X., and Zhou, Z. Machine learning: accelerating materials development for energy storage and conversion. *InfoMat*, 2020, 2(3), 553–576.

40. Nasab, M. A., Zand, M., Hatami, A., Nikoukar, F., Padmanaban, S., and Kimiai, A. H. A hybrid scheme for fault locating for transmission lines with TCSC. In *2022 International Conference on Protection and Automation of Power Systems (IPAPS)* (Vol. 16, pp. 1–10). IEEE, 2022.

41. Dashtaki, A. A., Hakimi, S. M., Hasankhani, A., Derakhshani, G., and Abdi, B. Optimal management algorithm of microgrid connected to the distribution network considering renewable energy system uncertainties. *International Journal of Electrical Power & Energy Systems*, 2023, 145, 108633.

42. Borozan, S., Giannelos, S., and Strbac, G. Strategic network expansion planning with electric vehicle smart charging concepts as investment options. *Advances in Applied Energy*, 2022, 5, 100077.

10 IoT Contribution in Construct of Green Energy

Anjan Kumar Sahoo
College of Engineering Bhubaneswar

CONTENTS

ABSTRACT

The demand for electrical energy continues to rise, and more applications rely on it. As a result, it is a requirement that electrical energy be produced, distributed, and used as efficiently as feasible. Furthermore, recent problems with nuclear power facilities have prompted calls for other environmentally friendly energy generation solutions. Two key technologies will play important roles in solving parts of those future difficulties, among numerous possibilities. One option is to switch from traditional, fossil-based energy sources to green energy sources for electrical power generation. The other one is implementation of Internet of Things (IoT) in green energy sources. The use of renewable energy resources has been increased since 2015 United Nation Climate Change Conference. Analysis and optimal operating status maintenance using a remote monitoring system is needed in the case of photovoltaic (PV) power generation and unstable wind. The implementation of various open IoT platforms like Arduino, Raspberry Pi, etc. can definitely improve the performance of green energy. The popularity and usability of green energy can be improved by inclusion of IoT. The chapter represents the recent contributions of IoT in the green energy area. The researchers can get the technical idea to continue their research in the application of IoT.

10.1 INTRODUCTION

The main disadvantages of renewable energy sources are their fluctuating production due to unpredictable environmental behaviour and their difficulty in operating on a predetermined timetable. The power generation can be easily managed by the method of collection and then analysis. The status of power generation needs to be monitored continuously. The data obtained can be easily anticipated by ensuring optimal preservation of upcoming energy production. This greater stability also aids grid dependability and flexibility.

In this chapter, the author suggests implementation approaches for building an energy monitoring system that is cost-effective and based on open Internet of Things (IoT) software and hardware platforms. And LoRa, which supports long-distance networks with low power, is implemented using an inexpensive solution that does not require a telco's base station. Because of its simplicity, low development costs, and wide range of applications, the monitoring method can be implemented to the upcoming energy IoT system.

10.2 LORA AND IoT MONITORING SYSTEM

With chip manufacturers investing heavily in the market, the IoT industry is bringing a lot of technology and solutions to market, and it is developing at an exponential rate. It is not, however, without challenges. Ensure that the "things" or end nodes can connect to the internet is one among the most difficult components of building the IoT. Today's IoT devices communicate via a multitude of technologies, but none of them are suitable for their intended function and use. Wi-Fi is now ubiquitous, but it uses a lot of energy and sends a lot of data. While this is excellent, it isn't the best solution for IoT devices with limited energy or those only need to transfer small amounts of data [1, 2]. LoRa stands for Long Range Radio. It is a novel wireless protocol that is intended for long-range and limited communications. It is mostly used for M2M and IoT networks. Through a serial interface, the Arduino-based IoT node connects to the controller of an energy device. It collects the information about battery status, present current, voltage, and temperature. The LoRa modem is serially connected to Arduino as shown in Figure 10.1. Send Api Function of LoRa protocol stack is called by the embedded software. The sleep function at the power level is utilized by the IoT node to provide a low-power mode and is set to wake up on a regular basis. In accordance with regulatory specifications, some LoRa settings such as the maximum output power, channel centre frequency, coding rate, channel bandwidth, and spreading factor are chosen. The received energy data as well as automation, radio, and timestamp data are sent to the Raspberry Pi via serial interface by the LoRa modem. The data received via the LoRa modem is saved in MongoDB, a NoSQL database that supports a structure that is ideal for large-scale big data [4].

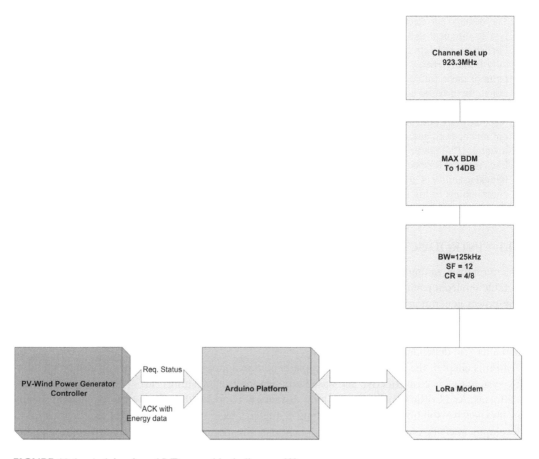

FIGURE 10.1 Arduino-based IoT sensor block diagram [3].

10.3   HYBRID MICROGRID WITH IoT

The hybrid smart grid provides new opportunities for solar-powered microgrids to be regulated and accessible via IoT. In Reference [3], Deepak Kumar and Bisht have presented the hybrid smart grid with IoT.

A microgrid system is a hybrid smart grid that has been improved to include solar power producers, hybrid ultra-capacitor (HUC), a centralized server that regulates the load based on the power that is used, home battery banks. The current central stations that operate electrical grids update their cycle at every 15-minute. The load is analysed and generators are changed at every 15 minutes, by the central station. The sources of green energy like wind turbines, solar panels, etc. do not have the same output that is predictable with traditional energy sources, rendering solar panels susceptible to surges. Surges induced by unexpected loads, such as household appliances, are not borne by solar panels and thus impossible to predict. If the power grid falls below a certain threshold power, the entire system may fail. Hence, as the green energy sources are changeable, they are incompatible with existing power grid systems.

We can use battery storage to minimize the system failures in unexpected outcome scenarios like the ones described above. This produces deep discharge during surges, reducing the life cycle of the battery. If the battery is connected to a HUC, the HUC will carry the power, protecting the battery from any surge. Because of its self-sustaining nature, the cost of constructing a microgrid could be covered in a couple of months. A user can also exchange leftover electricity from the previous day with the neighbours who are linked to the main grid. The information will be sent through IoT [5]. On a regular basis, data are synced from smaller hybrid microgrids' local cache to the main grid's database. The real-time information of every linked end points are obtained from analytics engine. The standby power is ensured by the operators during emergency.

Battery bank, solar panels, and a HUC comprise a hybrid microgrid system. The battery and HUC bank store solar energy. A hybrid microgrid technology keeps the battery charge at 100%. HUC bank is responsible for any load changes.

Renewable energy generation is crucial for guaranteeing the country's energy security and is also a viable strategy for combating changes in the global climate. It accelerates equipment development and enables more cost-effective expansion. Its growth will undoubtedly result in energy technology innovation and will aid in the establishment of new energy strategic sectors [6].

The authors discuss IoT sensors [7] that were charged using an energy-harvesting method, as well as popular products that were stored at the linked gateways. The proposed technique rewarded harvesting technology vendors financially while simultaneously enhancing quality of service (QoS) and system utility. To reduce on-grid energy use, an optimal solution [8] used green power and coupled caching and content forwarding.

The authors of Reference [9] provided an excellent examination of the microgrid management method using IoT. The microgrid management technique utilizes green energy while minimizing energy demand from grid. Green power sources, such as solar power from PV panels and wind power from aluminium air ventilators (wind turbine), can be stored in a battery and used to power loads more efficiently when operating off-grid. A clever energy-saving method has also been included, utilizing Passive Infrared Sensor and Light-Dependent Resistor to handle basic loads such as the different types of lights and fans. It includes an intelligent scheduling algorithm that uses concept of IoT to regulate a scalable load based on the consumer's needs. One is from microgrid and the other is from power grid. The basic loads can be efficiently powered by renewable energy, while the additional interruptible loads are directly connected to the grid. As a result, both energy use and utility bills are lowered. During certain seasons, such as rainy or overcast seasons, solar panels produce less energy, that is not sufficient to power the base type loads. In such instances, a relay is used to connect the microgrid's base loads to the grid supply. This process is entirely automated. The advantages of energy management become obvious when the conventional and fully automated modes are contrasted.

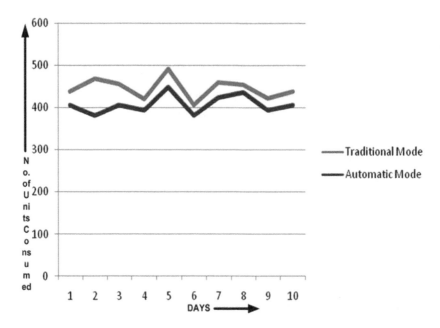

FIGURE 10.2 Performance comparison of different modes.

The conventional and fully automated modes are examined. The results are displayed in Figure 10.2 as a graph. The comparison of units consumed in watts for both modes in terms of loads is done. Because of the integration of renewable energy and scheduled loads via IoT, traditional modes consume significantly more units than autonomous modes. Renewable energy improves system efficiency by reducing the quantity of energy consumed.

10.4   HYBRID GREEN ENERGY HARVESTING USING IoT

In Reference [10], the authors built a smart hybrid green energy harvesting, storage system, and an IoT monitoring system. The energy storage device was used to save energy from water flow. It may be used for all the applications with low voltage also. The largest quantity of energy which this prototype may produce per day is around 100 Wh. The quantity of energy harvested and used can be monitored by using an IoT-based energy monitoring system. The component for capturing energy was tested in a variety of situations to discover the parameters that result in the greatest quantity of energy harvested.

Energy storage devices are frequently used to use energy harvested at a later time. Capacitors, supercapacitors, and batteries can all be used to store energy [11–14]. Energy storage devices are used in a wide range of applications, including renewable energy-harvesting systems [15–20]. When establishing energy storage systems for specific uses, however, various aspects must be considered. Expense, size, safety, reliability, life span, charging/discharging, and general administration are all factors to consider [21, 22].

10.5   CONCLUSION

This chapter defines a complete, self-sustaining, open-source green energy system. In this chapter, the author aimed to illustrate the role of IoT in the use of renewable energy by providing a few real-time studies in detail. Initially, we thoroughly investigated a few current survey studies on the usage of IoT in green energy development. Based on this information, a research gap was recognized and emphasized by the author on the impact of IoT on current technologies. There are three types of green technology strategies: resource management, resource allocation, and green energy utilization. Furthermore, the breakthroughs and uses of green technology in next-generation communications systems were explored in this chapter. Renewable energy technology solutions, on the other hand, confront additional challenges in balancing energy usage and performance of the network. These difficulties and upcoming research objectives were effectively recognized for the development of an ideal green technology strategy.

REFERENCES

1. Semtech Corporation, "SX1272/73-860 MHz to 1020 MHz Low Power Long Range Transceiver," www.semtech.com, March 2015.
2. LoRa Alliance Technical Committee, "LoRaWAN™ Regional Parameters," LoRa™ Alliance, July 2016.
3. Deepak Kumar Aagri and Altanai Bisht, "Export and Import of Renewable Energy by Hybrid MicroGrid via IoT," *IEEE*, 2018.
4. Hong Sun Hag and Cho Kyung Soon, "Full Stack Platform Design with MongoDB." *Journal of the Institute of Electronics and Information Engineers*, vol. 53, no. 12, pp. 152–158, 2016.
5. Qinghai Ou, "Application of Internet of Things in Smart Grid Power Transmission," *Third FTRA International Conference on Mobile, Ubiquitous, and Intelligent Computing (MUSIC)*, Sep 2012.
6. Bai Hong, Shi Wenhui, and Chen He, "Renewable Energy Technology Development Status and Future Key Technology Statement in China," *IEEE 3rd International Conference on Renewable Energy and Power Engineering*, 2020.
7. J. Yao and N. Ansari, "Caching in Energy Harvesting Aided Internet of Things: A Game-Theoretic Approach," *IEEE Internet Things Journal*, vol. 6, no. 2, pp. 3194–3201, 2019.
8. A. Khreishah, H. Bany Salameh, I. Khalil, and A. Gharaibeh, "Renewable Energy-Aware Joint Caching and Routing for Green Communication Networks," *IEEE Systems Journal*, vol. 12, no. 1, pp. 768–777, 2018.
9. C. Nayanatara, S. Divya, and E. K. Mahalakshmi, "Micro-Grid Management Strategy with the Integration of Renewable Energy Using IoT," *International Conference On Computation Of Power, Energy, Information and Communication*, 2018.
10. Hazlee Azil Illias, Nabilah Syuhada Ishak, Hazlie Mokhlis, and Md Zahir Hossain, "IoT-based Hybrid Renewable Energy Harvesting System from Water Flow," *IEEE International Conference on Power and Energy (PECon)*, 2020.
11. N. Yan, B. Zhang, W. Li, and S. Ma, "Hybrid Energy Storage Capacity Allocation Method for Active Distribution Network Considering Demand Side Response," *IEEE Transactions on Applied Superconductivity*, vol. 29, pp. 1–4, 2019.
12. Y. Xu, C. Li, Z. Wang, N. Zhang, and B. Peng, "Load Frequency Control of a Novel Renewable Energy Integrated Micro-Grid Containing Pumped Hydropower Energy Storage," *IEEE Access*, vol. 6, pp. 29067–29077, 2018.
13. J. D. Watson, N. R. Watson, and I. Lestas, "Optimized Dispatch of Energy Storage Systems in Unbalanced Distribution Networks," *IEEE Transactions on Sustainable Energy*, vol. 9, pp. 639–650, 2018.
14. J. Fang, Y. Tang, H. Li, and X. Li, "A Battery/Ultracapacitor Hybrid Energy Storage System for Implementing the Power Management of Virtual Synchronous Generators," *IEEE Transactions on Power Electronics*, vol. 33, pp. 2820–2824, 2018.
15. G. Giaconi, D. Gündüz, and H. V. Poor, "Smart Meter Privacy with Renewable Energy and an Energy Storage Device," *IEEE Transactions on Information Forensics and Security*, vol. 13, pp. 129–142, 2018.
16. Y. Teng, Z. Wang, Y. Li, Q. Ma, Q. Hui, and S. Li, "Multi-Energy Storage System Model Based on Electricity Heat and Hydrogen Coordinated Optimization for Power Grid Flexibility," *CSEE Journal of Power and Energy Systems*, vol. 5, pp. 266–274, 2019.
17. N. Nagill, S. Reddy K., R. Kumar, and B. K. Panigrahi, "Feasibility Analysis of Heterogeneous Energy Storage Technology for Cloud Energy Storage with Distributed Generation," *The Journal of Engineering,* vol. 2019, pp. 4970–4974, 2019.
18. C. Lin, D. Deng, C. Kuo, and Y. Liang, "Optimal Charging Control of Energy Storage and Electric Vehicle of an Individual in the Internet of Energy with Energy Trading," *IEEE Transactions on Industrial Informatics*, vol. 14, pp. 2570–2578, 2018.

19. H. F. Habib, C. R. Lashway, and O. A. Mohammed, "A Review of Communication Failure Impacts on Adaptive Microgrid Protection Schemes and the Use of Energy Storage as a Contingency," *IEEE Transactions on Industry Applications*, vol. 54, pp. 1194–1207, 2018.
20. H. Bitaraf and S. Rahman, "Reducing Curtailed Wind Energy Through Energy Storage and Demand Response," *IEEE Transactions on Sustainable Energy*, vol. 9, pp. 228–236, 2018.
21. R. H. Byrne, T. A. Nguyen, D. A. Copp, B. R. Chalamala, and I. Gyuk, "Energy Management and Optimization Methods for Grid Energy Storage Systems," *IEEE Access*, vol. 6, pp. 13231–13260, 2018.
22. J. Wu, X. Xing, X. Liu, J. M. Guerrero, and Z. Chen, "Energy Management Strategy for Grid-Tied Microgrids Considering the Energy Storage Efficiency," *IEEE Transactions on Industrial Electronics,* vol. 65, pp. 9539–9549, 2018.

11 Smart IoT System-Based Performance Improvement of DC Power Distribution within Commercial Buildings Using Adaptive Nonlinear Ascendant Mode Control Strategy

P. Ramesh
University College of Engineering

CONTENTS

ABSTRACT

The energy demand based on the increasing population and industrialization has incited the prerequisite for generating electrical energy from renewable sources. The inclination for this research is more grounded because of the usually melting away trademark resources and the overall common issues realized by fossil fuels. Renewable energy sources are perceived as one of the best answers for the developing energy emergency. Of these, photovoltaic sources have been the most encouraging future energy source because of their advantages, such as pollution-free, enduring, cost-sparing, low maintenance, and noise-free, environment-friendly properties. The solar source will become more successful if integrated into the existing power distribution infrastructure, which necessitates efficient Direct Current (DC)-Alternating Current (AC) conversion. Day by day, on the other side, many of the electrical loads are fitting to DC input, especially in the consumer products such as *Compact Fluorescent* Lamps and Light-Emitting Diode lights, television sets, fans, refrigerators, air conditioners, laptops, personal computers, and other electronics and communication gadgets used in workplaces and homes. Hence, there is a transition towards DC distribution inside buildings to improve the performance of the controllers used in between the Electrical Energy Sources and the Electrical and Electronic Loads. Examples of Commercial Buildings include Banks, Software, Communication, and Networking companies and Malls. Most of them use the above DC loads. This work mainly aims at improving the performance of controllers used in between AC/DC energy sources and the DC loads. Adaptive Nonlinear Ascendant Mode (ANAM) control strategy is proposed for this work. ANAM control strategy is developed, which takes care of the parameters such as power flow reliability, system efficiency, power-voltage curve, steady-state and transient response, overshoot control, and Total Harmonic Distortion. An experimental setup representing the Simulink model has been designed for a commercial building configuration. The proposed controller is applied for both AC/DC and DC/DC converters and the performance of the proposed controller is validated through simulation by checking the above parameters. The simulation and the experimental results confirm that the recommended configuration sources are more reliable and efficient than the existing Proportional Integrated Derivative and sliding mode control methods within both steady-state and transient performance.

11.1 INTRODUCTION: BACKGROUND AND DRIVING FORCES

The performance of the existing power system is increasing day by day with the integration of Renewable Energy Source (RES) which ensures better performance and efficiency. In recent days, solar, wind turbine, and fuel cells are used for RES. The Direct Current (DC) in the building power distribution framework has attributed to many circumstances, like the quick development of photovoltaic (PV) framework establishment [1] and the emergence of batteries in the building field [2]. Eventually, the expanding utilization of loads in the market with a DC source on a variable frequency drive motor (variable frequency drive) and Light-Emitting Diode lighting systems are increased [3]. From PV to direct power DC equipment, the flexibility of power supply in the building's power distribution and framework [4], from the PV back to Alternating Current (AC), can lessen the energy transfer misfortunes from the DC.

TABLE 11.1
Different Types of DC Loads in Commercial Buildings

S. No.	Different Types of DC Loads
1	Lighting system
	(Volt 12V with 3W, 5W, 7W, 9W, 15W, 40W, 100W, 500W and 1000W)
2	Television (12V with 10W [24 inch], 20W [32 inch] and 50W [50 inch])
3	Video audio monitors (12V with 40W, 100W and 250W)
4	Computer (48V with 200W)
5	Elevator (75V with 2000W)
6	Automatic gate opening systems (12V with 100W)
7	Mixer grinder (12V with 600W)
8	Air cooler (12V with 36W, 48W, 75W, 125W and 200W)
9	Washer and dryers (24V with 750W, 1000W and 1500W)
10	Water heater (24V with 300W, 400W and 500W)
11	Water purifier (36V with 20W, 50W and 100W)
12	Ceiling fans (12V with 35W)

Nowadays, commercial buildings consume 61% of the total electrical energy of the nation [5], and the essential loads happen to be from lighting frameworks. Table 11.1 demonstrates various sorts of DC loads which are utilized in commercial buildings. In any case, the current in commercial building power distribution framework depends on AC and DC energy from RES. RES is expected to change over from DC to AC and AC to DC, to be appropriated to the load [6, 7]. If DC power is taken to the DC bus directly, the power conversion productivity can be fundamentally improved. Along these lines, it is the need of great importance to building up a DC distribution framework to adjust to DC renewable energy and DC loads [8].

The most usually utilized control techniques for AC to DC and DC to DC converters are pulse width modulation (PWM) current and voltage mode control methods, the Proportional Integrated (PI) and the Proportional Integrated Derivative (PID)-based Pulse-Width Modulation strategies. For example, these traditional control strategies, P, PI, and PID don't perform acceptably under huge load and power system parameter differences. The existing Sliding Mode Method [33] performs effectively under load variation conditions but doesn't represent transient conditions. Therefore, in this work, an Adaptive Nonlinear Ascendant Mode (ANAM) control method is introduced to accommodate transient, steady-state conditions and verify performance parameters like Total Harmonic Distortion (THD), peak time and peak overshoot time.

11.2   RESEARCH BACKGROUND

PV power generation has been a practical option compared to conventional fossil-fuel power sources [9], which are mostly renewable and induces greenhouse gas emissions. PV power generation is introduced in urban zones to diminish energy transmission venture and lower energy misfortunes [10, 11]. The PV boards are generally associated with the Maximum Power Point Tracking (MPPT) controller, which builds up the energy output. A few techniques have been proposed all through this MPPT controller [12–14], and the Perturbation and Observation strategy is famously utilized [15].

There are several technical issues involved in integrating a conventional electric grid with decentralized DC power generation [16, 17]. Any PV generators will include the need for large amounts of auxiliary services, which are relatively expensive [18, 19] and have intermediate nature. Microgrid technology is implemented to increase renewable energy use, especially in urban applications [20, 21]. A microgrid is a collection of energy sources (both conventional and renewable), energy storage systems and power loads, which can be found in a unit with utility grids. PV power generation can be streamlined by energy storage and ready to be consumed by local loads.

Thus, the economic advantages of PV power irregular effect on the utility grid are altogether decreased [22–24]. Depending upon the widespread bus character, microgrids can be named AC microgrids, DC microgrids and hybrid AC-to-DC microgrids. Of late, analysts have progressively found that DC microgrids have higher potential because of their similarity with DC hardware, for example, PV generators, battery storage and upcoming DC loads [25–27]. Previous researchers proposed a few improvement techniques for the energy management of commercial buildings [28]. They are frequently found on examining load varieties for an indicated period on the client-side, analysing and inferring the example of behaviour to structure the productive energy management framework [29, 30].

This has been tended to by the mix of nearby RES and storage gadgets [31]. Different optimization models for coordinating interest-side administration calculations into the connection between smart building and utility grid have been discussed [32]. The DC distribution voltage is used in commercial buildings [33] using the sliding mode control (SMC) method. However, there is a downside to switching power in the event of switching when the SMC power quality is substandard. In Reference [34], the authors introduce a fuzzy logic method to control the application of the Building Power Management System.

11.3 MATERIALS AND METHODS

Figure 11.1 shows the proposed topology of the combined power source consisting of PV panels, AC-DC converters and DC-DC converters to connect AC and DC loads, respectively. The converter is controlled by an ANAM controller method. Here, power converters will be managed under an ANAM for interfering with resources and a universal DC bus. Based on the reference signals from the PV panel and AC supply, the ANAM generates the constant $380V_{DC}$ [6, 35] through power converters. This model controls the supervision of building energy management and ensures precise operation and the operation of its mechanism under various operating conditions.

11.3.1 MODELLING OF PV CELL

This work utilizes a PV model as appeared in Figure 11.2 and this model gives progressively exact outcomes. The model comprises a source current (I_{SC}), a diode and series resistance (R_S). It includes the impacts of short-circuit current (I_{SC}) and diode saturation current (I_O). It utilizes a single diode for idealization factor (n) and is set to accomplish the best I–V response.

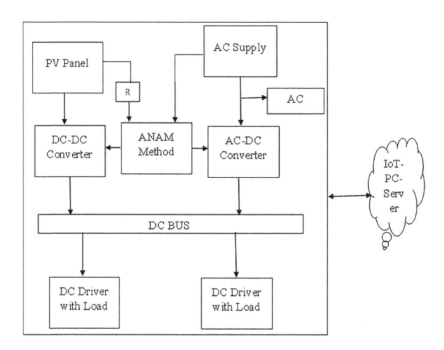

FIGURE 11.1   Block diagram of proposed system.

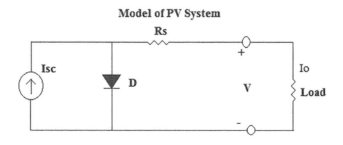

FIGURE 11.2   Model of PV system.

The relation of reverse saturation current of the diode, cell current I and short-circuit current I_{SC} is expressed through equation (11.1) [36].

$$I = I_{SC} - I_O\left[e^q\left(\frac{V + I_{RS}}{nKT}\right) - 1\right],$$ (11.1)

where

I = Response of solar cell current
V = Response of solar cell voltage
T = Solar cell temperature
K = Boltzmann constant
q = Elementary charge ratio
n = Ideality factor of diode.

Equation (11.2) discusses the mathematical operation of short-circuit current (I_{SC}) based on the rate of cell temperature.

$$I_{SC} \mid T = I_{SC} \mid T_{ref}[1 + \alpha(T - T_{ref})],$$ (11.2)

where

I_{SC} = Rate of short-circuit current
T_{ref} = Value of reference for temperature
α = Short-circuit current's temperature coefficient.
The short-circuit current (I_{SC}) is proportional to the intensity of irradiance.
I_{SC} at a given irradiance (G) is given as

$$I_{SC} \mid G = \frac{G}{G_0}I_{SC} \mid G,$$ (11.3)

where

G_0 = Represent the irradiance value.

Based on different sunlight intensities (in kW/m²), the I–V characteristic is shown in Figure 11.3.

11.3.2 DC-DC BOOST CONVERTER

The functional working diagram of the DC-DC boost converter is shown in Figure 11.4. This converter is used to convert the input voltage into a variable DC output voltage.

11.3.2.1 Boost Converter Circuit

Figure 11.5 shows the circuit diagram of the DC-DC boost converter. This circuit is used to convert the electrical voltage from one level to another based on switching action to improve the response of charging and discharging energy from inductors and capacitors.

Numerous control methods are used for controlling switch (Q1). Low-cost and straightforward controller structures are always in demand. In this work, ANAM control method is used to improve the performance of the DC-DC boost converter. The ANAM provides voltage regulation, electrical isolation, noise reduction and transient protection between an input and output side.

11.3.2.2 Controller Design and Modes of Operation

The DC-DC boost converter's output voltage must be controlled to keep the input voltage and load variations constant despite being steady. The proposed ANAM function is designed to adjust the time-varying proportional area of the step according to the control of ANAM.

$$S_S = K_1(t) \times (u_\infty - u_0),$$ (11.4)

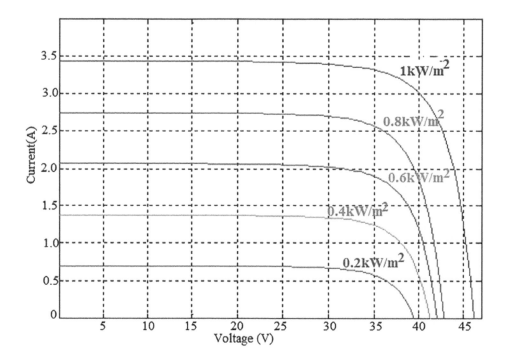

FIGURE 11.3   I-V characteristics of a PV system.

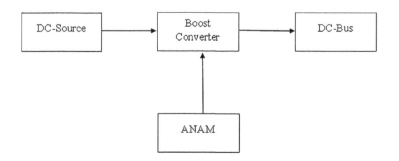

FIGURE 11.4   Block diagram of DC-DC converter.

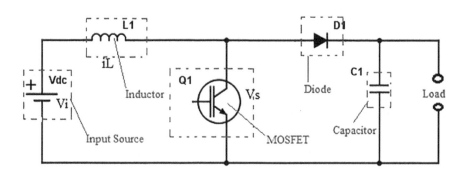

FIGURE 11.5   Circuit diagram of boost converter.

where

S_S = State space

μ_{∞} = Desired output voltage

μ_o = Actual output

K_1 = Switching interval of positive.

$K_1(t) > 0$ is a constant or time-varying parameter of ANAM according to the transition interval. It must be appropriately selected and modelled to meet the position of reaching/hitting the sliding coefficient stability condition. When $K_1(t)$ is considered as a constant in this control approach, its derivative is zero at any given point. Therefore, the differential sliding function is expressed by equation (11.5).

$$S_S \mid d_S = -K_1(t)u_o, \tag{11.5}$$

where

d_S = first order trending path.

This results in the equivalent control described by

$$U_{eq} = 1 - \frac{u_O}{i_L} \tag{11.6}$$

where

i_L = Load current

U_{eq} = Result of equivalent model

Then the corresponding trending law is defined by

$$S_S = -K_1(t)(u_\infty - u_O). \tag{11.7}$$

For the overall SMC of the converter, the overall control technique must have the option to control the framework to follow the yield of the reference esteem, which is characterized by the accompanying formula:

$$u = \begin{cases} u^+, & \text{if } S_S > 0 \\ u^-, & \text{if } S_S < 0 \end{cases}, \tag{11.8}$$

where

u^+ and u^- = Control inputs.

Replacing the mean state-space model of the system as a law of trend, the PWM generation dynamic duty cycles can be obtained directly:

$$0 < d(t) = 1 - \frac{K_1 u_i + \sqrt{K_1^2 u_i^2 + K u_O(u_\infty - u_O)}}{2K_1 u_O}, \tag{11.9}$$

where

$$k = \frac{4L}{R}(CRK_1^2 - k_1 k_2), \tag{11.10}$$

where

L = Inductor value

R = Load resistance

C = Capacitor value

k = Switching intervals.

When the error between the reference output and the actual output voltage is 0, equation (11.9) becomes:

$$d(t) = 1 - \frac{u_i}{u_\infty}. \tag{11.11}$$

The inherent general loose relationship exists between the input and output voltage for any DC-DC boost converters. It can be seen that the controller is independent of the current of the inductor and only requires load resistance when designing the sliding coefficients. There are two different operating modes based on the inductor current response, such as Continuous Current Mode (CCM) and Discontinuous Current Mode (DCM). CCM is used in this work because it gives a low signal-to-noise ratio compared with DCM.

11.3.2.2.1 Continuous Current Mode

When the inductor current is continuously charged and discharged during the switching period, it does not reach zero instantly, and it is called CCM operation, as shown in Figure 11.6. In CCM, each switching period T_S consists of two parts, namely $D_1 T_S$ and $D_2 T_S$, $D_{1+}D_2$, where D_1 is the duty cycle and $D_2 = 1 - D_1$.

During $D_1 T_S$, the inductor current linearly increases and during $D_2 T_S$, the inductor current linearly decreases. When the inductor current changes from a positive to a negative value (zero-crossing) and again ramps back to a positive value in each switching cycle, it is called the Forced Continuous Current Mode of Operation (FCCM). Inductor current in the FCCM of the process is shown in Figure 11.7. As compared with CCM, the FCCM reaches zero value, but CCM did not enter zero value.

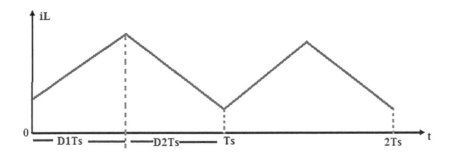

FIGURE 11.6 Inductor current response – CCM operation.

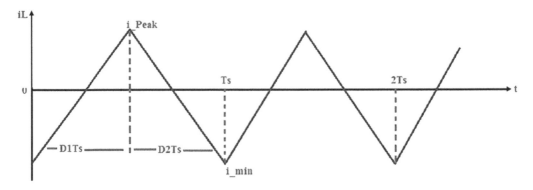

FIGURE.11.7 Inductor current response FCCM operation.

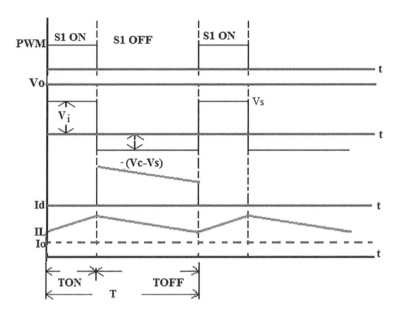

FIGURE 11.8   The pulse characteristics of CCM mode.

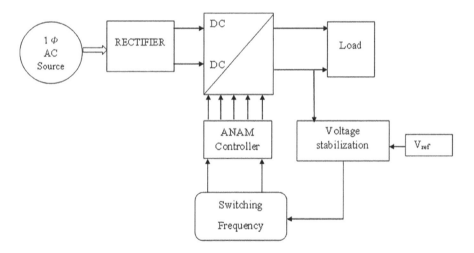

FIGURE 11.9   AC-DC converter-block diagram.

The gate triggering signal, output voltage variation, diode current and inductor current are depicted in Figure 11.8 for CCM operation of a boost converter.

11.3.3   AC-DC CONVERTER

The functional block diagram of the ANAM-based AC-DC converter is depicted in Figure 11.9. The ANAM control system gives accurate results against various components such as voltage stabilizer, maintains unity power factor and minimizes switching losses.

11.3.3.1   Buck-Boost Converter Circuit

The circuit outline of the ANAM-based Buck-Boost converter is shown in Figure 11.10. It comprises a DC-interface capacitor, a diode with arrangement association and a DC reactor. The proposed

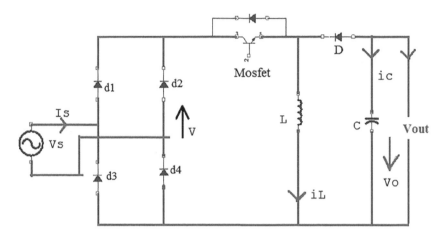

FIGURE 11.10   Circuit diagram Buck-Boost converter.

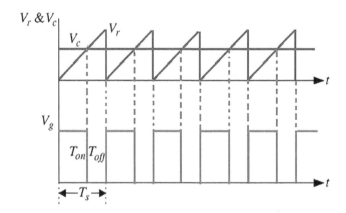

FIGURE 11.11   Generation of PWM.

converter is pertinent to both step-up and step-down applications. In this configuration, the energy storage/transfer element of the inductor and the output voltage characteristics of the step-up and step-down functions can be easily controlled by a suitable modification scheme of the power semi-conductor switch.

11.3.3.2   Switching Pulse Generation of Buck-Boost Converter

The pulse generation of MOSFET is developed by having a variable amplitude V_C with a DC reference signal, a consistent amplitude V_R and exchanging frequency F_S. The extent among V_C and Vr is known as the duty cycle, and $d = V_C/Vr$, which is portrayed as the total on-schedule and off-timespan $T_S = T_{on} + T_{off}$ extent. The ordinary yield voltage is changed by changing the variable V_C to control the obligation cycle D (Figure 11.11).

11.3.3.3   Modes of Operation of Buck-Boost Converter

The operation of Buck-Boost converter relies upon the equivalent circuits of the switching gadget and the diode D. The transformation gadget relies upon the condition of the equipment and the diode. Each chopping cycle comprises a few distinctive sub-modes.

Mode1: Charging Mode

When in charging mode, the MOSFET is there in ON stage, so the diode D is altered as reverse bias with the goal that the graceful voltage shows up over the inductor when the power is on. The inductor's current should ascend to the i_L and move through the info side, as shown in Figure 11.12a. As such, the loop terminals isolate the supply and yield capacitor flexibly. In this manner, the capacitor

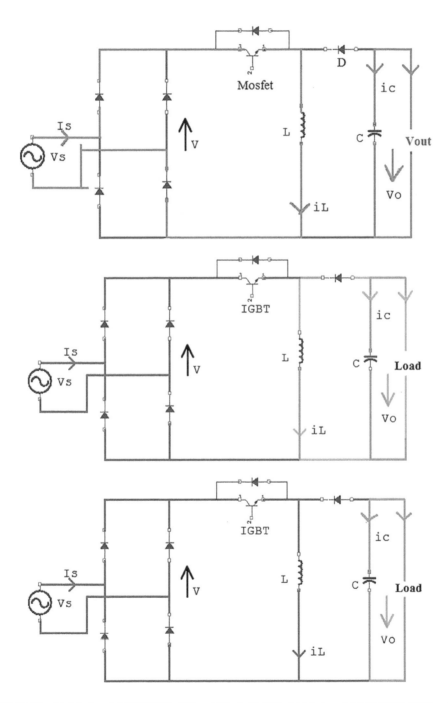

FIGURE 11.12 (a). Mode – 1 MOSFET is ON and Diode D OFF. (b) Mode – 2 MOSFET is OFF and Diode D ON. (c) Mode – 3 MOSFET is OFF and Diode D OFF.

charge is gathered from the past period, which must be included in the load. Gray lines on the image additionally show (Figure 11.12a) current ways for load flows during mode 1.

Mode 2: Discharging Mode

The discharge mode is corresponding to the charge mode. Exactly when MOSFET is turned off, the diode D will be forward-biased and the structure is moved to this mode. The inductor current i_L drops through the yield side and travels through the yield capacitor C and the DC load as shown in Figure 11.12b. The inductor's voltage upsets its polarity and forward-biased diode D, which has stored the inductor current. The imperativeness in the inductor current will drop until the switching device finishes the following obligation cycle or until the inductor current drops to zero again (mode 3). In this mode, the capacitor is charged with diminishing inductor current.

Mode 3

Mode 3, the switching device is still off. Notwithstanding, the circuit conditions may make the inductor current i_L drop to zero with the objective that the diode D gets opposite biased. This mode continues until the switching part is turned on again. The system will remain in the discharge mode for a period dictated by the structure parameters, if this period does not surpass the off time of the switching gadget. If it is expected that the current i_L streams regularly, the activity of mode-3 disappears. Figure 11.12c shows an equivalent circuit during this mode of activity.

The V–I characteristics of Voltage and Current Response for each mode are shown in Figure 11.13.

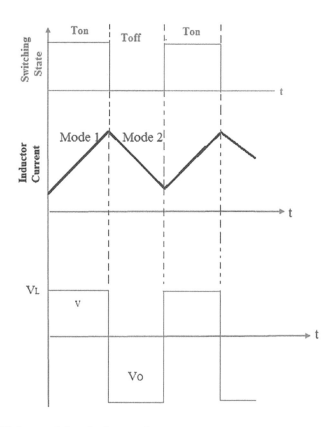

FIGURE 11.13   V-I characteristics of voltage and current response.

11.4 OPTIMIZATION AND POWER MANAGEMENT ANALYSIS OF CONVERTERS USING ADAPTIVE NONLINEAR ASCENDANT MODE CONTROL STRATEGY

Power or power management is the main requirement of a power converter system. When a shifting energy source side is unbalanced or load side changes are adaptively optimized, the system is implemented to steady the overall performance of the system with minimal power loss, which affects the system. This work proposes an adaptive nonlinear sliding control method in which optimal control, such as power management control, can be a natural and system architecture horizon for this type of design. More importantly, despite the complexity of the proposed system due to many nonlinear constraints, the new result is obtained based on the classical theory of optimal control, allowing real-time results and solving the optimal control problem. More specifically, ANAM is first used to find that the power management problem has a beneficial solution.

11.4.1 ANAM – ALGORITHM STEPS

Step 1: The first step is to fundamentally characterize the size of the population (S) and iteration (j).

Step 2: Generate people randomly, where $j = 1, 2, 3...$ Individuals carried by S. Set the maximum number of iterations parameter.

Step 3: Enumerate and evaluate the ideal ability of the individual's ultimate goal depending on its voltage variety and full load.

Step4: Calculate parameters by considering the three components cited below.

   i.   Individual variation based on other loads.

   ii.   Input power factor.

   iii.   The frequency of the switching converter.

Step 5: Calculating $T_i + 1$ during the test depends on this certainty and the client. The defence is not precise y' (for example, 10% resilience failure).

Step 6: By implementing errors, control the load changes as follows:

$$\Delta s = \gamma.V_{out.e}, \tag{11.12}$$

Where $V_{out.e}$ knowledge function.

Step 7: Adjust the unique load as follows:

$$s = s + \Delta s \tag{11.13}$$

Step 8: If the progress is not passed in the issuing state, go to Step 3 at that time.

Step 9: Individual loads can be upgraded to new locations individually.

Step 10: If the end criterion is fulfilled, then go to Step 3 and repeat the appropriate method.

Step 11: Finding the best mind arrangement in the investigation space, at that point, if the end criteria is compliant.

The working flow chart of the proposed ANAM-based DC power management system is shown in Figure 11.14. Regulate the PWM control signal of the DC-DC boost converter until the condition $(dI/dV) + (I/V) = 0$ is fulfilled. The transient response, THD and overall system efficiency are used to evaluate the performance of the proposed system.

Peak Time: The peak time is the time required for the reaction to show up at the essential peak of the overshoot. The boundary of peak time was resolved as it is used at which the peak regard occurs. It is meant by $t = tp$, in which the first derivate of the reaction is zero.

Peak Overshoot: Peak overshoot assesses the reaction of steady-state error. The numerical check of peak overshoot is computed as equation (11.14).

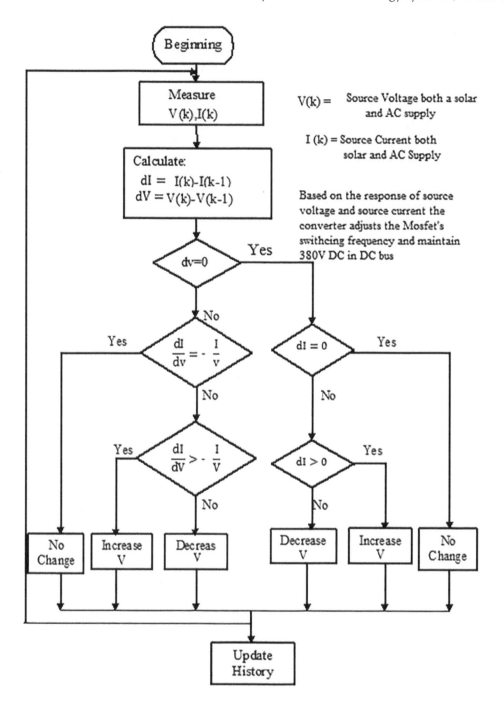

FIGURE 11.14   Flow chart of ANAM.

$$M_p = C(t_p) - C(\infty) \tag{11.14}$$

where

M_p = Maximum voltage response of peak overshoot
$C(t_p)$ = Response of peak value with respect to time
$C(\infty)$ = Final peak response.

Steady-State Error: Steady-state error is the difference between the perfect last output obtained and the genuine one when the system shows up at a steady state when its direction may be depended upon to continue if the structure is undisturbed.

$$S_S = R_S - RV_S \tag{11.15}$$

where

$S_S V_S$=Response of steady-state error,

RV_S=Reference state response and is same as Running state response.

Total Harmonic Distortion: THD is an estimation of the harmonic distortion present in a sign and is characterized as the proportion of the whole of the powers of every single consonant segment to the power of the fundamental frequency.

$$\text{THD} = \frac{1}{V_1}\left[\sum_{n=2,3,\ldots}^{\infty} V_n^2\right]^{1/2} \tag{11.16}$$

11.5   IoT DATA CONTROL SYSTEM

In remote areas, there is a requirement for continuous monitoring of the power electronics converter framework to guarantee that steady yield is guaranteed. The flow research depicts the hardware and software plan of building energy checking frameworks in a remote zone. The observing framework is outfitted with THD and transient investigation with a remote module for data transmission. The data obtained are shown on an IoT platform, and applicable triggers are set appropriately. Accurately when a DC power board framework is interfaced with the internet, it creates an excellent Internet Protocol (IP). Along these lines, in this system, IP is made by controllers. The page is made out of the objective that when IP is given in the Uniform Resource Locator (URL), the control page, as in Figure 11.15, shows the system observing on the site page; thus, a customer can see the status of the proposed structure.

The flow of execution in Figure 11.15 starts by checking internet affiliation. Check whether the internet association is proper and the IP address appears in the serial terminal. On the off chance that there is an issue with the internet, it will exhibit blunder. In the wake of checking the internet association, the server port arrangement is affirmed. On the off chance that there is an issue, the slip-up message appears in a serial terminal. On the off chance that the server port game plan is legitimate and the server runs, Transmission Control Protocol (TCP) connection gets related, and Remote Procedure Call (RPC) is set up. A customer can give decisions in webserver to control machines. RPC orders inside the software summon the microcontroller exercises. When the TCP affiliation is set up, the Hypertext Transfer Protocol (HTTP) server starts running. By then, the HTML 5 code similarly begins. At long last, when the IP is given in the URL, the establishment of HTML 5 code runs, and the site page is appeared. When a customer gives flags, the related RPC begins, and the microcontroller would play out the necessary action as per the message given by the customer.

11.5.1   IoT DATA COMMUNICATION

Remote monitoring systems are intended to monitor intricate amenities in industries, power plants, network operations centres, airports and spacecraft with some grade of automation. In this study, a smart remote monitoring system is proposed using IoT for monitoring the ordered parameters of DC power management. This study tries to describe the design and implementation of an interconnected mechanism of DC power management and the measurement of the reliable parameters.

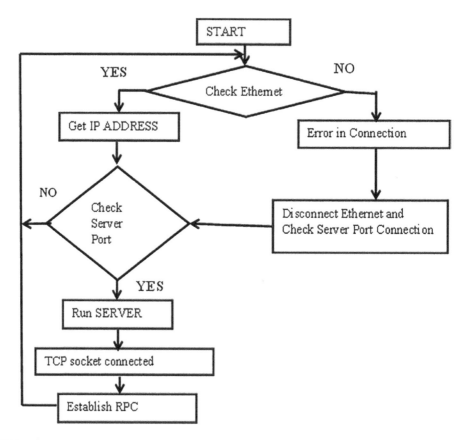

FIGURE 11.15   Flow chart of IoT communication.

The settings of DC power management are stored in local workstations and shared with the cloud to display the web link parameters. The proposed system structure is a combination of customized IoT devices with an information system for data aggregation.

IoT is an open-source application and Application Programming Interface to scrutinize and form information from things using the HTTP protocol over Local Area Network. IoT engages the creation of sensor logging applications, zone following applications and a relational association of things with declarations. The client ought to make the record first. The announcement contains channels that are discrete for various activities. The channel has fields that are distinctive for different boundaries in the checking framework. In the wake of out of boundary scenario, the cloud has worked in capacities that speak to the qualities as graphs. Figure 11.16 shows the flowchart for IoT data communication and the steps are listed below.

Stage 1: Start.
Stage 2: Microcontroller senses the power utilization of converter through the current sensor and voltage divider, transient response and THD levels.
Stage 3: IoT driver gets the microcontroller output information through a serial port and shows the acquired data on the web page through python content.
Stage 4: IoT driver sends the monitoring information on to the cloud.
Stage 5: The cloud shows the information as a diagram, clearly explainable to the whole client.
Stage 6: Stop.

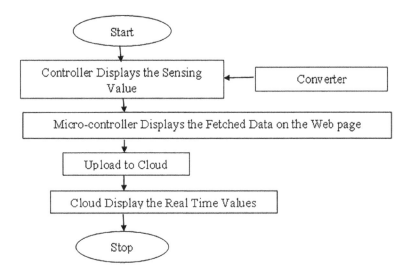

FIGURE 11.16   Flow chart of cloud data communication.

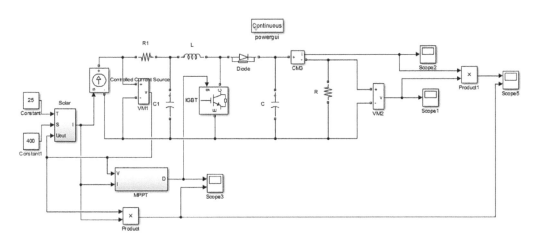

FIGURE 11.17   Simulink model solar with DC-DC converter.

11.6   RESULTS AND DISCUSSION

This section discusses the performance analysis of the proposed solar-based DC power management system. The simulation of the proposed system is implemented using MATLAB® software. The simulation circuit and obtained results are discussed in the following subsections.

11.6.1   Performance Analysis of Solar-Based DC-DC Converter

The Simulink model of the proposed solar-based DC-DC converter is shown in Figure 11.17, and the simulation parameters are listed in Table 11.2. The simulation results and performance analysis of DC-DC converter with solar sources are dealt as follows:

The simulation result of solar voltage is shown in Figure 11.18. The maximum peak of the solar voltage is 200V.

TABLE 11.2
Simulation Parameters of DC-DC Converter with Solar

Software Tool	MATLAB 2016a
Renewable power generation source	Solar PV Array
Total capacity	20 kWp
Indifference time	105–450 seconds
Startup power	40W
Nominal voltage	635Vdc
Short-circuit current, Isc (A)	24A
Power Conditioning Unit Parameters	
DC-DC converter	380V
Rated voltage	211V
Resistance	0.02 Ohm
Inductance	10 µh
Capacitance	200 µf

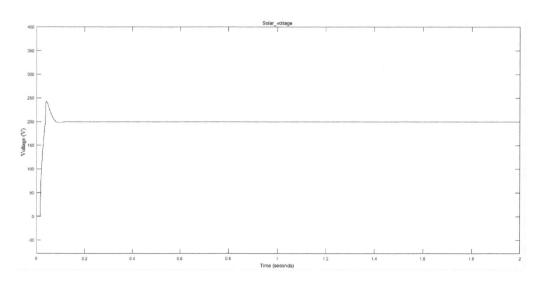

FIGURE 11.18   Input voltage from solar.

The simulation response of the proposed DC-DC converter is shown in Figure 11.19. The maximum peak of the DC-DC converter response is 380 V.

The simulation response of Load Voltage and Load Current for DC-DC converter is shown in Figure 11.20. The maximum peak of the DC-DC converter response is 48 V (Figure 11.20a), and the load current response is shown in Figure 11.20b.

11.6.2   PERFORMANCE ANALYSIS OF AC-DC CONVERTER

Figure 11.21 shows the simulation model of the proposed AC-DC converter, and the simulation parameters are listed in Table 11.3.

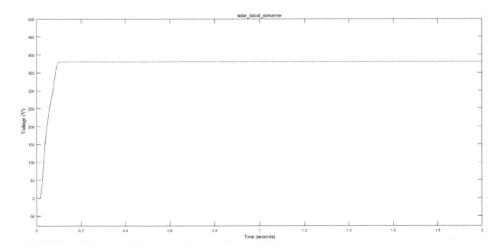

FIGURE 11.19 Simulation response of DC-DC converter.

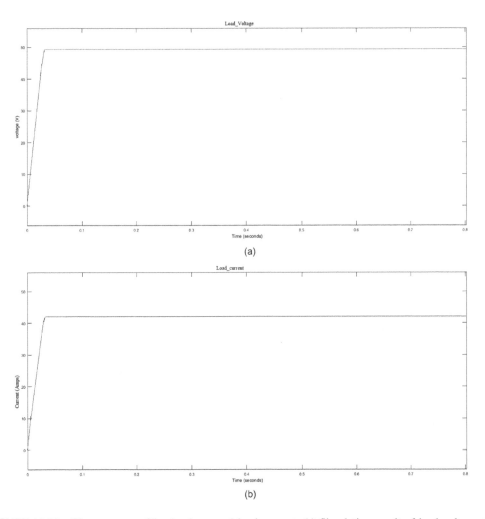

FIGURE 11.20 The response of load voltage and load current. (a) Simulation result of load voltage. (b) Simulation result of load current.

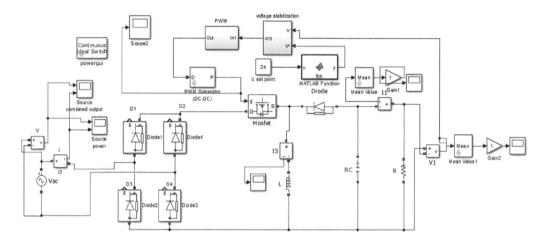

FIGURE 11.21   Simulink model of proposed AC-DC converter.

TABLE 11.3
Simulation Parameters of AC-DC Converter

Parameters	Values
AC source	20 kVA
V_{in} (RMS)	230 ±10% V
V_{out}	380V DC
Maximum load	2000W
MOSFET switching frequency	5 kHz
Input power factor	0.9715
Inductor	100e−4 H
Capacitor	400e−8 Farad
Diode (forward voltage)	0.8V
MOSFET (internal diode resistance)	1e−6 ohms

The AC source voltage and source current are shown in Figure 11.22. These waveforms clearly state that both current and voltages drawn from the AC source are in phase, and hence, the power factor is unity-based irrespective of load variations.

The AC-DC converter voltage response across the DC bus from the conversion of the solar voltage is shown in Figure 11.23. The maximum peak of the AC-DC converter response is 380 V.

The load voltage of AC-DC converter from the conversion of the source voltage is shown in Figure 11.24. The maximum peak of the DC-DC converter response is 48 V (Figure 11.24a) and the load current response is demonstrated in Figure11.24b.

The simulation of the THD response of the proposed energy management system for Input AC Voltage the proposed system is shown in Figure 11.25. By using ANAM, THD is 3.31% with an equivalent frequency of 50 Hz.

The prototype model of proposed building energy management is shown in Figure 11.26. In this work, solar power is used as a primary source. By using ANAM, the demand energy problem is significantly controlled, and this power management system is monitored through the IoT webpage.

The IoT webpage, which is the manipulator module, is where the data are converted into the actual values using multiplication factors and then the values are displayed.

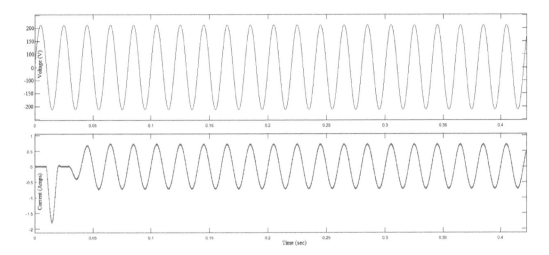

FIGURE 11.22 AC source voltage and source current.

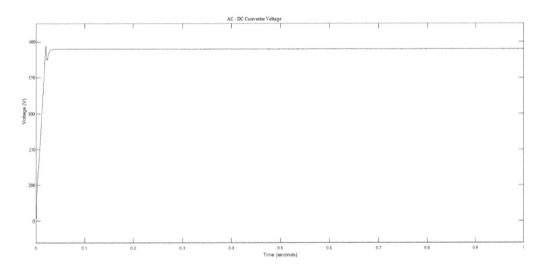

FIGURE 11.23 AC-DC converter voltage.

Table 11.4 discusses the simulation results of Peak Time, Peak Overshoot Time, Recovery Time and Steady-State Error. These results are obtained using 2000 W loads.

The simulation results of Peak Time, Peak Overshoot Time, Recovery Time and Steady-State error are shown in Figure 11.27. As compared with existing fuzzy logic and SMC methods, the proposed Adaptive Nonlinear Sliding Method achieves the best result against all parameters. The peak time of sliding is 0.9417 seconds, the peak time of fuzzy logic is 0.6010 seconds, and the peak time of ANAM (AC-DC) is 0.12 seconds and that of ANAM (DC-DC) is 0.13 seconds. As compared with existing methods, the peak time response is improved by up to 21.6% by using ANAM. The peak overshoot time of sliding mode is 1.522 seconds, the peak overshoot time of fuzzy logic is 1.145 seconds, and the peak overshoot time of ANAM (AC-DC) is 0.15 seconds and that of ANAM (DC-DC) is 0.123 seconds. As compared with fuzzy logic, the peak overshoot response is improved up to 28% by using ANAM. The steady-state error of sliding mode is 10%, the steady-state error of fuzzy logic is 8% and the steady-state error of ANAM is 6%.

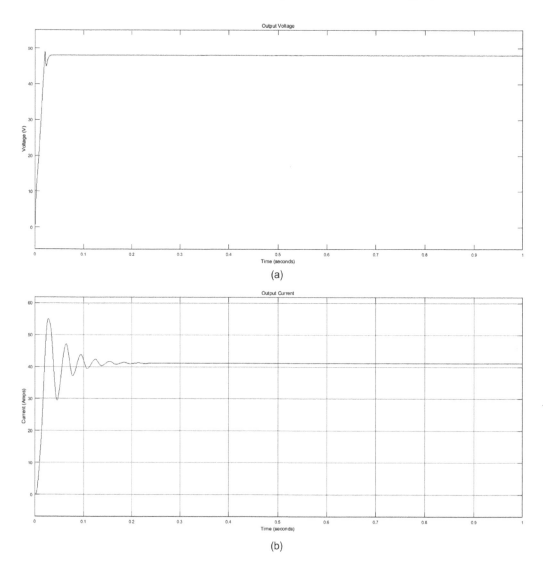

(a)

(b)

FIGURE 11.24 The response of load voltage and load current with AC-DC converter. (a) Load voltage with AC-DC converter. (b) Load current with AC-DC converter.

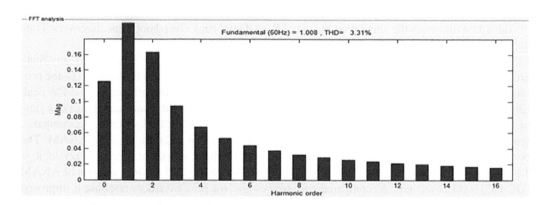

FIGURE 11.25 THD analysis of input AC voltage the proposed system.

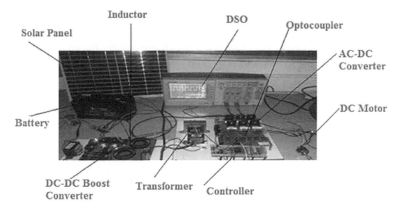

FIGURE 11.26   Prototype model of proposed system.

TABLE 11.4
Performance Analysis of Control System Parameters

Methods	Sliding DC-DC	Fuzzy Mode DC-DC	Mode ANAM AC-DC	DC-DC
Peak time (seconds)	0.9417	0.6010	0.12	0.13
Peak overshoot time (seconds)	1.522	1.145	0.15	0.123
Recovery time (seconds)	0.66	0.56	0.2	0.18
Steady state error (%)	10	8	6	4
THD (%)	13.5	8.2	3.3	–

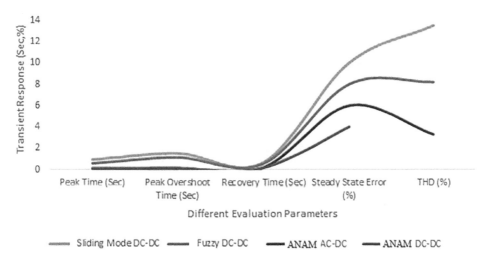

FIGURE 11.27   Performance evaluation of proposed system.

The sliding mode obtains a recovery time of 0.66 seconds, the peak overshoot time of fuzzy logic is 0.56 seconds, and the peak overshoot time of ANAM (AC-DC) is 0.20 seconds and that of ANAM (DC-DC) is 0.18 seconds. As compared with sliding mode, the recovery time response is improved by up to 30% by using ANAM. The THD of sliding mode is 13.5%, the THD of fuzzy logic is 8.2% and the THD of ANAM is 3.3%.

11.7 CONCLUSION

This work proposes an ANAM method that can be applied for commercial building DC devices. The aim of maintaining constant DC bus voltage using DC bus signalling will be realized, considering different operations. The proposed control empowers the highest usage of PV power during different working states of the building power management network. DC bus voltage levels are utilized to arrange the sources and storages in the framework. It goes as a control contribution for the working mode switching during various working conditions. Using a decentralized control method, the system becomes more flexible and expandable, which can integrate more microgrids without changing the control method. System simulations will be carried out to validate the proposed control methods for the distributed integration of PV and energy storage in commercial buildings. The proposed controller has also been applied to an AC-DC converter, and the performance of the proposed controller is validated through simulation. The acquired simulation and test results affirm that the suggested arrangement sources are more robust and effective than the existing source setup. As compared with the current system, the proposed method achieves the best results; for example, peak time is 0.12 seconds, peak overshoot time is 0.15 seconds, recovery time is 0.20 seconds, the steady-state error is 6% and THD is 3.31%. In the future, the system will introduce deep learning methods to improve the power quality issues of a solar-based commercial building application system.

REFERENCES

1. Sharip, Mohd R. M., Haidar, Ahmed M. A., and Jimel, Aaron C., "Optimum Configuration of Solar PV Topologies for DC Micro grid," *International Journal of Photoenergy*, 2019, Article ID 2657265, 13 pages, 2019. doi:10.1155/2019/2657265
2. Rosales-Asensio, Enrique, de Simón-Martín, Miguel, Borge-Diez, David, Blanes-Peiró, Jorge Juan, and Colmenar-Santos, Antonio, "Microgrids with Energy Storage Systems as a Means to Increase Power Resilience: An Application to Office Buildings," *Energy*, 2019, 172(C), 1005–1015.
3. Sundareswaran, Kinattingal, Ark Kumar, Kevin, Raman Venkateswaran, Payyalore, and Sathya, P., "Solar Photovoltaic Fed Dual Input LED Lighting System with Constant Illumination Control," *Frontiers in Energy*, 2016. doi:10.10.1007/s11708-016-0420-z
4. Lai, Elisa, Muir, Stewart, and Erboy Ruff, Yasemin, "Off-Grid Appliance Performance Testing: Results and Trends for Early-Stage Market Development," *Energy Efficiency*, 2019. doi:10.1007/s12053-019-09793-z
5. Vishwanath, Arun, Chandan, Vikas, and Saurav, Kumar, "An IoT Based Data Driven Pre-Cooling Solution for Electricity Cost Savings in Commercial Buildings," *IEEE Internet of Things Journal*, pp. 1–1, 2019. doi:10.1109/JIOT.2019.2897988
6. Arunkumar, G., Devaraj, Elangovan, Sanjeevikumar, P., Holm-Nielsen, Jens, Leonowicz, Zbigniew, and Joseph, Peter, "DC Grid for Domestic Electrification," *Energies*, 2019. doi:10.3390/en12112157
7. Ullah, S., Haidar, A.M.A., and Zen, H. "Assessment of Technical and Financial Benefits of AC and DC Microgrids based on Solar Photovoltaic," *Electrical Engineering*, 2020. doi:10.1007/s00202-020-00950-7
8. Kitson, J., Williamson, S.J., Harper, P.W., Mcmahon, Chris, Rosenberg, G., Tierney, M., Bell, K., and Gautam, B., "Modelling of an Expandable, Reconfigurable, Renewable DC Microgrid for Off-Grid Communities," *Energy*, 160, 2018. doi:10.1016/j.energy.2018.06.219
9. Comello, S., Reichelstein, S., and Sahoo, A. "The Road Ahead for Solar PV Power," *Renewable and Sustainable Energy Reviews*, 2018, 92, 744–756.
10. Chuang, Jules, Lien, Hsing-Lung, Den, Walter, Iskandar, Luvian, and Liao, Pei-Hsuan, "The Relationship Between Electricity Emission Factor and Renewable Energy Certificate: The Free Rider and Outsider Effect," *Sustainable Environment Research*, 2018, 28. doi:10.1016/j.serj.2018.05.004

11. Zhou, Yuanbing and Chen, Xing, "Mechanism of CO_2 Emission Reduction by Global Energy Interconnection," *Global Energy Interconnection*, 2018, 1(4), 409–419.

12. Zandi, Z. and Mazinan, A., "Maximum Power Point Tracking of the Solar Power Plants in Shadow Mode Through Artificial Neural Network," *Complex & Intelligent Systems*, 2019. doi:10.1007/s40747-019-0096-1

13. Mossad, Mohamed I., Osama Abed El-Raouf, M., Alahmar, Mahmoud, and Fahd, A. Banakher, "Maximum Power Point Tracking of PV System-Based Cuckoo Search Algorithm; Review and Comparison," *Energy Procedia*, 2019, 162, 117–126. doi:10.1016/j.egypro.2019.04.013

14. Schuss, Christian, Fabritius, Tapio, Eichberger, Bernd, and Rahkonen, Timo, "Moving Photovoltaic Installations: Impacts of the Sampling Rate on Maximum Power Point Tracking Algorithms," *IEEE Transactions on Instrumentation and Measurement*, 2019, pp. 1–9. doi:10.1109/TIM.2019.2901979

15. Kamran, M., Mudassar, M., Fazal, M.R., Asghar, M.U., Bilal, M., and Asghar, R., 2020. "Implementation of Improved Perturb & Observe MPPT Technique with Confined Search Space for Standalone Photovoltaic System," *Journal of King Saud University-Engineering Sciences*, 32(7),432–441.16. Anzalchi, A. and Sarwat, A., "Overview of Technical Specifications for Grid-Connected Photovoltaic Systems," *Energy Conversion and Management*, 2017, 152, 312–327.

17. Jo, J.H., Aldeman, M.R., and Loomis, D.G., "Optimum Penetration of Regional Utility-Scale Renewable Energy Systems," *Renewable Energy*, 2018, 118, 328–334.

18. Vezzoli, Carlo, Ceschin, Fabrizio, Osanjo, Lilac, M'Rithaa, Mugendi, Moalosi, Richie, Nakazibwe, Venny, and Diehl, Jan Carel, "Distributed/Decentralised Renewable Energy Systems," In: *Designing Sustainable Energy for All. Green Energy and Technology*. Springer, Cham, 2018. doi:10.1007/978-3-319-70223-0_2

19. Carli, Raffaele and Dotoli, Mariagrazia, "Decentralized Control for Residential Energy Management of Smart Users' Microgrid with Renewable Energy Exchange," *IEEE/CAA Journal of Automatica Sinica*, 2019, 6, 641–656.

20. Nasir, Mashood, Khan, Hassan, Hussain, Arif, Mateen, Laeeq, and Zaffar, Nauman, "Solar PV-Based Scalable DC Microgrid for Rural Electrification in Developing Regions," *IEEE Transactions on Sustainable Energy*, 2017, 1–1. doi:10.1109/TSTE.2017.2736160

21. Hasan, A.S.M., Chowdhury, D., and Khan, M.Z.R., "Performance Analysis of a Scalable DC Microgrid Offering Solar Power Based Energy Access and Efficient Control for Domestic Loads. *arXiv preprint arXiv:1801.00907*, 2018.

22. Ban, Mingfei, Yu, Jilai, Shahidehpour, M., and Guo, Danyang, "Optimal Sizing of PV and Battery-Based Energy Storage in an Off-Grid and Nanogrid Supplying Batteries to a Battery Swapping Station," *Journal of Modern Power Systems and Clean Energy*, 2018, 7. doi:10.1007/s40565-018-0428-y

23. Kumar, Shailendra and Singh, Bhim, "Self-Normalized Estimator-Based Control for Power Management in Residential Grid Synchronized PV-BES Microgrid," *IEEE Transactions on Industrial Informatics*, 2019, 1–1. doi:10.1109/TII.2019.2907750

24. Maranda, Witold, "Analysis of Self-Consumption of Energy from Grid-Connected Photovoltaic System for Various Load Scenarios with Short-Term Buffering," *SN Applied Sciences*, 2019, 1. doi:10.1007/s42452-019-0432-5

25. Manikanta, M., Kumar, U. Ranjith, and Lakshmi, Ch Prasanna, "Energy Management System for Sudden Load Varying with PV-Battery & Diesel Generator Hybrid System (August 27, 2019)," *International Journal for Modern Trends in Science and Technology*, August 2019, 05(08).

26. Worighi, Imane, Geury, Thomas, El Baghdadi, Mohamed, Van Mierlo, Joeri, Hegazy, Omar, and Maach, Abdelilah, "Optimal Design of Hybrid PV-Battery System in Residential Buildings: End-User Economics, and PV Penetration," *Applied Sciences*, 2019, 9, 1022. doi:10.3390/app9051022

27. Upasani, Mayuri and Patil, Sangita B., "Grid Connected Solar Photovoltaic System with Battery Storage for Energy Management," *2018 2nd International Conference on Inventive Systems and Control (ICISC)* (2018): 438–443.

28. Yu, H.J.J., "A Prospective Economic Assessment of Residential PV Self-Consumption with Batteries and Its Systemic Effects: The French Case in 2030," *Energy Policy*, 2018, 113, 673–687.

29. Omar, Moien A. and Mahmoud, Marwan M., "Design and Simulation of a PV System Operating in Grid-Connected and Stand-Alone Modes for Areas of Daily Grid Blackouts," *International Journal of Photoenergy*, 2019, Article ID 5216583, 9 pages. doi:10.1155/2019/5216583

30. Fara, Laurentiu and Craciunescu, Dan, "Output Analysis of Stand-Alone PV Systems: Modeling, Simulation and Control," *Energy Procedia*, 2017, 112, 595–605. doi:10.1016/j.egypro.2017.03.1125

31. Longo, Michela, Foiadelli, Federica and Yaïci, Wahiba, "Simulation and Optimization Study of the Integration of Distributed Generation and Electric Vehicles in Smart Residential District," *International Journal of Energy and Environmental Engineering*, September 2019, 10(3), 271–285.

32. Cheng, Lefeng, Zhang, Zhiyi, Jiang, Haorong, Tao, Yu, Wang, Wenrui, Xu, Wei Feng, Hua, Jinxiu, "Local Energy Management and Optimization: A Novel Energy Universal Service Bus System Based on Energy Internet Technologies," *Energies*, 2018. doi:10.20944/preprints201805.0094.v1

33. Rehman, Abdul, Ashraf, Muhammad, and Bhatti, Aamer, "Fixed Frequency Sliding Mode Control of Power Converters for Improved Dynamic Response in DC Micro-Grids," *Energies*, 2018, 11, 2799. doi:10.3390/en11102799

34. L. Hernández, José, Sanz, Roberto, Corredera, Álvaro, Palomar, Ricardo, and Lacave, Isabel, "A Fuzzy-Based Building Energy Management System for Energy Efficiency," *Buildings*, 2018, 8, 14. doi:10.3390/buildings8020014

35. Ryu, M.H., Kim, H.S., Baek, J.W., Kim, H.G., and Jung, J.H., "Effective Test Bed of 380-V DC Distribution System Using Isolated Power Converters," *IEEE Transactions on Industrial Electronics*, 2017, 62, 4525–4536.

36. Hatti, Mustapha, Renewable Energy for Smart and Sustainable Cities. Artificial Intelligence in Renewable Energetic Systems' IC-AIRES2018, held on 24–26 November 2018, at the High School of Commerce, ESC-Koléa in Tipaza, Algeria.

12 Artificial Intelligence Methods for Hybrid Renewable Energy System

P. Srividya
RV College of Engineering

CONTENTS

ABSTRACT

This chapter presents a review on application of artificial intelligence (AI) approaches for hybrid energy systems. As the traditional grid power electricity is not reachable to people in remote and rural areas, they are more dependent on fossil fuels. Fossil fuels are already in shortage around the globe. This necessitates for hybrid energy system which aims at integrating various types of generation, storage and consumption technologies into a single system in order to generate power round the clock with lower cost and higher reliability. Uncertainty of renewable energy sources and the ability of AI techniques to process the inputs intelligently have provoked the usage of AI in energy sectors. At present, AI is being used in the energy sectors for smart electricity generation, transmission and energy consumption. It makes integration of energy sources into the grid easier. It enhances the constancy, effectiveness, robustness and security of energy systems. AI models can do predictions in shorter time and can use the values for future predictions as well. Hence, the various AI techniques to improvise the performance of the hybrid system will be studied in the chapter.

DOI: 10.1201/9781003331117-12

12.1 INTRODUCTION

Energy plays a significant role in the economic development of a country. It is being used for industrialization and urbanization of a country. According to the global reports, the demand for energy raises approximately by 2% per annum. Fossil fuels being the major source of energy are also the main source of environmental pollution. The depletion of fossil fuels due to its over dependence has raised the demand for alternate energy sources like rain, wind, geothermal, biomass, generating energy from solid waste, etc. Renewable sources of energies are abundant in nature, are pollution free and sustainable. The awareness among people about the abnormal weather changes and the drastic increase in the cost of traditional energy sources has pushed many nations to provide alternate energy approach that disseminate renewable energy (RE) systems. Also the grid power is not reachable to remote places. This has also played a role in switching over to alternative sources of energies in these areas. Although RE sources are considered as promising power sources, they are unpredictable and are weather-dependent. This necessitates the integration of RE sources like solar and wind along with the conventional energy source to form a hybrid system [1] as illustrated in Figure 12.1. Hybrid scheme creates reliable and environmental friendly electricity. The main constraint of the hybrid system is its unpredictable nature and size. The system should be sized optimally to meet the load demands with minimum cost. If the constraint is met, the system can be used to increase production of the electrical power and to supply the rural areas. If the prediction about the RE is done precisely, the efficiency of hybrid system will increase.

The hybrid energy system shown in Figure 12.1 consists of the following:

a. PV panel array that converts solar energy into direct current.
b. Wind turbine that converts mechanical energy into electrical energy.
c. Fuel cells as a backup source of power, battery bank to store excess energy and supply energy when required.
d. DC to DC converters to step up the DC voltage to a higher value.
e. DC to AC converters to generate AC signal from DC signal.
f. Regulator to ensure power supply to the load.

12.2 RENEWABLE ENERGY SOURCES

Energy being an essential requirement for various sectors can be derived from two different sources – renewable and nonrenewable sources. Coal, natural gas and petroleum form the sources of nonrenewable energy resources, whereas energy generated from wind, biomass, solar, sea waves, rain and hydro-generation forms the RE resources. According to ministry of new and RE, about

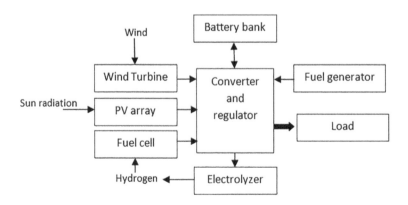

FIGURE 12.1 Hybrid renewable energy systems.

80% of the energy demand is supplied from fossil fuels and 5% from nuclear sources and the rest by nonrenewable energy sources. The statistics shows that the main sectors that shows slow switchover to RE sources includes transport and power generation.

12.3 APPLICATION OF ARTIFICIAL INTELLIGENCE (AI) TO HYBRID ENERGY SYSTEMS

Application of AI for energy systems brings a shift in the energy sector from traditional power generation to a more cost-effective system. AI can be used in management of electricity, smart grids, energy generation, loss prediction in the electrical grid, forecasting the load, equipment failure prediction and many more. Prerequisites for usage of AI in the energy system are the energy sector digitalization and the availability of large set of data. AI makes the energy sector more efficient and secure by analyzing and evaluating the huge volume of data available. The algorithms used to model, analyze and predict the performance of the hybrid systems are complex involving differential equations that demands huge computation power and time. It needs long-term meteorological inputs like temperature, wind, radiation, etc. Measuring these qualities has several shortcomings like unavailability of data at most of the locations, even if it is available, quality of data might be inferior. AI overcomes these difficulties. It uses techniques to learn critical information patterns hidden in a multidimensional information domain.

12.3.1 AI FOR POWER GRID AND SMART GRID

Decentralization and digitization of the power grid have increased the difficulty in managing the large number of grid participants and in grid balancing as it demands analysis of huge data. AI helps in processing this data quickly and as efficiently as possible. Smart grids allow transportation of electricity and data. The rise in the demand for wind and solar plants for power generation has also raised the demand for intelligent power generation and consumption. AI plays a vital role in evaluating, analyzing and controlling the consumers, producers and storage facilities that are connected to the grid. AI very particularly focuses on integration of electro-mobility. AI assists these by monitoring, coordinating and detecting the anomalies involved in generation, utilization and transmission in real time and later also develops appropriate solutions to rectify it. AI also coordinates maintenance work and finds the optimal time for network and individual system maintenance. This reduces the cost and disturbances caused to the network operation.

12.3.2 AI IN ELECTRICITY TRADING

AI when applied to power trading helps in improving the forecasts. Systematic evaluation of weather data and other historical data is possible using AI. Stability of the grid and security of the grid increase due to better forecasts. AI also facilitates in accelerating the integration of renewable. In energy industry, machine learning (ML) and neural networks improve forecasts. With application of AI, there is already a decline in the demand for control reserve, although the volatile power generators have a greater share in the market.

12.4 HYBRID RENEWABLE ENERGY SYSTEMS (HRESS) WITH MACHINE LEARNING

ML can be efficiently used in generation and demand sectors for both grid-connected and stand-alone sources based on the necessities and the nature of the obstacle to enhance the performance. Figure 12.2 shows the various sectors like power prediction, demand forecasting, RE management systems and others where ML methods are used. Usages of ML methods in various HRESs are as follows:

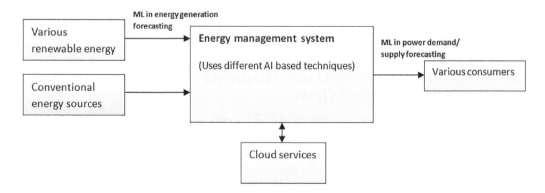

FIGURE 12.2 Advanced technologies usage in HRES [1].

i. Forecasting energy generation that is predicting the output that can be generated from RE sources. This is done using historical data. The only challenge involved will be due to the dependence of the source on environmental conditions. Different neural networks can be used in predicting the output from RE sources.

ii. Geographical location, sizing and configuration of renewable plants play a vital role in HRES. Plant location is based on weather conditions, establishment cost and availability of energy sources like wind, solar radiation, etc. ML assists in all such decision makings.

iii. RE-integrated smart grid is a new generation of power plant that optimizes the grid sectors from generation up to distribution including storage. Expansion of power grid and the need to make it intelligent and efficient are the driving forces to use AI and ML in problems faced by power grids like fault detection, grid operation, data management by grid and others.

iv. ML helps in renovating or in finding alternate materials required in building RE sources like solar cells, batteries and others.

12.5   RENEWABLE ENERGY FORECASTING APPROACHES

Energy forecasting is a method of estimating the energy required for various demands in order to maintain adequate supply. There are two approaches toward energy forecasting – top-down and bottom-up approaches. In top-down approach, prediction is done at the top most level, whereas in bottom top approach, it is done at bottom most level. Bottom-up approach is most suited for hybrid systems and shown in Figure 12.3.

12.5.1   Prediction of Solar Energy

Photovoltaic or photothermal energies are the main forms of solar energies. Although photovoltaic energy does not generate greenhouse gas, it is unpredictable in nature as it depends on diverse environmental factors like air pressure, ambient temperature, wind direction, wind speed and others. ML can be used in predicting the solar irradiance at a particular place on hourly, daily or on monthly basis.

12.5.2   Prediction of Wind Energy

The wind energy is obtained by the conversion of kinetic energy of air into electricity. This energy is also dependent on environmental factors like solar radiation, wind power and other ambient conditions. Hence, prediction of this energy becomes significant. Prediction can be done using physical

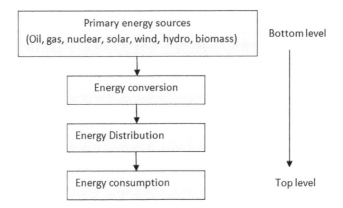

FIGURE 12.3   Forecasting – bottom-up approach.

approach or using ML approach based on historical data analysis. Accurate prediction still remains a challenge due to the irregular wind speed over a time period. The accuracy in forecasting can be increased by adopting Artificial Neural Network (ANN)-based techniques like Convolutional Neural Network (CNN).

12.5.3 PREDICTION OF HYDROPOWER ENERGY

The hydropower energy is obtained by the conversion of kinetic energy of water into electricity. Hydro energy has higher reliability and performance compared to other sources. It also has lower maintenance cost and is capable of adjusting according to the changes occurring in the load. This energy is dependent on the following factors – size of the turbine and the volume of water passing through it. This necessitates for the prediction of the turbine size and the volume of water that can pass through it. But the relation between the two is non-linear and complex in nature. Using AI and ML techniques like support vector machine and genetic algorithm, optimization and prediction are possible.

12.5.4 PREDICTION OF BIOMASS ENERGY

By harnessing biological sources, biomass energy can be obtained. It forms a part of the carbon cycle where the transfer occurs from the ambience into the plants, from plants into the soil and from soil into the ambience when the plant decays. The bio-energy finds utility in many sectors including transportation and heating. Most of the biomass is produced in the agricultural fields, forests and in the waste lands. Biomass prediction can be done precisely using certain AI-ML techniques like linear regression models.

The drawbacks that exist in the predictions for these forms of energy are the lack of availability of data sets, equipment to measure them at some locations and standard techniques that can be adopted in measurements.

12.6 NEURAL NETWORK TECHNIQUES APPLIED IN THE PREDICTION OF RENEWABLE ENERGY

Energy prediction is a two-step process – the first step involves data acquisition and processing, and the second step is ML algorithm application for energy prediction. Data acquisition and processing involve passing the data through processing module to normalize and remove the unwanted data. This forms the data preprocessing stage, which is not required for all the data training techniques.

TABLE 12.1
Different Parameters Involved in Prediction of Renewable Energy [1]

Input Parameter	Predicting Solar Energy	Predicting Wind Energy	Predicting Hydro Energy
Time	Yes	Yes	Yes
Season	Yes	Yes	Yes
Location	Yes	Yes	Yes
Solar power	Yes	–	–
Solar radiance	Yes	–	–
Diffuse irradiance	Yes	–	–
Temperature	Yes	Yes	–
Wind power	–	Yes	–
Wind speed	–	Yes	–
Water flow	–	–	Yes
Hydraulic head	–	–	Yes

The data are then clustered to create a training data set and analyzed to establish the correlation of data in order to get the exact delay involved in the forecasting models. Input parameter required for different types of energy is different and is summarized in Table 12.1.

Prediction of solar energy is based on parameters like time, season, sun's altitude, latitude/longitude, solar power and ambience temperature. Wind energy is based on time, season, location of wind turbines, wind power, and its direction and speed. Hydro power prediction depends on time, season, location of hydro power plants, water flow, pressure and humidity. For biomass energy prediction, usually image-based techniques are used. In prediction of RE, methods like the statistical method uses collected data, ML techniques like ANN, numerical methods based on weather forecast/ satellite images or combination of all the above can be used. Neural network-based approaches are the most commonly used ML models in prediction. It has the capability to learn, memorize and build the connections among non-linear data. The neural network-based approaches are fault-tolerant and are capable to estimate for any continuous non-linear function with good accuracy. Many variations of the neural network like CNN, recurrent neural network (RNN) [2] or multi-layer perceptron (MLP) also prove to be efficient in prediction. ML-based training and prediction stages are shown in Figure 12.4.

12.6.1 MLP MODELS

MLP is also called as feed-forward neural network, and it possess the most simple neural network architecture. All the neurons of a particular layer have a forward connection with every neuron on the subsequent layer. The activation function is present in each neuron. The layered structure includes hidden, input and output layers that are capable of approximating the future values by accepting certain input. The MLP model is a fully interconnected network in which each network is assigned a weight W_{ij}. The network is then trained by assigning small values to the weights. The activation function generates the output by acting on the received inputs. The optimal combination for the output is achieved by optimizing the weights using learning method and back-propagation algorithm.

12.6.2 CNN MODELS

These models are particularly specialized in processing huge matrix of data. They have the capability of predicting the future behavior of the power generation using the past information on the renewable sources of energies like wind and solar energy. The CNN models are themselves capable

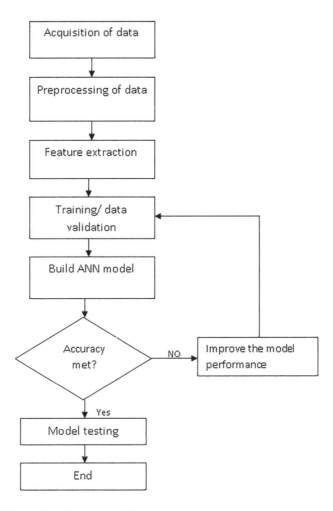

FIGURE 12.4   ML-based training and prediction stages.

of processing the data. Hence, additional feature extraction methods are not required. CNN model generally employs four different layers – input, convolution, pooling and output layers that are capable of performing input feeding, feature extraction and output prediction.

Input feeding is performed by the input layer. It specifies the input time series data that are fed to the weights for processing. Feature extraction is performed by convolution and pooling layers. The convolution layer which can be a single layer or multiple layers establishes the relationship between the time series input data. The pooling layer then decreases the dimension of the data associated to the target variable and generates feature maps as output. The output layer then combines the feature maps to the fully connected network layers through flattening layer. This layer helps in flattening the feature maps into columns to easily insert the data into the network layer. The fully connected network layer then predicts the future value depending on the features present in the input and the target variables. The block diagram of a basic CNN model to forecast time series data is shown in Figure 12.5.

12.6.3 RNN MODELS

RNN models have infinite impulse response, whereas CNN has finite impulse response. RNN models has directed cyclic graph which cannot be unrolled. It shares parameters with neurons present

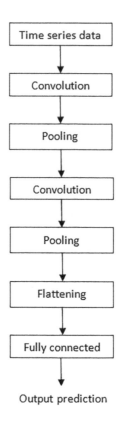

FIGURE 12.5　Block diagram of a basic CNN model to forecast time series data.

in different layers in order to predict better results. It can be used in predicting time series and is capable to hold the previous information to yield the output. This helps it in predicting the future values. As the model can remember information, it proves to be efficient in solar and wind power predictions.

RNN can also have additional stored states that can be under the direct control of the neural network. The storage can also be replaced by a different network or graph if it incorporates time delays or has feedback loops. These states are called gated memory or gated states and forms a part of long short-term memory networks (LSTMs).

LSTM model is upgraded version of RNN that helps to capture the time series dependency that exists in prediction applications. The prominent feature of LSTM is that it can remember information passed through the network for longer time. It is more efficient in RE forecasting purposes.

12.7　LEARNING ALGORITHMS FOR ANN TRAINING

A popular learning algorithm used in ANN is back-propagation algorithm. This algorithm trains the neural network using chain rule method and is illustrated in Figure 12.6. ANN training using back-propagation algorithm involves two different phases:

1. Propagation phase involving forward pass in which the input propagates from input layer to the output layer and backward pass in which the outputs are taken from the output layer

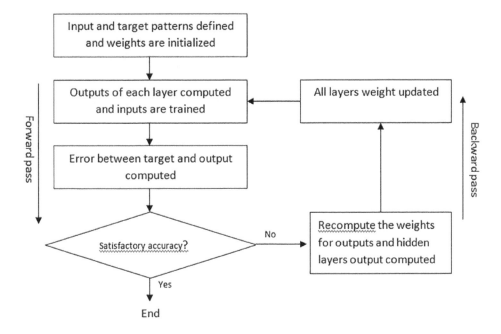

FIGURE 12.6   Back-propagation algorithm.

and compared with the target. The difference between the two is considered as error. The error propagates from the outer nodes to the inner nodes.

2. Weight updation phase which involves minimization of the error to an acceptable level by updating the weights until satisfactory results are obtained by using the gradient in gradient descent algorithm. Gradients are obtained by multiplying the input stimulus with the output delta and the gradients are then added to previous weights to obtain new values for the weights.

12.8   CONCLUSION

Although fossil fuels are the main source of energy, they form the major component in polluting the environment and are also depleting at a faster rate. As an alternate to this, RE sources like solar, wind, hydro and biomass can be used. But these sources are dependent, unreliable and unpredictable in nature. This forces the integration of RE sources with fossil fuels to form a hybrid energy system. Accurate prediction still remains a challenge in hybrid energy system and demands for the development of a predictable model. Over the last few decades, the hybrid energy systems have witnessed the usage of AI techniques in smart generation of electricity, RE prediction, emergency response and in quick energy delivery.

This chapter suggests ANN-based techniques to develop a predictable model for hybrid system. ANN has become popular as an analytical alternative technique to mathematical model because of its capacity to express the non-linear nature of input and output variables. They are adaptive and learn from examples, perform predictions at higher speed and handle data that are noisy and not complete. Without any prior knowledge on the nature of the primary process, they have the capacity to make approximations of non-linear functions with minimum parameters.

REFERENCES

1. Md Mijanur Rahman, Mohammad Shakeri, Sieh Kiong Tiong, Fatema Khatun, Nowshad Amin, Jagadeesh Pasupuleti, Mohammad Kamrul Hasan, "Prospective Methodologies in Hybrid Renewable Energy Systems for Energy Prediction Using Artificial Neural Networks," *Sustainability*, 13(4), 2393, 2021.
2. Saravanan Krishnan, Ramesh Kesavan, B. Surendiran, G. S. Mahalakshmi, *Handbook of Artificial Intelligence in Bio Medical Engineering*, CRC Press, 2021.

13 Bidirectional Converter Topology for Onboard Battery Charger for Electric Vehicles

Shreyaa Parvath, P. Mirasree,
R. Nivedhithaa, and G. Kanimozhi
Vellore Institute of Technology

CONTENTS

Abstract

In this chapter, a multifunctional onboard battery charger (OBC) for electric vehicles has been modeled that facilitates charging of low-voltage batteries and high-voltage batteries simultaneously without additional switching circuits or bulky capacitors. The active power decoupling circuit that obliterates second-order ripples doubles as the charging circuit for the low-voltage battery. In addition, thin-film dc-link capacitors used in the proposed converter replace larger capacitor banks in conventional converter topology. To facilitate simultaneous charging of both low-voltage and high-voltage batteries, a dc-dc converter is included that can operate in buck, boost and buck-boost modes. The transformer core is shared by the dc-dc converter and the low-voltage charging circuit, thus reducing the overall build of the converter. A prototype model of the multifunctional converter is designed and simulated using MATLAB®/Simulink.

13.1 INTRODUCTION

Over the last few years, with the depletion of fossil fuels and the pollution levels rising in most parts of the world, various sectors have been working to achieve more sustainable and environment-friendly alternatives, including the transportation sector. As a result, notable growth has been observed in the production of plug-in electric vehicles on both national and global scale [1,2]. In this fast-paced world, people require electric vehicles with batteries that can be charged in a small

DOI: 10.1201/9781003331117-13

duration of time. The battery charger is one of the most crucial components of the plug-in electric vehicle, as it mainly determines the charging time. This has created the necessity to have an effective and long-lasting onboard battery charger (OBC) [3]. An OBC with high efficiency and high power density can be achieved by adjusting the components and other factors in different topologies [4]. In single-phase battery chargers, a voltage ripple is observed at the dc-link that causes fluctuation in the power. Large capacitor banks can be used to smoothen out the ripple but due to their bulky size, it poses a crucial hindrance to obtaining a high power density for the charger [5]. The utilization of active power decoupling (APD) circuits is the most common approach adopted to absorb the second-order ripples [6].

A converter with high power density would require a compact but efficient design that can prove to be a problem when multifunctional chargers are required. Integrated power units can help decrease the weight, space and cost of a charger and hence cause an increment in the overall power density of the charger [7–9], which means the traction battery can be employed to charge the auxiliary battery and both the batteries can also be charged simultaneously. Moreover, in the case of multifunctional chargers, the number of components used is also high, ultimately resulting in more losses. In References [10,11], an integrated OBC model with APD function and capability to charge an auxiliary battery is proposed with reduced dc-link capacitance that ultimately leads to reduced converter size. Similarly, in References [12,13], an OBC with the capability to simultaneously charge auxiliary and traction batteries is proposed. However, these configurations operate only under unidirectional power flow conditions that are not ideal for the vehicle-to-grid (V2G) mode of operation.

Therefore, a multifunctional OBC, i.e., one that can operate in multiple modes and facilitate charging of the axillary and traction battery simultaneously without any additional switching circuits or bulky capacitors to make sure that the system is small in size, is proposed.

13.2 WORKING PRINCIPLE OF THE OBC

The circuit configuration of the OBC is displayed in Figure 13.1. The rectifier with APD function converts the ac grid input to dc output as required for the charger application. In APD method the ripple power is absorbed at twice the grid frequency by using active switches and energy storage units such as capacitors [14,15]. The electrostatic fields of the capacitor store the ripple energy. This produces lower voltage stresses but requires a higher capacitance value. Alternatively, two capacitors C_1 and C_2 of the same value connected in series can be used to absorb the ripple power.

13.3 MODES OF OPERATION

Mode 1 – Grid-to-Vehicle (G2V) Mode

In this mode, the traction battery is charged from the grid through the front-end rectifier and the buck-boost converter. The ac-dc converter portrayed in Figure 13.2 operates as an APD circuit. The circuit is analyzed in open-loop conditions. The bidirectional buck-boost converter helps in

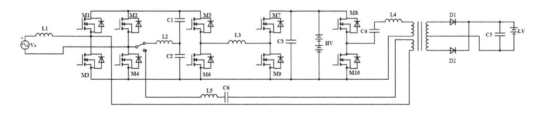

FIGURE 13.1 Circuit diagram of the charger.

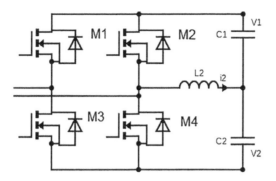

FIGURE 13.2 Active power decoupling circuit.

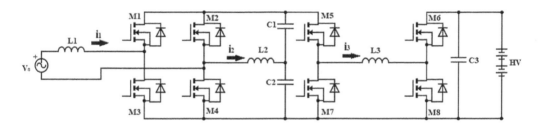

FIGURE 13.3 Charging circuit for Mode 1 operation (G2V).

controlling the current and voltage of the traction battery [16]. Figure 13.3 displays the circuit involved in Mode 1 operation.

The ac-dc stage acts as a bridge between the grid and the converter. The values of the components involved are calculated based on the following equations:

$$P_{\text{in}} = V_s I_s + V_s I_s \cos(2\omega t) = P_{\text{dc}} + P_{\text{ac}}, \tag{13.1}$$

where V_s is the grid Root Mean Square (RMS) voltage, I_s is the grid RMS current, C_f is the capacitance of filter capacitor, P_{dc} is the power fed to the ac-dc crter and P_{ac} is the ac component of the ripple power. The input inductance L_1 has been neglected. The APD circuit decouples the two components of P_{in}.

$$V_{C_1}(t) = V_{dC_1} + V_C \sin(\omega t + \varphi) \tag{13.2}$$

$$V_{C_2}(t) = V_{dC_2} - V_C \sin(\omega t + \varphi). \tag{13.3}$$

From equations (13.2) and (13.3),

$$i_{C_1}(t) = \omega C_f V_C \cos(\omega t + \varphi) \tag{13.4}$$

$$i_{C_2}(t) = -\omega C_f V_C \cos(\omega t + \varphi). \tag{13.5}$$

The instantaneous power of the APD circuit, P_{APD}, is given by

$$P_{\text{APD}} = i_{C_1}(t) V_{C_1}(t) + i_{C_2}(t) V_{C_2}(t) \tag{13.6}$$

$$P_{\text{APD}} = \omega C_f V_{C_2} \sin(2\omega t + 2\varphi). \tag{13.7}$$

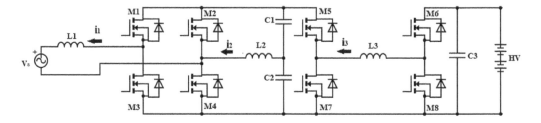

FIGURE 13.4   V2G mode of operation.

The ripple power is absorbed by the capacitors C_1 and C_2. At half the dc-link voltage ($V_c = \dfrac{V_{dc}}{2}$), the ripple power of the APD circuit is at its highest.

Mode 2 – Vehicle-to-Grid Mode (V2G)

Figure 13.4 displays the circuit for Mode 2 where the energy is fed back to the grid. The ac-dc converter functions as an inverter and helps in regulating the grid current and dc-link voltage. The bidirectional buck-boost converter helps in controlling the discharging current from the battery [17].

Mode 3 – High-Power Low-Voltage Charging (HP-LVC) Mode

In Mode 3, the circuit is disconnected from the grid and the buck-boost converter. The auxiliary battery is charged by HP-LVC circuit. The circuit diagram for Mode 3 is illustrated in Figure 13.5. The traction battery charges the auxiliary battery when the vehicle is disconnected from the grid. This is achieved by means of LLC resonant converter. The switches M_1, M_2, M_3 and M_4 act as the primary end of the transformer. The resonant converter filters out harmonics of the dc input to give sinusoidal ac voltage [18]. But this mode of charging works only during traction battery gets charged.

Mode 4 – Low-Power Low-Voltage Charging (LP-LVC) Mode

Conditions may arise where the auxiliary battery gets depleted even while the traction battery is being charged. Taking this constraint into concern, an additional low-power circuit of 500 W is added, which facilitates simultaneous charging of traction and auxiliary batteries.

In Mode 4, both traction and auxiliary batteries are charged simultaneously by an ac-dc converter, dc-dc converter and the LP-LVC circuit. Once the auxiliary battery is charged, the charger switches to G2V operation. Figure 13.6 displays the circuit for Mode 4.

The theoretical waveforms of current through inductor L_3 during G2V and V2G operation are given in Figure 13.7.

13.4   DESIGN SPECIFICATIONS

The value of C_f can be determined as

$$C_f = \frac{4P_{in}}{V_{dc}\omega},$$

(13.8)

where $C_f = C_1 = C_2$.

Hence, the equivalent capacitance of the ac-dc circuit is

$$C_{eq} = \frac{C_f}{2}.$$

(13.9)

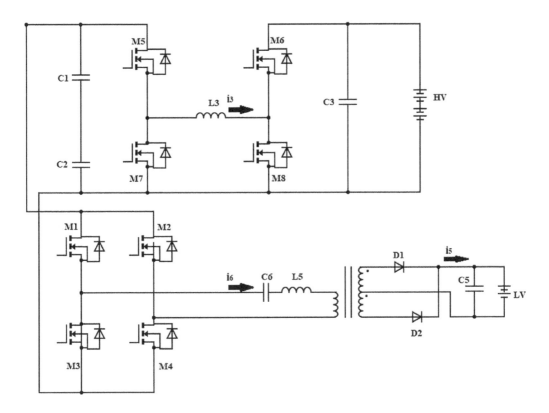

FIGURE 13.5 Charging circuit for Mode 3 operation.

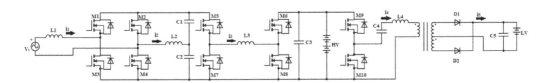

FIGURE 13.6 Charging circuit for Mode 4 operation.

From equation (9.8), the required capacitance is calculated. The dc-link capacitance is given by

$$C = P_{in}\omega\, V_{dc}\Delta V_{dc}, \tag{13.10}$$

where ΔV_{dc} is the peak-to-peak ripple voltage. Since the APD circuit is functioning in continuous mode, some reactive power is consumed by L_r (filter inductance). To take this into account, α is used.

$$\alpha = \omega 2L_2 2C_f \tag{13.11}$$

The dc-link voltage needs to be maintained at a constant value for charging and discharging the traction battery operates over a wide range of voltage (250–430 V) to fulfill all the roles of the traction battery. A 1 kW HP-LVC is considered. To determine the duty cycle (D),

$$V_o = \frac{2V_sN_sD}{N_p}. \tag{13.12}$$

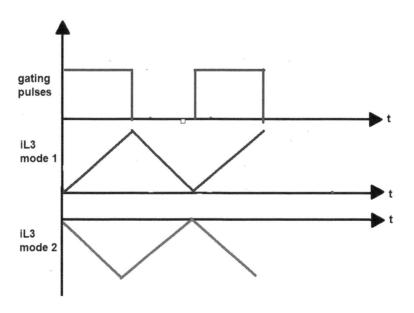

FIGURE 13.7   Theoretical waveform of current through inductor L_3 during V2G and G2V operation.

$$\text{Transformer turns ratio}(n) = \frac{N_s}{N_p}\alpha\frac{V_o}{V_s} = 25, \tag{13.13}$$

where N_p denotes the number of turns in the primary winding of the transformer and N_s denotes the number of turns in the secondary winding of the transformer. Average inductor current is given by

$$i_4 = \frac{V_o}{R}. \tag{13.14}$$

Assuming 40% current ripple,

$$\Delta i_4 = 0.4 \times i_4.$$

The value of inductance is given by

$$L_5 = \frac{V_o(0.5 - D)}{\Delta\, i_4\, f}. \tag{13.15}$$

Assuming 5% voltage ripple,

$$\Delta V_o = 0.05 \times V_o, \tag{13.16}$$

$$\frac{\Delta V_o}{V_o} = \frac{1 - 2D}{32 f^2 L_5 C}. \tag{13.17}$$

13.5   SIMULATION RESULTS

In this section, a comprehensive analysis of the results observed in the converter simulation is discussed. During Mode 1 operation, the charging and discharging are controlled by reversing the direction of grid current. Table 13.1 shows the design specifications of the converter.

TABLE 13.1

Design Specifications of the Proposed Converter

Parameters	Value
Grid voltage	230 V RMS
Grid frequency	50 Hz
Switching frequency	100 kHz
dc-link voltage	350 V
Grid inductor (L_s)	4 µH
Traction battery voltage	250–430 V
Power of traction battery charger	2 kW
Auxiliary battery voltage	12–14 V
Power of HP-LVC	1 kW
Power of LP-LVC	0.5 kW
Transformer ratio	25:1
Resonant inductance, L_4	68 µH
Resonant capacitance, C_5	74 nF
Resonant inductance, L_5	34 µH
Resonant capacitance, C_6	37 nF

13.5.1 Mode 1 and Mode 2 Operation

Figure 13.8a and b shows the current and voltage waveforms of the traction battery during G2V operation. During G2V operation, a constant value of 350 V is maintained at the dc-link, which is represented in Figure 13.8c.

The direction of current through the inductor changes with change in mode of operation. With reversal in power flow, the current state is also reversed, which is represented through current across inductor L_3 in Figure 13.8d. The dc-link voltage is acquired across the capacitors C_1 and C_2 and with half the voltage across each capacitor as represented in Figure 13.8e and f. During Mode 2 operation, the waveform will be reversed owing to reverse current direction.

Mode 2 corresponds to V2G operation where the power stored in traction battery is transferred to the grid. Since the traction battery is discharged to feed the grid, the System on Chip (SoC) of the battery gradually decreases as displayed in Figure 13.9a. The current through the circuit is reversed, which is illustrated through the direction of current through L_3 in Figure 13.9b.

13.5.2 Mode 3 – HP-LVC Circuit

During Mode 3 operation, the traction battery discharges to charge the auxiliary battery. The voltage and current waveforms of the auxiliary battery during Mode 3 operation are represented in Figure 13.10a and b. The discharge of the traction battery is indicated through the decrease in SoC in Figure 13.10c. The current through resonant inductor L_4 is represented in Figure 13.10d.

13.5.3 Mode 4 – LP-LVC Circuit

During Mode 4 operation the half-bridge LLC resonant converter is turned ON. The switching frequency of the switches is maintained at 100 kHz and the high voltage (HV) charging and LP-LVC are operated simultaneously. Figure 13.11a and b displays the required output current and voltage values across the auxiliary battery.

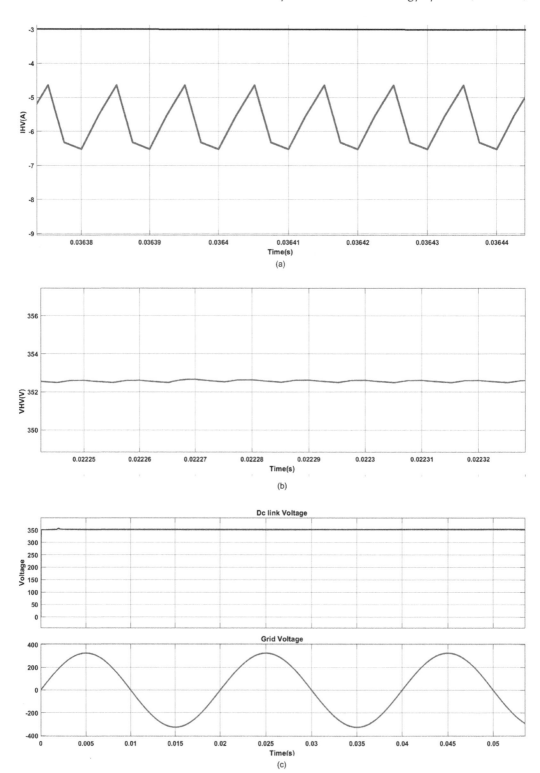

FIGURE 13.8 Waveforms during Mode 1 operation: (a) current waveform for traction battery, (b) output voltage of traction battery, (c) dc-link voltage and source voltage.

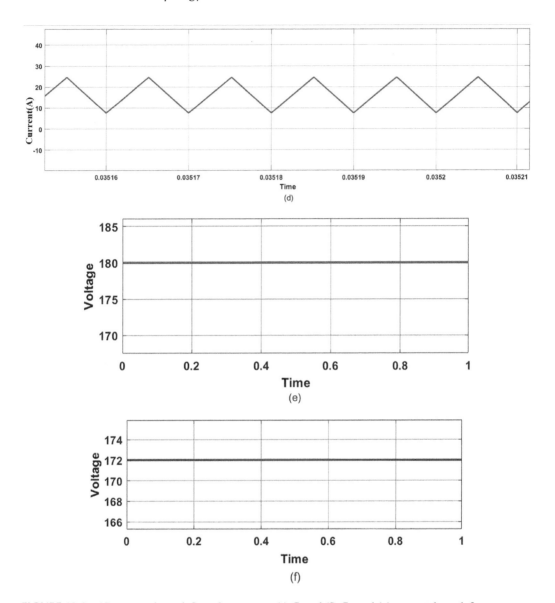

FIGURE 13.8   (d) current through L_3, voltage across (e) C_1 and (f) C_2, and (g) current through L_2.

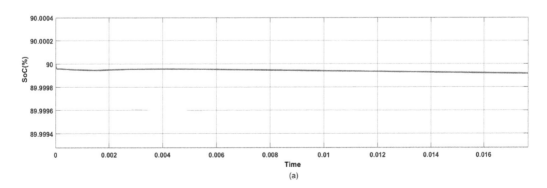

FIGURE 13.9   Related waveforms during Mode 2 operation: (a) SoC of high-voltage battery.

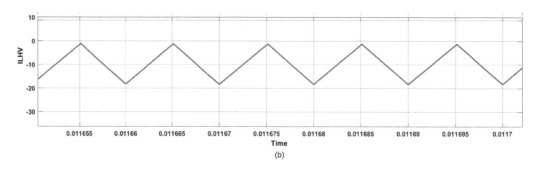

FIGURE 13.9   (b) current through L_3.

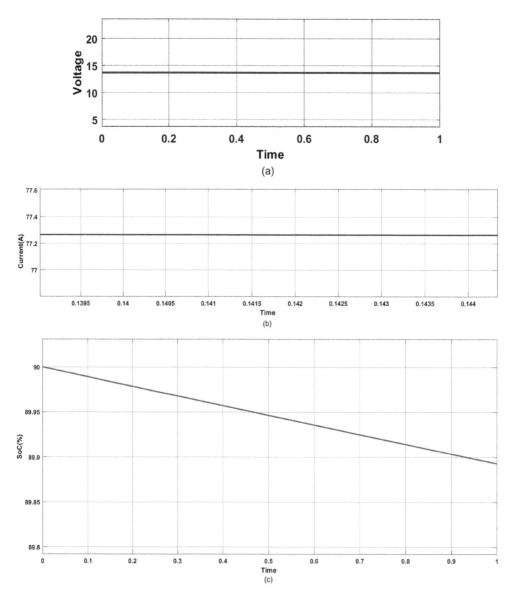

FIGURE 13.10   Waveforms during Mode 3 operation: (a) voltage in the auxiliary battery, (b) current waveform for the auxiliary battery, (c) SoC of the traction battery, and (d) current across resonant inductor.

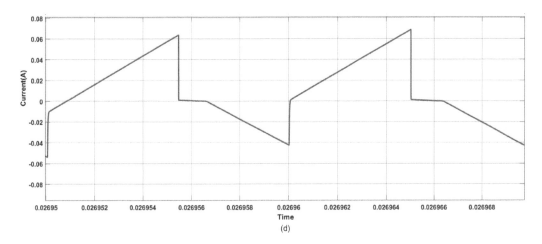

FIGURE 13.10 (d) current across resonant inductor.

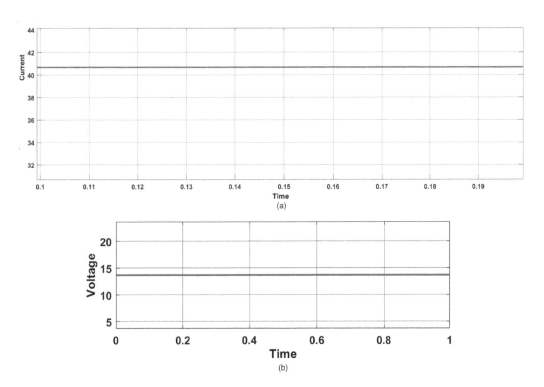

FIGURE 13.11 Waveforms for Mode 4 operation: (a) current across the auxiliary battery and (b) voltage across the auxiliary battery.

Figure 13.12a illustrates the variation in output voltage levels of the high-voltage battery during Mode 1 operation when the initial SoC of the battery is varied. The expected output is achieved at higher SoC levels. While the output voltage varies, the current shows negligible variation with decreasing SoC levels. Figure 13.12b shows the consequent variation in the efficiency of the converter. The general efficiency of the converter during Mode 1 operation is estimated to be 93.75% and that during Mode 2 operation is estimated to be 91.98%.

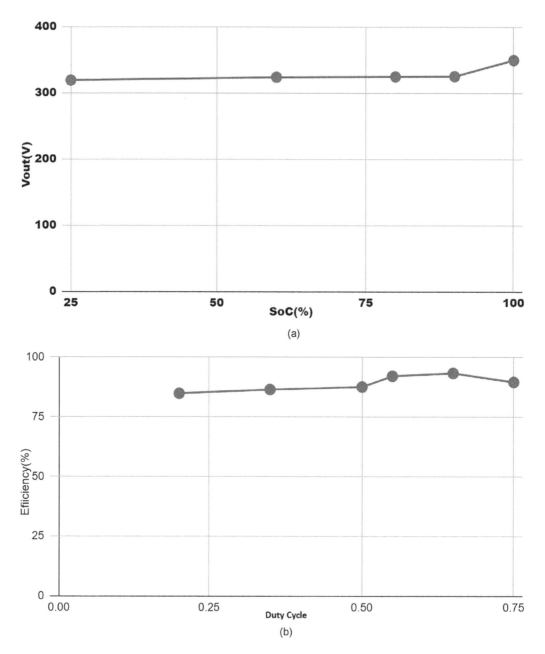

FIGURE 13.12   (a) Measured voltage during varying SoC levels of high-voltage battery during Mode 1 operation. (b) Efficiency of the converter during Mode 1 operation with varying duty cycles.

13.6   CONCLUSION

A multifunctional OBC has been proposed in this chapter. The topology of the converter is such that it facilitates the reuse of components, with a decrease in the overall build of the converter and consequently the cost. The front-end converter functions as a rectifier, inverter and APD circuit. The LP-LVC circuit facilitates the charging of traction and auxiliary batteries simultaneously. The APD capability is achieved using the auxiliary charging circuit, hence avoiding the usage of bulky capacitors. The highest efficiency for G2V operation is 93.75% and for V2G operation is 91.89%.

REFERENCES

1. D. P. Tuttle and R. Baldick. "The evolution of plug-in electric vehicle-grid interactions." *IEEE Transactions on Smart Grid*, vol. 3, no. 1, pp. 500–505, 2012.

2. J. C. Gómez and M. M. Morcos. "Impact of EV battery chargers on the power quality of distribution systems." *IEEE Power Engineering Review*, vol. 22, no. 10, pp. 63–63, 2002.

3. M. Yilmaz and P. T. Krein. "Review of battery charger topologies, charging power levels, and infrastructure for plug-in electric and hybrid vehicles." *IEEE Transactions on Power Electronics*, vol. 28, no. 5, pp. 2151–2169, 2012.

4. K. Clement-Nyns, E. Haesen, and J. Driesen. "The impact of charging plug-in hybrid electric vehicles on a residential distribution grid." *IEEE Transactions on Power Systems*, vol. 25, no. 1, pp. 371–380, 2009.

5. Y. -S. Kim, C. -Y. Oh, W. -Y. Sung, and B. K. Lee. "Topology and control scheme of OBC–LDC integrated power unit for electric vehicles." *IEEE Transactions on Power Electronics*, vol. 32, no. 3, pp. 1731–1743, 2017. doi: 10.1109/TPEL.2016.2555622.

6. Y. Sun, Y. Liu, M. Su, W. Xiong, and J. Yang. "Review of active power decoupling topologies in single-phase systems." *IEEE Transaction on Power Electronics*, vol. 31, no. 7, pp. 4778–4794, 2016.

7. S. Kim and F. Kang. "Multifunctional on-board battery charger for plug-in electric vehicles." *IEEE Transactions on Industrial Electronics*, vol. 62, no. 6, pp. 3460–3472, 2014.

8. H. V. Nguyen, D.-D. To, and D.-C. Lee. "Onboard battery chargers for plug-in electric vehicles with dual functional circuit for low-voltage battery charging and active power decoupling." *IEEE Access*, vol. 6, no. 1, pp. 70212–70222, 2018.

9. H. V. Nguyen, S. Lee, and D. C. Lee. "Reduction of dc-link capacitance in single-phase non-isolated onboard battery chargers." *Journal of Power Electronics*, vol. 19, no. 2, pp. 394–402, 2019.

10. R. Hou and A. Emadi. "Applied integrated active filter auxiliary power module for electrified vehicles with single-phase onboard chargers." *IEEE Transactions on Power Electronics*, vol. 32, no. 3, pp. 1860–1871, 2017.

11. R. Hou and A. Emadi. "A primary full-integrated active filter auxiliary power module in electrified vehicles with single-phase onboard chargers." *IEEE Transactions on Power Electronics*, vol. 32, no. 11, pp. 8393–8405, 2017.

12. D.-H. Kim, M.-J. Kim, and B.-K. Lee. "An integrated battery charger with high power density and efficiency for electric vehicles." *IEEE Transactions on Power Electronics*, vol. 32, no. 6, pp. 4553–4565, 2017.

13. H. V. Nguyen, D. -D. To, and D. -C. Lee. "Onboard battery chargers for plug-in electric vehicles with dual functional circuit for low-voltage battery charging and active power decoupling." *IEEE Access*, vol. 6, pp. 70212–70222, 2018. doi: 10.1109/ACCESS.2018.2876645.

14. T. Kar and G. Kanimozhi. "Modeling and analysis of modified bridgeless pseudo boost converter." *International Journal of Control Theory and its Applications*, vol. 9, no. 7, pp. 3315–3326, 2016.

15. S. Bhagyashri, and G. Kanimozhi. "EV charging through three-port wireless power transfer." *International Journal of Recent Technology and Engineering*, vol. 8(1), pp. 2252–2258, 2019.

16. G. Amritha, G. Kanimozhi, C. Umayal, and S. Ravi. "Bidirectional resonant DC-DC converter for microgrid applications." *The International Journal of Recent Technology and Engineering (IJRTE)*, vol. 8, no. 5, pp. 5338–5345, 2020.

17. H. -S. Lee and J. -J. Yun. "High-efficiency bidirectional Buck–Boost converter for photovoltaic and energy storage systems in a smart grid." *IEEE Transactions on Power Electronics*, vol. 34, no. 5, pp. 4316–4328, 2019. doi: 10.1109/TPEL.2018.2860059.

18. X. Zhou et al. "A high efficiency high power-density LLC DC-DC converter for Electric Vehicles (EVs) on-board low voltage DC-DC converter (LDC) application." *2020 IEEE Applied Power Electronics Conference and Exposition (APEC)*, pp. 1339–1346, 2020. doi: 10.1109/APEC39645.2020.9124278.

14 Design and Analysis of Split-Source Inverter for Photovoltaic Systems

S.S. Harshad, S. Devi, R. Seyezhai, N. Harish,
R. Bharath Vishal, and V. Barath
Sri Sivasubramaniya Nadar College of Engineering

CONTENTS

ABSTRACT

Many DC–AC inverters have been developed recently for photovoltaic (PV) applications. Among the available inverters, split-source inverter (SSI) is gaining popularity due to its single-stage operation. The SSI has numerous advantages over the conventional inverters, such as the elimination of additional switching state for stepping up input voltage, and also, it has minimized voltage spikes, switch voltage stresses and passive component count along with continuous DC link voltage and input current. This chapter presents a unidirectional DC–AC configuration of SSI that is different from its conventional configuration. It has a higher efficiency, lower Total Harmonic Distortion (THD) factor and high power density in comparison with the conventional configuration. The proposed inverter is compared with other voltage-source inverters, namely the impedance and quasi-impedance source inverters. The comparison is performed in terms of its efficiency and THD. The simulation study is carried out using MATLAB®/SIMULINK, and from the results, the SSI is found to have a better voltage gain along with higher power density, thereby making it suitable for PV systems.

14.1   INTRODUCTION

An inverter is basically a device which converts a DC voltage into an AC voltage using switching devices. There are many ways for this process, starting from traditional voltage-source inverter, which is used to give a constant output voltage, but due to its limitations on reliability, efficiency and operations, single-stage inverters have been used widely during recent years to replace the conventional voltage-source inverters (VSIs) that require an additional stage [1]. A new system was

DOI: 10.1201/9781003331117-14

developed using impedance network for coupling the converter to the source, which is known as Z-source inverter (ZSI) [2,3].

The ZSI was developed to have single-stage conversion from DC to AC. The ZSI is capable of performing both buck and boost operations with the help of four normal states and a shoot-through state, and hence, the intermediate energy conversion stage is not necessary. The impedance network enables the inverter to operate with either voltage source or current source. It can also be used for both inductive and capacitive loads. In spite of the above-mentioned advantages, the inverter suffers from discontinuous input current, thereby affecting the quality of the input of the inverter.

The quasi-Z-source inverter (qZSI) was developed to inherit the advantages of the ZSI along with continuous input current [4]. The ripple in the source current ripple is minimized by having continuous input current, and this helps in enhancing the quality of the input current waveform. However, the voltage gain (G), i.e., the ratio of load voltage (V_o) to source voltage (V_{in}) is low and hence, for applications that require high-voltage gain conversions, there is a need of an alternative inverter topology with high gain values.

The three-phase high-gain inverter topology called split-source inverter (SSI) is proposed in Reference [5], which has continuity in source current and DC link voltage along with reduced voltage stress with increase in gain values. The SSI has high-voltage gain values along with reduced number of active components that is equal to VSI. It does not require additional switching states to boost the input voltage. This chapter compares the performance of single-phase SSI using modified sinusoidal pulse width modulation with ZSI and qZSI. The simulation results of ZSI, qZSI and SSI presented in this chapter are obtained using MATLAB/SIMULINK.

14.2 TOPOLOGY STUDY OF INVERTERS

14.2.1 VOLTAGE-SOURCE INVERTER

VSI converts a constant DC voltage into an AC voltage with desired magnitude and frequency. Figure 14.1 depicts the circuit topology of VSI. It is mainly used for grid interfacing of distributed

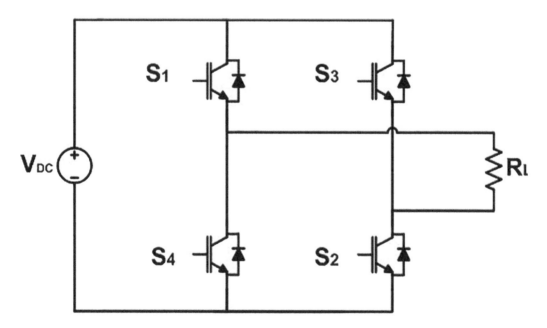

FIGURE 14.1   Circuit diagram of traditional VSI.

generation systems where it requires a two-stage conversion process consisting of a single input source. In VSI, metal oxide semiconductor field effect transistors (MOSFETs) are generally employed because it has high commutation speed, and therefore, it can deliver the output with higher efficiency. But, the output voltage of VSI is rich in lower order harmonics. Hence, to avoid these harmonics, many other single- stage conversion techniques are adopted and are discussed in the following sections.

The limitations of traditional VSI are as follows:

- It acts as a buck converter.
- It is vulnerable to noise and gets damaged if the circuit gets short-circuited or open-circuited.
- The output range is also limited.
- There can be mis-gating because of electromagnetic interference noise leading to shoot-through problem, thus reducing the inverter's reliability.

14.2.2 Z-Source Inverter

ZSI consists of a combination of inductor (L) and capacitor (C) network connected between the source and inverter to increase the reliability and efficiency. The ZSI can act as a buck-boost inverter. The output voltage can be varied by proper selection of modulation index and shoot-through duty ratio. Apart from the two active and zero states, ZSI consists of a shoot-through state additionally. The operation at active state is similar to a normal VSI, whereas, in the zero state, the load terminal is open-circuited through upper or lower legs. In shoot-through state, the impedance network is short-circuited to store the energy and the same is dissipated during the active state providing a boost. Figure 14.2 represents the topology of ZSI.

The output voltage of ZSI is given by equation (14.1) [2].

$$V_o = \frac{B \times M \times V_{dc}}{2},$$ (14.1)

where B represents buck-boost factor, M is the modulation index and V_{dc} is the DC link voltage.

For maximum boosting, the duty ratio is given by equation (14.2).

$$D = 1 - M$$ (14.2)

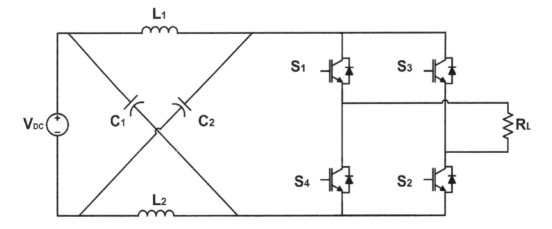

FIGURE 14.2 Circuit diagram of ZSI.

The limitations of ZSI are as follows:

- The source current is discontinuous.
- This system has lower reliability.

14.2.3 QUASI-Z-SOURCE INVERTER

The qZSI similar to ZSI boosts the input voltage using a shoot-through state. During this state, the impedance network stores the magnetic energy without short-circuiting the capacitors. This energy is delivered to the load to boost the output voltage. It has continuous and constant input current.

In shoot-through state, the diode is turned OFF and the inductors are energized through capacitors. In the null state, the load is open-circuited, i.e., there is no connection between the inverter circuit and the impedance circuit and, the capacitors are charged in this state. Finally, during the active state, the input voltage and the inductors provide energy to the inverter side. Figure 14.3 depicts the circuit diagram of qZSI.

Design Equations for qZSI: The impedance parameters are designed using equations (14.3) and (14.4) [3].

$$L_1 = L_2 = V_c T_0 \,/\, \Delta I_L \tag{14.3}$$

$$C_1 = C_2 = I_{av} T_0 / \Delta V_c \tag{14.4}$$

where L_1 and L_2 are inductance, C_1 and C_2 are capacitance, ΔI_L is the current ripple in L_1 and L_2, ΔV_c is the voltage ripple across C_1 and C_2 and T_0 is the shoot-through period.

The limitations of ZSI are as follows:

- It has reduced voltage gain.
- It has increased voltage stress across the component.

14.2.4 SINGLE-PHASE SPLIT-SOURCE INVERTER (SSSI)

Even though improved topologies like Z-source and quasi-Z-source provide a better boost, there are some problems associated with it. In order to overcome those problems and to obtain a better performance of the inverter, an SSSI is designed [6,7].

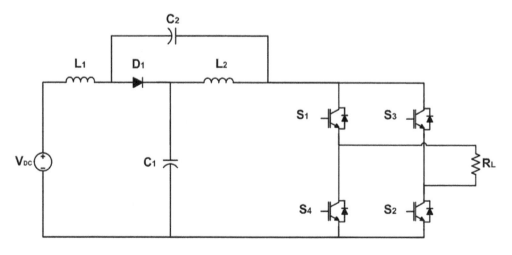

FIGURE 14.3   Circuit diagram of qZSI.

In SSSI, common cathode configuration reduces the inductance in the commutation path. The passive components used in this design include an inductor and capacitor on the DC side of the inverter. In addition, using only conventional switching states, it can boost the input voltage and also, with continuous DC link voltage, the voltage spikes across the switches can be minimized using decoupling capacitors. Figure 14.4 depicts the circuit diagram of SSSI.

Modes of Operation

There are four switching states in this inverter consisting of two active states (A_1 and A_2) and two zero states (Z_1 and Z_2). The inductor gets charged during A_1, A_2 and Z_1 states and during Z_2 it gets discharged, and charges the capacitor through the anti-parallel diodes (D_a and D_b). The various modes of operation are shown in Figure 14.5a–d.

14.3   SIMULATION OF DIFFERENT TOPOLOGIES

The comparison of the above-mentioned topologies was carried out using MATLAB/SIMULINK.

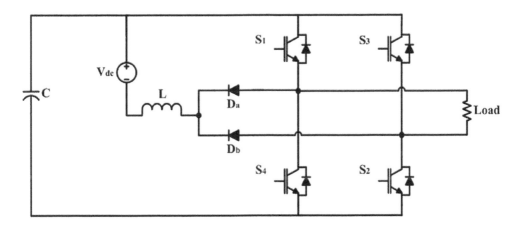

FIGURE 14.4   Circuit diagram of SSI.

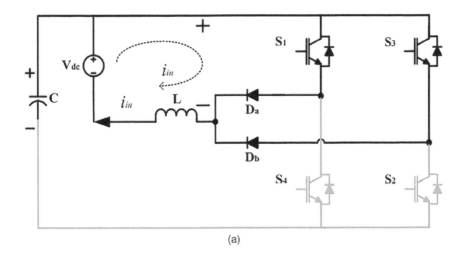

(a)

FIGURE 14.5   (a) Various modes of operation of SSSI.

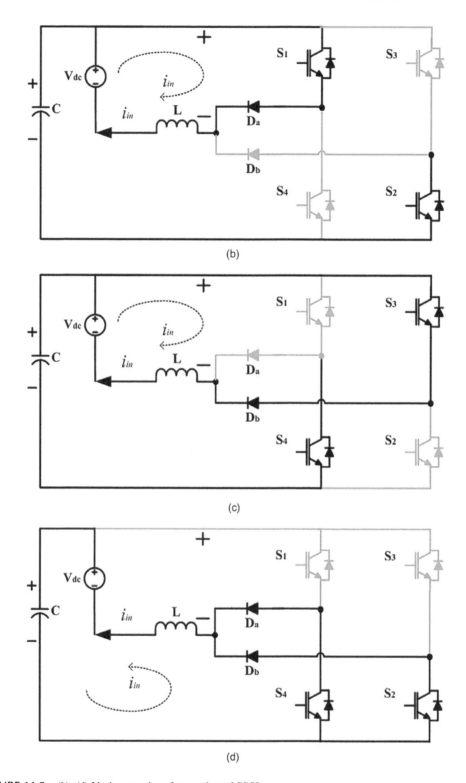

FIGURE 14.5   (b)–(d) Various modes of operation of SSSI.

14.3.1 GATE PULSE GENERATION FOR VARIOUS TOPOLOGIES

a. Simple Boost Control (SBC) Method

The SBC uses two constant signals as reference to generate shoot-through for ZSI and qZSI. The SBC control implemented for ZSI and qZSI is shown in Figure 14.6.

b. Modified Sinusoidal Pulse Width Modulation (MSPWM)

The MSPWM is shown in Figure 14.7. For SSSI, the pulses are generated by comparing the modified sinusoidal waveform with the triangular carrier waveform.

The simulation parameters for SSSI are designed using [7]. The simulation parameters for ZSI, qZSI and SSSI are given in Table 14.1.

Using the parameters mentioned in Table 14.1, ZSI was simulated; Figure 14.8 represents the voltage (V_o) and current (I_o) waveforms ZSI output.

From Figure 14.8, it is seen that the value of V_o is 33 V and the value of I_o is 1.25 A with a Total Harmonic Distortion (THD) value of V_o 4.09%.

Figure 14.9 depicts the V_o and I_o waveforms of qZSI.

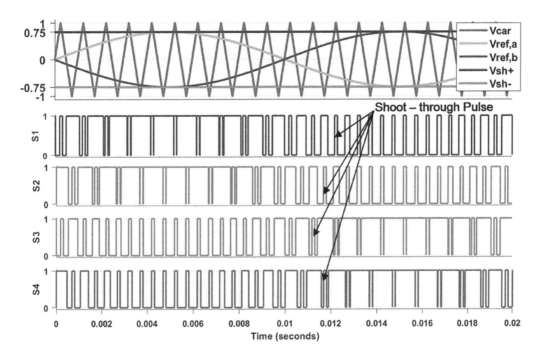

FIGURE 14.6   SBC control for ZSI and qZSI.

TABLE 14.1
Simulation Parameters for Inverter Topologies

Parameters	Values
Input voltage	16.5 V
Switching frequency	15 kHz
Modulation index	0.75 for SBC; 0.8 for MSPWM
Load resistance	30 Ω

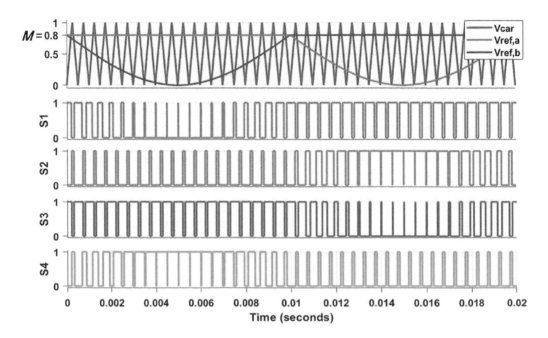

FIGURE 14.7   Pulse generation using modified SPWM.

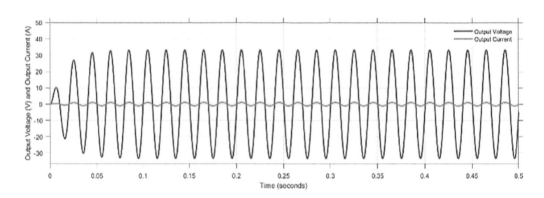

FIGURE 14.8   Voltage and current waveforms of ZSI output.

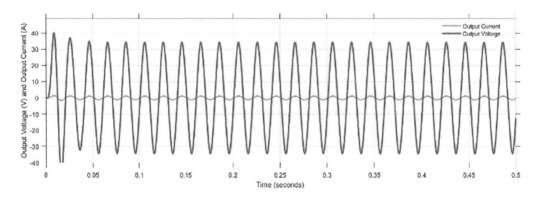

FIGURE 14.9 V_o and I_o waveforms of qZSI.

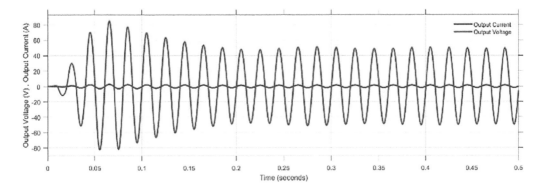

FIGURE 14.10 V_o and I_o waveforms of SSSI.

TABLE 14.2
PV Panel Specifications

Parameters	Values
Open circuit voltage (V_{oc})	21.24 V
Short circuit current (I_{sc})	2.55 A
Rated power	37.08 W
Voltage at maximum power	16.56 V
Current at maximum power	2.25 A
Total number of cells in series (N_s)	36
Total number of cells in parallel (N_p)	1

From Figure 14.9, it is seen that for qZSI, the value of V_o is 33 V and that of I_o is 1.25 A and the load voltage THD value is 4.65%.

Figure 14.10 represents the V_o and I_o of SSSI.

From Figure 14.10, it is seen that the value of V_o and I_o is 49 V and 1.65 A, respectively, and the THD value of V_o is 4.92%.

A photovoltaic (PV) panel with specifications given in Table 14.2 was modelled using MATLAB/SIMULINK [8] and has been integrated with SSSI to study the performance.

The short circuit current (I_{sc}) vs open circuit voltage (V_{oc}) and output power (P_o) vs output voltage (V_o) characteristics of PV model obtained are shown in Figure 14.11a and b, respectively.

Figure 14.12 represents the V_o and I_o waveforms of SSSI interfaced with PV.

From Figure 14.12, it is seen that the values of V_o and I_o are 49 V and 1.5 A, respectively, and the THD value of V_o is 4.97%.

14.4   COMPARISON AND RESULTS

The above-mentioned topologies are compared on the basis of component count, gain value and filter size [9]. Table 14.3 gives the comparison of various topologies.

From Table 14.3, it is found that the SSSI has the highest voltage gain along with reduced numbers of L and C in comparison with VSI, ZSI and qZSI.

Figure 14.13 represents the variation in V_o for different values of M for ZSI, qZSI and SSSI.

From Figure 14.13, it can be seen that for the same value of M, the voltage gain obtained with SSSI is greater than that obtained with ZSI and qZSI.

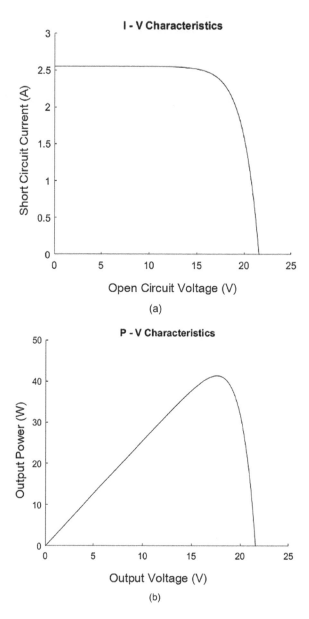

FIGURE 14.11   (a) and (b) PV model characteristics.

TABLE 14.3
Comparison of Various Topologies

Parameters	VSI	ZSI	qZSI	SSSI
Switch count	4	4	4	4
Inductor count	0	2	2	1
Capacitor count	0	2	2	1
Diode count	0	0	1	2
Voltage gain	-	$\dfrac{1}{2M-1}$	$\dfrac{1}{2M-1}$	$\dfrac{M}{1-M}$

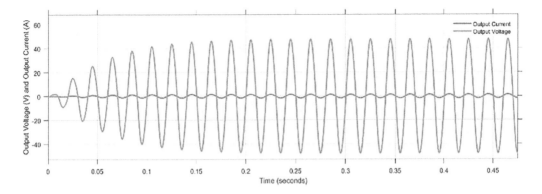

FIGURE 14.12   Output voltage and current waveforms of SSSI interfaced with PV.

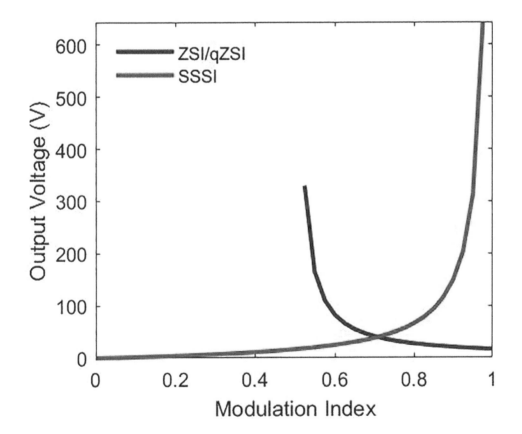

FIGURE 14.13   Plot of output voltage vs modulation index.

14.5   CONCLUSION

In comparison with conventional VSI, ZSI and qZSI, the SSSI has continuous DC link voltage and also the need for shoot-through states in order to boost the input voltage is eliminated. This results in decreased switching losses. In terms of voltage gain, SSSI provides a higher value than ZSI/qZSI, and by choosing an optimum switching frequency, the filter size can also be reduced, providing lower THD value, thus making it ideal for PV applications.

REFERENCES

1. M. H. Rashid, *Power Electronics Hand Book Devices, Circuits, and Applications*, Third Edition. Elsevier Inc., Amsterdam, Netherlands, 2011. ISBN 978-0-12-382036-5.
2. F. Z. Peng, "Z-source inverter," *IEEE Transactions on Industry Applications*, vol. 39, no. 2, pp. 504–510, 2003.
3. M. Murali, P. Deshpande, B. Virupurwala, and P. Bhavsar, "Simulation and fabrication of single phase Z-source inverter for resistive load," *UPB Scientific Bulletin, Series C*, vol. 78, no. 1, pp. 112–124, 2016. ISSN 2286-3540.
4. J. Anderson and F. Peng, "A class of quasi-Z-source inverters," in *2008 IEEE Industry Applications Society Annual Meeting*, pp. 1–7, 2008.
5. A. Abdelhakim, P. Mattavelli, and G. Spiazzi, "Three-phase split-source inverter (SSI): Analysis and modulation," *IEEE Transactions on Power Electronics*, vol. 31, no. 11, pp. 7451–7461, 2016.
6. O. G. Londhe and S. L. Shaikh, "Analysis of the single-phase split-source inverter by using different DC AC topology," *IJERT*, vol. 9, issue 7, pp. 1–3, 2020.
7. P. Davari, F. Blaabjerg, A. Abdelhakim, and P. Mattavelli, "Performance evaluation of the single-phase split-source inverter using an alternative DC–AC configuration," *IEEE Transactions on Industrial Electronics*, vol. 65, no. 1, pp. 363–373, 2018.
8. N. Pandiarajan and R. Muthu, "Mathematical modelling of photovoltaic module with Simulink," *International Conference on Electrical Energy Systems (ICEES 2011)*, Newport Beach, pp. 258–263, 2011.
9. U. Devaraj, S. Ramalingam, and D. Sambasivam, "Evaluation of modulation strategies for single-phase quasi-Z-source inverter," *Journal of the Institution of Engineers (India): Series B*, vol. 100, no. 2, pp. 1–9, 2018.

15 Electric Vehicles and Smart Grid
Mini Review

R. Atul Thiyagarajan, A. Adhvaidh Maharaajan,
Aditya Basawaraj Shiggavi, V. Muralidharan,
and C. Sankar Ram
Vellore Institute of Technology

Aayush Karthikeyan
University of Calgary

K.T.M.U. Hemapala
University of Moratuwa

V. Berlin Hency and O.V. Gnana Swathika
Vellore Institute of Technology

CONTENTS

ABSTRACT

This research paper is a literature review of the integration of electric vehicles (EVs) with smart grids. An increase in the production of EVs results in power fluctuations in grid systems. The EV-grid integration requires massive communication technologies including collection, transmission, and distribution of real-time data. Modern vehicle-grid techniques could bring sustainable changes in electromobility, resulting in improvement of transport facilities and reducing pollution.

15.1 INTRODUCTION: BACKGROUND AND DRIVING FORCES

EV and smart grid, and how they are elevating the efficient functionality environmentally as well as economically are discussed in this chapter. The electrical vehicle fleet has turned into mass storage of energy, which in turn allows the integration of renewable energy sources. Introducing

DOI: 10.1201/9781003331117-15

Vehicle-2-Grid (V2G) functionality in the mere future and their impact and limitation of implementing through modeling of hybrid generating source to charge or discharge EVs efficiently by the energy management system is important. A vehicle installed with V2G functionality offers multifunctional features [1]. There are different types of electric vehicles (EVs) like Plug-in Electrical Vehicle (PHEV), Battery Electrical Vehicle (BEV), and Hybrid Electrical Vehicle (HEV). The major barrier to implementing EV vehicles is insufficient charging facilities in developing countries like India, and the implementation of the V2G concept. Energy security and renewable energy are important aspects of this technology, and it offers a lot of potential to tackle global warming [2]. The challenges and issues related to V2G transfer and what are the impacts of V2G on the grid and computational analyses of the economic and modeling of V2G technology are discussed. The ultimate objective of V2G is to give maximum power and satisfaction to the consumer where EV owners can charge their vehicles at low cost and discharge the stored energy in the vehicle's battery when the market price shoots up; this provides the individual owning an EV the benefit of being connected to the grid. Emission of pollutants such as CO_2, CO, etc., can be reduced. The EV can be used as external storage devices for the grid. In this work, in the field of degradation of battery, bidirectional power flow, a configuration for V2G, and the critical aspects involved with V2G are analyzed [3].

The important problems that are being faced in this present situation and how electrification of transport vehicles plays an important role in negating the abovementioned environmental threats are also discussed. The critical indicators are based totally on air, climate, water, nature, and human affect. And, those attributes play an important function in modernizing the electrification technique. Technical experts nowadays consider these attributes and implement the respective solution so that the percentage of damage caused due to these attributes can be reduced and also can be prevented shortly [4]. The challenges of renewable energy demand a balanced ecosystem. Among many choices, lithium-ion batteries are widely used for storage. Here, the authors review almost 400+ published research papers initially. The management system includes batteries, supercapacitors, fuel cells, hybrid storage, power, temperature, and heat management. A monitoring system contains current and voltage monitoring systems and the paper discusses some critical topics such as temperature and safety management and recycling processes [5]. Nowadays, the grids are getting transformed into an instrumented, interconnected, and intelligent energy distribution system. A large amount of technology and technical experts are working hard to make the electric grid smarter. Improvising the grid affects major fields such as security, optimization, modeling, electronics, analytics, and physical science. This paper connects all these fields and the technical challenges of smart grid from the research perspective are discussed. External support from the government, industries, and the public is essential to design a smart grid. The technical aspects and limitations of these commodities are combined [6].

Electric-drive vehicles, whether by electric charging or liquid fueling, have major power sources containing electric utilities. The electric utilities are so formidable that they can use the EV as a battery and the normal vehicle as a hybrid vehicle. These electric utilities are also very useful in comparing the power ratings to use them in environmental development strategies. These ratings and cost economy can reduce the burden on EV owners and they can be used easily to advertise vehicle electrification and can also be the starting point for large-scale intermittent renewable energy resources [7]. We can call a grid an SG if it has two-way communication and bidirectional power flow with self-healing capability using advanced controllers making the entire grid smarter. Here, the consumers can become "prosumers," i.e., they can both consume power from the grid and produce and supply power to the grid. So, an individual owning an EV can become a prosumer by supplying the stored energy in the battery to the grid. One of the major challenges faced in a smart grid is data management. A huge amount of complex and unstructured data is to be managed and processed to get a better insight into the collected data. This paper demonstrates a case study of the IEEE-13 bus system [8].

The drivers of smart grid (SG), the evolution of SG, smart microgrids and their standards, and ongoing research in the field of SG are discussed. Major drivers of SG include the development of

smart cities across the globe, advancements in network infrastructure, favorable regulatory policies, and an increase in the adoption of electric vehicles (EV), artificial intelligence (AI), robotics,
Internet of Things (IoT), and automation. The three main ingredients of SG are information technology (IT), power system, and communication technology. Other ingredients include interoperability, E-storage facilities, demand response, agents, transmission, penetration of renewables, system
security, sensing and measuring units, and efficiency and reliability. The evolution of SG: electro-
mechanical meters (till the late 1970s), automatic meter readers—one-way communication devices
(1980–2000), smart meters—two-way communication devices (2000–2020), and advanced metering infrastructure (AMI)—that supports SG operations (from 2020). Microgrids are also known
as island grids that can meet the local demand and supply excess power to the main grid. It also
smoothens the intermittent power from the renewables [9].

The future development of EVs in China could be more affordable compared to other countries.
The features include the process to be clean and green, efficient and economical, and compatible
with the customers. The country concentrates on modernizing its whole grid system by constructing an ultra-high-voltage (UHV) power grid, integrating modern renewable energy sources into the
power grid directly, constraining the products and developing them into various voltage ratings, and
introducing many new techniques like superconductivity, energy storage, and other comprehensive
defensive technologies; some other measures and operation planning on the smart grid also play an
important role [10].

15.2 EV CHARGING

In recent years, adoption of EVs increased rapidly due to the penetration of renewable energy
sources in the grid. Adoption of EVs has its own advantages and disadvantages. For example, the
increasing and uncoordinated EV charging will add some burden on the grid. If EVs' charging and
discharging are coordinated with the energy generated from renewable energy sources (RES) and
of power system constraints, the vehicle batteries can be efficiently used as a power storage system
and provide flexible load for power grids. The business model of charging tariff of EVs cannot be
flat and cannot be sustained for a long period, as EVs have been increasing rapidly. This paper looks
at the impact of dynamic pricing on EV charging. The four main charging strategies are: residential tariff, commercial tariff, commercial tariff with rooftop solar panels, and solar energy with
dynamic pricing [11].

The rollout of the EVs compared to the decarbonization of power sector can bring environmental benefits in terms of CO_2 emission reduction and air quality. This process has a major limitation
in that it may directly increase the demand from the power grid. These further developments also
consider the driver capabilities and the traffic congestion and also other factors. Here, they produce
a random utility model that easily integrates with the activity-based demand module. Unlike other
models, this model captures the advantages of tactical charging methods when integrated with the
smart grid. These values at the end of the model presentation provide insights into the value placed
by individual public attributes [12]. The first-of-a-kind integrated high-fidelity grid and transport
model was used to identify effective ways for widespread electrification. The model includes three
control variables and the model focuses on the on-road passenger mobility in the entire city. Based
on the usage and customer details, the scenario of manufacturing an EV is divided into four types
of mobility factors. The results of this model include the technical accomplishments and progress
of the XFC requirement used here. As a limitation, new synergies and interconnections with the
electrical systems may arise. The model needs to meet the main requirement: the coordination of
charging stations, and the grid [13].

In the EV market, the charging time reduction plays an important role in the customer market,
and charging time should be reduced to 10–15 minutes. But today's technology makes this process very complicated and the cost of manufacturing a battery becomes costlier. In this paper, an
electrochemical model is designed and tested with a high-rate charge. This model also explains

the limitations of high electrode density cells and poor electrolyte transport. This model explains to us the future needs of the enabling 4C and 6C charging technologies. So, fast charging cannot be efficiently improvised by increasing electrode porosity and negative to positive ratio [14]. The management methods of the charging process in large numbers have a direct impact and can be reduced massively. By using efficient mathematical methods, the charging behavior of the EV can be managed on a large scale. An efficient algorithm is being developed here so that we can calculate the load shift potentials defined as the range of all charging curves and the factors and reviews that are being received by the company via customers. The analyzed results from Germany show that the data load shifting based on a mobility study is necessary for extradition [15].

The mathematical principles and practical implementation for the new and current development of the state of charges are discussed. The estimation methods are divided into four main types and for each method, there is a mathematical approach. As the %SOC is the important factor that directly corresponds to the battery performance, it is proportional to the EV performance. And, these data are also helpful to the regional control strategies to achieve power efficiency. Brief literature is also given in this paper to categorize or define the mathematical references of the SOC estimation [16]. A machine model is developed to estimate the charging demand of EVs. This paper deeply discusses the impact of historic traffic data and weather conditions so that the forecasting model is accurate. These data are based on the regions of South Korea. The model includes an unsupervised clustering model for the traffic model classification and a relational analysis to identify influential factors and also includes a decision tree model for each criterion. The case studies determine the demand of EV charging both in summer and winter so that the different load profiles for each case are presented clearly. This allows us to identify the load demand based on the historical data of traffic and seasons [17].

It is a comprehensive review of existing EV charging topologies, infrastructures, and charging levels. The two types of chargers are onboard and off-board chargers. There are two types of power flow, namely, unidirectional and bidirectional power flows. As the name suggests, unidirectional power flow chargers can only be used to charge the battery from the power system, it doesn't support V2G, whereas bidirectional chargers can be used to supply the energy stored in the battery back to the grid. One of the main challenges faced by onboard chargers includes less power rating due to constraints like weight, space, and cost. Electric drives are to be used to cater to this issue. There are two types of onboard chargers, namely, conductive and inductive types. Inductive chargers support wireless charging, but it takes a long time to charge. For a better charging rate, and space, off-board chargers are the best [18]. This deals with the energy loss problems in the distribution network associated with the EV association. The charging overlap is the main limitation of the present grid technology, which is not being able to control the overload. The model particle swarm optimization process makes the power interactions between the vehicle and grid much smoother. The main objective is to evenly distribute the charging and discharging periods to the owners of the vehicles. From the results, we can infer that the energy loss can be reduced to 25% and there are more uniform load values in the distributed network [19].

The article discusses the power reduction techniques in distributed systems and that particular segment has gained more importance in recent years. By considering these flaws in the distribution network, this paper provides a two-stage EV charging planning and network reconfiguration methodology to assess the problems of minimizing the power loss in both LV and HV distribution networks. This proposed method has been directly addressed in the medium voltage distribution networks. The results indicate that there is a 63.64% power loss reduction considering the best possible situation [20].

The advantages and disadvantages of estimating SOC methods introduced by ANFI systems of a traditional battery performance model are discussed. The model presented is a battery residual capacity model with very generalized ability, adaptability, and accuracy. Based on the charging and discharging rate of the vehicle, the SOC parameters are determined. These results are formulated by using the MATLAB® platform. They have high practical values and as a result there is a

3% difference between the theoretical and experimental values [21]. To estimate the discharging rate of the battery, the model is developed based on the performance of lithium-ion cells. These experiments concentrate more on the environment's ancillary services. And, the degradation data are very useful to test the product's reliability and other factors, which can improve the EV market. The results also show us that in a timescale of 10 years, a formidable number of experiments can be performed and simultaneously ensure the development of the industrial market [22].

EVs have their own advantages when compared with classical combustion engines. So, they can be regarded as an efficient mode of transport. The main impact of EV penetration is the recharging of batteries, which adversely affects the power system infrastructure. Since the cost of a distribution transformer is high in the distribution network, we need to maintain the system properly. Proper scheduling of charging EVs should be implemented to prevent a sustained overload and extend the life of the transformer. This paper proposes a smart charging method using which the cost of charging is minimized with optimized charging power concerning the real-time tariffs. Dynamic change in the rate of charging to demand then will lead to a decongestant in the network. If there is a decongestant in the network then, the network is less prone to overloads and equipment damages [23]. The framework to estimate and evaluate the reliability of the power circuits in chargers is considered to be the major factor affecting the utility load in the grid system soon. The general infrastructure of EV charging systems is defined to have different ratings. A suitable DC-DC topology has to be chosen to illustrate the reliability issues. By using these methods, the failure between the converters can be minimized so that there is a smooth transition. The framework developed has the potential to serve as a template for reliability evaluation and comparison of future developments in EVs [24].

15.3   VEHICLE TO GRID AND GRID TO VEHICLE

The development and increase in the number of EV manufacturers caused V2G and Grid-2-Vehicle (G2V) technologies to receive a lot of attention these days. These two techniques are very important in maintaining the load to a particular level without any hindrance or blackouts. The operation techniques include generation, distribution, and usage. The customer end is very much dependent on the reconditioning from the manufacturing industries about the load or exact amount of charge that the vehicle has consumed. These topologies also provide an area for a wide range of research [25]. The researchers have shown that the transportation fleet is one of the biggest sources of emission of CO_2. The introduction of the EVs as a substitute for oil- and gas-based automobiles has been the most efficient way to reduce the dependency on oil and gas and in turn eliminating the GHGs emissions. The penetration of the RESs and EVs are the green solutions to decrease the current environmental issues. Due to intermittent nature of RESs and high investment costs of developing EVs' infrastructure, tendency for using them is under the predicted estimations. The RESs act as a complementary system, which can solve a part of this problem. However, due to their intermittent and unpredictable nature, they cannot be efficient enough to support the grid in emergency. The introduction of the vehicle to grid technology as mobile energy and its integration with the RESs and smart grid are the most efficient ways to eliminate the problems in the demand and supply. V2G helps the power grid in peak load situations. The implementation of RESs and V2G will be more efficient and valuable to the environment and to the consumers [26].

Operating structures are based on strengthening the rules and status process for EV. They define the status parts of V2G and its development scales. EVs need to increase predominantly in numbers over the coming years. The V2G brings a great advantage to the power grid as well as owners of EVs and minimizes the economic status of the grid system. EV usually has less pollution than combustion engine vehicles used these days. In the near future, there can be emerging markets to increase renewable energy resources. It can bring big in economic status to the EV owners [27]. EV has suitable power control systems, advanced features to the grid, and proper segregation, in the storage of energy for unpredicted outages. The weather and growing process are about extensive gasses. Smart grid has a provision which pays the consumer for the energy for electric mobility. This aids in cheaper grid supervision, and sustainable and stable power, which in turn moves current peak capacity to baseload. If there exists excess energy from the EV withstanding more time and space, energy can be injected from the EV into the grid. Partitions are carried out by the two-way DC-to-DC chargers and inverters [28].

One of the main issues of EV owners is the degradation of battery capacity when used in V2G mode. The model with RES and EV energy storage system is utilized to minimize energy costs for the customers by studying the cost benefits of EV owners using various Feed-in Tariffs (FITs). The time of use (TOU) and fixed tariffs are the two different types of tariffs that are considered. In a few thresholds of FITs, the EV owners engaging in V2G are benefited. Usage of the EV is thoroughly monitored and surveyed, using which we can calculate the probability of EV on-road(running), whether the EV is parked or not, and the probability of plug-in. Different case studies are made to understand the optimization strategies in different research areas [29]. A new operation mode for an EV to help the power load management is being discussed here. The improved vehicle-for-grid is directly linked with the EV's operations under power load management control and smooth functioning of the power grid. The process of EVs injecting current harmonics and providing reactive power is the main factor in the (iv4g) process. In this process, the model as a whole is connected to the electrical installation of 230 V, 16 A, and 50 Hz ratings. The model results show that the proposed (Iv4G) mode performs well and can be easily related with respect to G2V and V2G operating modes [30].

The seizing of highly demanded hybrid PV systems is discussed here very briefly. Hybrid energy systems like PV-battery-diesel have been recognized as an effective solution for this current situation. Results show that utilizing the V2G parking lot can minimize the cost for effective sizing of the hybrid system by a massive 5.21%. The scheme relating to the demand side is gaining more importance. The heuristic operation algorithm plays an important role in reducing the iterative time-consuming process, which needs much more timing. All three possible cases that can reduce the effects on the controllable loads are being discussed here [31]. The energy loss problems in the distribution network are associated with the EV association. The charging overlap is the main limitation of the present grid technology, which is not being able to control the overload. The model particle swarm optimization process makes the power interactions between the vehicle and grid much smoother. The main objective is to evenly distribute the charging and discharging periods to the individual vehicle owners. The finding reveals that the energy loss can be reduced to 25% and there are more uniform load values in the distributed network [32].

Renewable sources and low carbon are becoming key factors in setting the needs and curbing emissions [33,34]. Slow movement of conventional to EVs majorly depends upon the needed load and peak values. In many developed countries, the governments concentrate more on electrifying the transport sector, which also acts as a base of competition for manufacturers. These processes succeed once there is a change in energy. These are also high technological advancements that can serve both public and private transport [35]. We also need loads and DERs that are able to function independently. There is a model for a stable power grid proposed in which PV-equipped homes and an EV capable of conducting the analysis of transient loads, load sharing, and faults are linked. Proposed controls take into account difficult conditions in EVs and PV by managing the power of the battery in EVs by changing the V and F values in the silenced mode. Maximum Power Point Tracking (MPPT) is changed for minimizing the chance of silent mode and a maximum charge battery. The control inertia is utilized in a unified control manner. Controller processes have a good chance for changing V and F at PCC under the simulated scenarios [36].

15.4 VEHICLE TO GRID AND GRID TO VEHICLE

The strengthening of the series and functional objectives for EVs is discussed. It conveys the importance of V2G and its development scales. EVs are intended to increase drastically over the next few years. The V2G brings a great advantage to the power grid as well as owners of EVs by giving spinning constraints and minimizing the economic value in power plants. EVs usually have lower environmental pollution than existing types of vehicles. In the near future, it can be increasingly supported by renewable energy. Hence, it could bring large economic value to the grid operator and the EV owners. In the long run, we will also benefit from the whole process [37].

The EV market is steadily growing, and it is important to identify the advanced charging techniques to prevent stress on the grid. A bidirectional converter reduces these limiting effects of vehicle charging but with diverse benefits and economical offers. To evaluate different combinations of V2G based on the increase in self-consumption and peak shaving , an EV fleet was developed by a mixed-integer linear model predictive control framework. The optimal framework is coupled to calculate the life of the EV battery. This path is analyzed with the sizes ranging 1 to 150 vehicles with massive changes like energy shift between the energy changes, 2-way schemes, and descriptive formulations. The deployment gives flexibility, leading to better battery life improvements [38]. EV that takes part in V2G gives services to the operators. Hence, an exact validation of where the energy is expected to enable necessary operations of the system is a must. It is also important to provide consumer satisfaction. It is important to find the feasibility of V2G opportunities for different cases. Using survey data, the value of V2G service provision is determined. The provision of the V2G service depends on both the energy in EVs and provisional probability [39]. Two main factors of electrification of vehicles are increased driving range and fast charging capabilities. But, there is always a competing nature in the power supply department. The paper deeply discusses the model-based engineering technique that can supplement empirically and iteratively designed cells. There are many frameworks to be optimized partitions configured in automated market cells that are generally given. There are many applications of methodologies to a common module pack that needs to be investigated thoroughly for a design of a BEV and PHEV. The study in the paper is complemented by the battery optimal layer design (BOLD) toolbox [40]. An extensive review of the economic aspects of the management and risk analysis of modern problems such as the zero-emission EV is considered to be the greatest or a more modern solution. With the government's constant support, the plug-in-EV (PHEV) can be made popular. In the future, when PHEVs are randomly connected to the power grid, they can bring new challenges to the power management department of the grid. This paper also shows the theoretical response of an EV connected to a grid over the years and validates the PHEVs market [41].

Distributed energy resources as modern technology in power systems could play a determinant role in reliability issues. This paper conveys the heuristic approach to solving the Reliability Constrained Unit Commitment (RCUC) problems. Some attributed reliability indicates the commitment problem for hybrid particle swarm. The total operational cost for this process improves as the range of acceptable performance limits decreases. A composed unit commitment can be obtained using his proposed method [42]. The challenges faced (RES) mainly throw light on the Multiple Agent System (MAS). An agent is a computational system that interacts with real-world systems with features like autonomy and reactiveness (Example: Robot). A MAS consists of multiple intelligent agents to solve complex problems. With the help of MAS, the time delay can be reduced in areas like transmission switching and protection, communication, analysis/prediction, and plant control management. Due to the distributed nature of the SG, MAS is widely used in SG. The power electronic converters used to connect RES to the grid are deeply analyzed. Distributed and overlapping transmission lines controlled by MAS are modeled for analyzing the distributed generators. And, few other simulations are made for analyzing the frequency variations in the grid [43]. The growing demand for electricity due to the advancing technology demands the grid to be very smart, efficient, reliable, quick (in control), secure, and eco-friendly. A grid with all the abovementioned features is called SG. Nearly 80% of the electricity is consumed by the cities and is responsible for greenhouse gas emissions (80%). The conventional power grid mainly focuses on the generation, transmission, and distribution of electricity. This paper describes the necessity of making the conventional grids "smarter." It's not possible to say how far research in SG is required to make the grid entirely smart. But recent trends in infrastructure advanced communication protocols, intelligent electronic devices (IED), etc., encourage the researchers in the field of SG [44]. An EV can both supply and consume power to and from the grid. If it consumes power, it naturally increases the load and reduces the stability, which directly affects the power quality, whereas in V2G mode, it enhances the stability in the grid. The intermittent nature of RES and the dynamic nature of EVs affect the overall quality. To cater to these problems, this paper describes the application of advanced charging technology and proposes a transient voltage stability margin index. For the estimated index value, we can evaluate how different loads like EV affect power quality and suggest solutions for it so that the instability injected by these loads can be removed immediately [45]. The main factor that affects EV owners is degradation of battery capacity when used in V2G mode. This paper proposes a model with RES, EV, to reduce energy costs for customers by studying cost benefits of EV owners using various FITs. Two different types of tariffs are considered in a few thresholds of FITs by the EV owners. Usage of EV is thoroughly monitored, using which we can calculate the probability of EV on-road(running), whether the EV is parked or not, and the probability of plug-in. Different case studies are done to understand the optimization strategies in different cases [46]. As smart grid optimizations are an important concept, optimal dispatch of microgrids has a significant impact on reducing fuel consumption, pollution, and costs. Scheduling of microgrids under grid-connected conditions is optimized, by considering the operational cost and the environmental protective cost of the microgrid system. Distributive parts in the grid model contains PV arrays, turbines, engines, microturbines, and EVs. Related to the changes and techniques of the system, an advanced optimization is the way to solve the model. This releasing part of microgrids having EVs has better low-cost and systematic advantages compared to models without EVs [47].

Due to the rapid growth of electrical vehicles, charging the EVs to grid causes an unexpected spike in load decrement, which in turn affects the health of the power system. Ways to increase productivity of the EVs with a plan by finding suitable drive techniques. The advantage of EV is that it can store energy, which in turn can distribute power to meet a peak load demand. MOMVO [Multi-Objective Multiverse optimization] algorithm is used to reduce the effects of charging and discharging opportunities in the main system and eliminate the cost associated with the operation on behalf of both the EV owner and the utility company. EVs are made to charge and discharge at places which are merged to a bus of an economical radial system. The partition of G2V and V2G methods gives many ways to supply the utility grid on the basis of checking the demand of the power system, leveling of load status, and maximum load cutoff. This reduces burden on the substations and lines [48].

15.5 EFFECTS IN VEHICLE ELECTRIFICATION

The impact and advantages that are brought about by the introduction of PHEV are discussed. Impacts as a consequence of the introduction of PHEVs are constant load demand, reduced reserve margins, and peak load demands. The advantage of using PHEVs has reduced carbon pollution in the environment. The analysis uses brief modeling supply and demand in the electrical sector. A fixed recharging schedule is applicable to all vehicles; schedule doesn't adjust to electric supply to meet the rising demand. The inventory of electrical supply is based on the results from various models (NEMS) that can mimic supply and demand differently. It does not model the transmission system and does not reflect the complexities of air emission regulations [49]. A droop controller uses modern standards that can be used with all series of EVs with respect to international standards like IEC 61851 and SAEJ1772. The field validation tested three ancillary services: congestion management, local voltage support, and frequency-controlled normal operating reserve. The total delay was subjected to 2–3 seconds on average, and never surpassed 4 seconds. An underestimating criterion in current magnitude was detected when there was a decrease in the charging rate of the EV, which may turn into a greater problem [50]. Electrical storage in grid-connected entities has high capability to aid the transfer toward a reliable renewable energy supply. The lithium-ion batteries are expected to play a vital role in the transition, due to their high energy density and potential capacity provided by PHEV. The disadvantage of using lithium-ion battery in the grid may result in further degradation. Using intelligent control of these batteries can minimize the rate of further degradation and their utilization is cost-effective. A cycle-life experiment was performed on lithium polymer cells in which the Depth of Discharge (DOD) has a significant effect on the life of a battery cycle. The closer to the end of the discharge, the faster the equivalent series resistance rises. Increase in the rate of Equivalent Series Resistance (ESR) depends on the DOD [51].

V2G uses EV to provide power for electric systems. Vehicle has 20 times the reliable power potential, <1/10th the utilization, and 1/10th the capital cost per kW. In short, EVs should be modified for high-value critical rules and regulation, which provide about 3% of the EV fleet. V2G supports for maximum power and storage for renewable electric generation. The long-term case for V2G boils down to a choice. The chances of V2G to travel along with these paths together more swiftly are profitable than planning either system in isolation [52].

The use of EV is modeled using various power system techniques and environmental reviews based on travel demand analysis. The systematic review of these diverse approaches defines the existing contrast in the modeling techniques. For a time-being review of demand management, activity-based modeling (ABM) is being used. However, for the current (ABM) model, there is a substantial limitation on EV-grid interactions, and these data can be used to identify the charging behavior [53]. Here, the stochastic simulation methodology is being used to collect the travel delay data and charging profiles for the number of EVs. The dependent structures determine the parameters and those data are being collected by the nonparametric copula function, an iterative module for the conditional functions that is being used at each level. These methods are very much different from the conventional charging prediction methods. These methods use real-world data as an input to the model. The disjoint trip level enhances the model's capabilities and usefulness. Such data being created here are much useful for the power aggregate prediction and demand and many more power system services [54].

The BEV prototypes and their advancement in a short span are discussed. In the next stage of being adopted to the BEV, CleanMPG claims that the United States is estimated to have a million charging stations. The beginning stages of the BEV charging method include connecting them to home networks and private chargers. However, many Asia-pacific countries lean toward the public charging network as there is a reduced number of private charging facilities. These stations are more appropriate to be installed in the urban areas. This paper also provides the data points and the detailed traffic details, so the time for charging EVs can be significantly reduced [55]. An intelligent unit commitment (UC) for the V2G application. Due to the advent of EVs and power electronics, individuals owning an EV can become a prosumer by supplying the stored energy in the battery to the grid. The EV perhaps is regarded as a small portable power plant. And in turn, it reduces the reliance on the small valuable units in the existing power plant to meet the ever-increasing demand, which reduces the operating costs. The UCs usually help electricity utilities or producers in managing the supply–load demand by determining when and which electricity generating unit is to be turned on/off. The complex unit commitment with V2G development issue is solved using a leveled hybrid particle swarm optimization. This balanced hybrid binary PSO uses binary codes to turn on/off the supply from the vehicles [56]. As discussed earlier, an EV can both supply and consume power to and from the grid. In the G2V mode, the power quality drops. This paper proposes a method to identify any aberrations in the overall power quality of the grid. The Monte Carlo method is used to study the impact of PHEV on power quality. This model considers the random plug-ins and plug-outs of EVs, location, charging strategy, overall market penetration, and charging durations of EVs. This study says that the modern plug-in type hybrid EVs (PHEV) have only a negligible effect on the power quality. This method can also be used to study the impact of BEV. As PHEVs are gaining popularity in the market, this paper mainly focuses on PHEVs. To analyze the impact of BEV based on power quality, a few modifications are to be made [57]. The main problem in electric power systems is distribution imbalance. One of the solutions to the problem is the energy storage system. The objective of the paper is to analyze the residential photovoltaic system for PHEV loads along with residential loads. A combination of two subsystems is cascaded by a DC connection. The primary subsystem consists of a current-controlled boost converter, MPPT, and a PV array to harvest the solar energy. The second subsystem consists of a current-controlled DC/AC inverter. According to the load profile, a power management algorithm is established to regulate the power flow between the grid and EV battery. This system is reliable and robust on the worst days [58]. Charging of EVs leads to unexpected spikes in load demands, which influence the secondary voltage quality. This paper proposes a way to mitigate the voltage drop concerns by using TOU pricing. This pricing scheme uses the off-peak generation for EV charging, thus delaying in improving grid support. TOU schedule assesses various factors for charging EV with off-peak rates between 8 pm and 3 am. Charging the EV when the demand is less will minimize the effect of secondary service voltages, thus enhancing the grid potential and customer benefits. Load penetration due to EV is also increasing rapidly. Therefore, integrating TOU schedule in EV charging will decrease the peak load demand in secondary service voltage [59].

Electrical vehicle usage is increasing rapidly. Due to the penetration of EVs, the electrical network will grow steadily, but there are major challenges ahead in the electrical networks. The major challenges in the penetration of EVs are rise in load consumption during peak demand, which in turn increases the cost for evolving a network to provide for the required load. This paper proposes a charging/discharging methodology with a two-price and voltage-based load management program to cope up with the economic and technical purposes due to the penetration of EVs. Being sensitive in the cost and voltage in the load management will help in the reduction of cost as well as load characteristics is thickened. Without network development, managing the penetration of EVs is impossible. Besides, due to the same level of penetration, the network capacity is fully occupied, sequentially minimizing the reliability which in turn has a significant impact on the network parameter [60].

15.6 CONCLUSION

The main objective of this paper is to review the works of literature on EVs and smart grid technologies, and also the recent trend in smart grid and EVs are thoroughly reviewed. This paper is subdivided into five sections: Introduction, EV charging, Vehicle-2-Grid(V2G) & Grid-2-Vehicle(G2V), modernizing the grid, and effects of vehicle electrification. Due to the growing demand for electricity, SG architectures are being adapted. We can turn a "plain vanilla grid" into an SG by adding features like two-way communication and bidirectional power flow with self-healing capability. The phenomenon of integrating EVs with the smart grid and smart grid with EVs is known as G2V, and V2G, respectively. Various methods of integrating the EVs with SGs and EV charging methods are reviewed to understand the optimization strategies.

REFERENCES

1. Arfeen, Zeeshan; Khairuddin, A.; Kausar, Mehreen; Humayun, Usman; and Attaullah, Khidrani. (2020). V2G facts and facets with modeling of electric vehicle fast-charging station-status and technological review. *Pakistan Journal of Scientific Research*, 1(2), 1–7.
2. Goel, Sonali; Sharma, Renu; and Rathore, Akshay. (2021). A review on barrier and challenges of electric vehicle in India and vehicle to grid optimisation. *Transportation Engineering*, 4, 100057. doi: 10.1016/j.treng.2021.100057.
3. Thakre, Mohan; Mahadik, Yogesh; and Yeole, Dipak. (2021). Potentially affect of a vehicle to grid on the electricity system. *IOP Conference Series: Materials Science and Engineering*, 1084, 012077. doi: 10.1088/1757-899X/1084/1/012077.
4. Environment and Climate Change Canada. (2019). Canadian Environmental Sustainability Indicators: Greenhouse Gas Emissions.
5. Hasan, Mohammad Kamrul; Mahmud Md; Ahasan Habib, A.K.M.; Motakabber, S.M.A; and Islam, Shayla. (2021). Review of electric vehicle energy storage and management system: Standards, issues, and challenges. *Journal of Energy Storage*, 41, 102940.

6. Rosenfield, M.G. (2010). The smart grid and key research technical challenges. *2010 Symposium on VLSI Technology*, pp. 3–8, IEEE.

7. Kempton, Willett and Letendre, Steven E. (1997). Electric vehicles as a new power source for electric utilities. *Transportation Research Part D: Transport and Environment*, 2(3), 157–175. doi: 10.1016/s1361-9209(97)00001-1.

8. Shahzad, Umair. (2021). Smart grid and electric vehicle: Overview and case study. *Journal of Electrical Engineering, Electronics, Control and Computer Science*, 8, 1–6.

9. Farhangi, H. (2010). The path of the smart grid. *IEEE Power and Energy Magazine*, 8(1), 18–28. doi: 10.1109/MPE.2009.934876.

10. Zhang, R.; Du, Y.; and Yuhong, L. (2010). New challenges to power system planning and operation of smart grid development in China. *International Conference on Power System Technology*, pp. 1–8, IEEE.

11. Chen, Qin and Folly, Komla. (2022). Impact of renewable energy and dynamic pricing on electric vehicles charging. *2022 30th Southern African Universities Power Engineering Conference (SAUPEC)*, Durban, South Africa. doi: 10.1109/SAUPEC55179.2022.9730690.

12. Daina, Nicolò; Sivakumar, Aruna; and Polak, John W. (2017). Electric vehicle charging choices: Modelling and implications for smart charging services. *Transportation Research Part C: Emerging Technologies*, 81, 36–56.

13. Meintz, Lipman; Palmintier, Muratori; Panossian, Jadun; Laarabi, Waraich; Desai, Meier; Moffat, Sheppard; Meintz, Andrew; Lipman, Tim; Palmintier, Bryan; Muratori, Matteo; Panossian, Nadia; Jadun, Paige; Laarabi, Haitam; Waraich, Rashid; Desai, Ranjit; Meier, Alexandra Von; Moffat, Keith; and Sheppard, Colin. (2021). Grid-Enhanced, Mobility-Integrated Network Infrastructures for Extreme Fast Charging (GEMINI-XFC).

14. Colclasure, Andrew M.; Dunlop, Alison R.; Trask, Stephen E.; Polzin, Bryant J.; Jansen, Andrew N.; and Smith, Kandler. (2019). Requirements for enabling extreme fast charging of high energy density Li-ion cells while avoiding lithium plating. *Journal of the Electrochemical Society*, 166(8), A1412.

15. Clement-Nyns, Kristien; Haesen, Edwin; and Driesen, Johan. (2009). The impact of charging plug-in hybrid electric vehicles on a residential distribution grid. *IEEE Transactions on Power Systems*, 25(1), 371–380.

16. Chang, W.Y., 2013. The state of charge estimating methods for battery: A review. *International Scholarly Research Notices*..

17. Arias, Mariz B., and Sungwoo Bae. (2016). Electric vehicle charging demand forecasting model based on big data technologies. *Applied Energy*, 183, 327–339.

18. Yilmaz, Mustafa and Krein, Philip T. (2013). Review of battery charger topologies, charging power levels, and infrastructure for plug-in electric and hybrid vehicles. *IEEE Transactions on Power Electronics*, 28, 2151–2169.

19. Bouhouras, Aggelos; Kothona, Despoina; Gkaidatzis, Paschalis; and Christoforidis, Georgios. (2022). Distribution network energy loss reduction under EV charging schedule. *International Journal of Energy Research*. doi: 10.1002/er.7727.

20. Kothona, D. and Bouhouras, A.S. (2022). A two-stage EV charging planning and network reconfiguration methodology towards power loss minimization in low and medium voltage distribution networks. *Energies*, 15(10), 3808..

21. Robat, Amin Rezaei Pish, and Salmasi, Farzad Rajaei. (2007). State of charge estimation for batteries in HEV using locally linear model tree (LOLIMOT). *2007 International Conference on Electrical Machines and Systems (ICEMS)*, IEEE.

22. Bhoir, Shubham; Caliandro, Priscilla; and Brivio, Claudio. (2021). Impact of V2G service provision on battery life. *Journal of Energy Storage*, 44, 103178. doi: 10.1016/j.est.2021.103178.

23. Visakh, Arjun and Selvan, M. (2022). Smart charging of electric vehicles to minimize the cost of charging and the rate of transformer aging in a residential distribution network. *Turkish Journal of Electrical Engineering and Computer Sciences*, 30, 927–942. doi: 10.55730/1300-0632.3819.

24. Ghavami, M.; Essakiappan, S.; and Singh, C. (2016). A framework for reliability evaluation of electric vehicle charging stations. *IEEE Power and Energy Society General Meeting (PESGM)*, pp. 1–5, IEEE.

25. Hannan, M. A.; Mollik, Md; Al-Shetwi, Ali; Abd Rahman, Muhamad; Mansor, Muhamad; Begum, Rawshan; Muttaqi, Kashem; and Dong, Z.Y. (2022). Vehicle to grid connected technologies and charging strategies: Operation, control, issues and recommendations. *Journal of Cleaner Production*, 339, 130587. doi: 10.1016/j.jclepro.2022.130587.

26. Bibak, Bijan and Tekiner-Moğulkoç, Hatice. (2020). A Comprehensive Analysis of Vehicle to Grid (V2G) Systems and Scholarly Literature on the Application of Such Systems. *Renewable Energy Focus*, 36. doi: 10.1016/j.ref.2020.10.001.

27. Solanke, Tirupati; Ramachandaramurthy, Vigna K.; Yong, Jia Ying; Pasupuleti, Jagadeesh; Kasinathan, Padmanathan; and Arul, R. (2020). A review of strategic charging–discharging control of grid-connected electric vehicles. *The Journal of Energy Storage*, 28, 101193. doi: 10.1016/j.est.2020.101193.

28. Arfeen, Zeeshan; Abdullah, Md Pauzi; Husain, Nusrat; Kausar, Mehreen; Iqbal, Najeeb; and Rashid, Muhammad. (2021). Exposure of electrically driven vehicles and smart vehicle-to-grid technology. *International Journal of Progressive Sciences and Technologies*, 30, 220–228. doi: 10.52155/ijpsat. v30.1.3859.

29. Mwasilu, Francis; Justo, Jackson; Kim, Eun-Kyung; Do, Ton; and Jung, Jin-Woo. (2014). Electric vehicles and smart grid interaction: A review on vehicle to grid and renewable energy sources integration. *Renewable and Sustainable Energy Reviews*, 34, 501–516. doi: 10.1016/j.rser.2014.03.031.

30. Manu, D.; Shorabh, S.G.; Swathika, O.G.; Umashankar, S.; and Tejaswi, P. (2022). Design and realization of smart energy management system for Standalone PV system. In *IOP Conference Series: Earth and Environmental Science* 1026(1), 012027.

31. Swathika, O.G.; Karthikeyan, K.; Subramaniam, U.; Hemapala, K.U.; and Bhaskar, S.M. (2022). Energy efficient outdoor lighting system design: Case study of IT campus. *IOP Conference Series: Earth and Environmental Science*, 1026(1), 012029.

32. Sujeeth, S.; and Swathika, O.G. (2018). IoT based automated protection and control of DC microgrids. *In 2018 2nd International Conference on Inventive Systems and Control (ICISC)*, pp. 1422–1426, IEEE.

33. Patel, A.; Swathika, O.V.; Subramaniam, U.; Babu, T.S.; Tripathi, A.; Nag, S.; Karthick, A.; and Muhibbullah, M. (2022). A practical approach for predicting power in a small-scale off-grid photovoltaic system using machine learning algorithms. *International Journal of Photoenergy* 2022, 1–21.

34. Odiyur Vathanam, G.S.; Kalyanasundaram, K.; Elavarasan, R.M.; Hussain Khahro, S.; Subramaniam, U.; Pugazhendhi, R.; Ramesh, M.; and Gopalakrishnan, R.M. (2021). A review on effective use of daylight harvesting using intelligent lighting control systems for sustainable office buildings in India. *Sustainability*, 13(9), 4973.

35. Swathika, O.V. and Hemapala, K.T.M.U. (2019). IOT based energy management system for standalone PV systems. *Journal of Electrical Engineering & Technology,* 14(5), 1811–1821.

36. Swathika, O.V. and Hemapala, K.T.M.U. (2019). IOT-based adaptive protection of microgrid. In *International Conference on Artificial Intelligence, Smart Grid and Smart City Applications* Springer, Cham, pp. 123–130.

37. Kumar, G.N. and Swathika, O.G. (2022). Chapter 19: AI applications to renewable energy an analysis. In: O.V. Gnana Swathika, K. Karthikeyan, and Sanjeevikumar Padmanaban (Eds.), *Smart Buildings Digitalization: IoT and Energy Efficient Smart Buildings Architecture and Applications*. CRC Press, Boca Raton, FL, p. 283.

38. Swathika, O.G. (2022). Chapter 5: IoT-based smart. In: O.V. Gnana Swathika, K. Karthikeyan, and Sanjeevikumar Padmanaban (Eds.), *Smart Buildings Digitalization: IoT and Energy Efficient Smart Buildings Architecture and Applications*. CRC Press, Boca Raton, FL, p. 57.

39. Lal, P.; Ananthakrishnan, V.; Swathika, O.G.; Gutha, N.K.; and Hency, V.B. (2022). Chapter 14: IoT-based smart health. In: O.V. Gnana Swathika, K. Karthikeyan, and Sanjeevikumar Padmanaban (Eds.), *Smart Buildings Digitalization: Case Studies on Data Centers and Automation*. CRC Press, Boca Raton, FL, p. 149.

40. Chowdhury, S.; Saha, K.D.; Sarkar, C.M.; and Swathika, O.G. (2022). IoT-based data collection platform for smart buildings. In: O.V. Gnana Swathika, K. Karthikeyan, and Sanjeevikumar Padmanaban (Eds.), *Smart Buildings Digitalization: Case Studies on Data Centers and Automation*. CRC Press, Boca Raton, FL, pp. 71–79.

41. Peng, Minghong; Liu, Lian; and Jiang, Chuanwen. (2012). A review on the economic dispatch and risk management of the large-scale plug-in electric vehicles (PHEVs)-penetrated power systems. *Renewable and Sustainable Energy Reviews*, 16(3), 1508–1515.

42. Ghanbarzadeh, Taraneh; Goleijani, Sassan; and Moghaddam, Mohsen Parsa. (2011). Reliability constrained unit commitment with electric vehicle to grid using hybrid particle swarm optimization and ant colony optimization. *2011 IEEE Power and Energy Society General Meeting*, IEEE.

43. Fuchs, F.; Dietz, R.; Garske, S.; Breithaupt, T.; Mertens, A.; and Hofmann, L. (2014). Challenges of grid integration of distributed generation in the interdisciplinary research project Smart Nord. *2014 IEEE 5th International Symposium on Power Electronics for Distributed Generation Systems (PEDG)*, pp. 1–7. doi: 10.1109/PEDG.2014.6878627.

44. Butt, Osama Majeed; Zulqarnain, Muhammad; and Tallal Majeed Butt. (2021). Recent advancement in smart grid technology: Future prospects in the electrical power network. *Ain Shams Engineering Journal*, 12(1). doi: 10.1016/j.asej.2020.05.004.

45. Zhang, Chong et al. (2014). Impacts of electric vehicles on the transient voltage stability of distribution network and the study of improvement measures. *2014 IEEE PES Asia-Pacific Power and Energy Engineering Conference (APPEEC)*, pp. 1–6, IEEE.

46. Sun, Y.; Yue, H.; Zhang, J.; and Booth, C. (2019). Minimization of residential energy cost considering energy storage system and EV with driving usage probabilities. *IEEE Transactions on Sustainable Energy*, 10(4), pp. 1752–1763. doi: 10.1109/TSTE.2018.2870561.

47. Lu, Xinhui; Zhou, Kaile; and Yang, Shanlin. (2017). Multi-objective optimal dispatch of microgrid containing electric vehicles. *Journal of Cleaner Production*, 165. doi: 10.1016/j.jclepro.2017.07.221.

48. Kasturi, Kumari; Nayak, Chinmay; and Nayak, Manas. (2019). Electric vehicles management enabling G2V and V2G in smart distribution system for maximizing profits using MOMVO. *International Transactions on Electrical Energy Systems*, 29, e12013. doi: 10.1002/2050-7038.12013.

49. Hadley, Stanton; and Tsvetkova, Alexandra. (2009). Potential impacts of plug-in hybrid electric vehicles on regional power generation. *The Electricity Journal*, 22, 56–68.

50. Knezovic, Katarina; Martinenas, Sergejus; Andersen, Peter; Zecchino, Antonio; and Marinelli, Mattia. (2016). Enhancing the role of electric vehicles in the power grid: Field validation of multiple ancillary services. *IEEE Transactions on Transportation Electrification*, 1–1. doi: 10.1109/TTE.2016.2616864.

51. Dogger, Jarno; Roossien, Bart; and Nieuwenhout, Frans. (2011). Characterization of Li-ion batteries for intelligent management of distributed grid-connected storage. *IEEE Transactions on Energy Conversion*, 26, 256–263. doi: 10.1109/TEC.2009.2032579.

52. Kempton, Willett and Tomić, Jasna. (2005). Vehicle-to-grid power implementation: From stabilizing the grid to supporting large-scale renewable energy. *Journal of Power Sources*, 144, 280–294. doi: 10.1016/j.jpowsour.2004.12.022.

53. Daina, Nicolò; Sivakumar, Aruna; and Polak, John W. (2017). Modelling electric vehicles use: A survey on the methods. *Renewable and Sustainable Energy Reviews*, 68, 447–460.

54. Brady, John and O'Mahony, Margaret (2016). Modelling charging profiles of electric vehicles based on real-world electric vehicle charging data. *Sustainable Cities and Society*, 26, 203–216.

55. Ip, Andy; Fong, Simon; and Liu, Elaine. (2010). Optimization for allocating BEV recharging stations in urban areas by using hierarchical clustering. *2010 6th International Conference on Advanced Information Management and Service (IMS)*, IEEE.

56. Saber, Ahmed Yousuf and Venayagamoorthy, Ganesh Kumar. (2010). Intelligent unit commitment with vehicle-to-grid: A cost-emission optimization. *Journal of Power Sources*, 195, 898–911.

57. Jiang, Chen Hui et al. (2014). Method to assess the power-quality impact of plug-in electric vehicles. *IEEE Transactions on Power Delivery*, 29, 958–965.

58. Gurkaynak, Yusuf and Khaligh, Alireza. (2009). Control and power management of a grid connected residential photovoltaic system with Plug-in Hybrid Electric Vehicle (PHEV) load. *Conference Proceedings - IEEE Applied Power Electronics Conference and Exposition – APEC*, pp. 2086–2091. doi: 10.1109/APEC.2009.4802962.

59. Dubey, Anamika; Santoso, Surya; Cloud, Matthew; and Waclawiak, Marek. (2015). IEEE power and energy technology systems journal determining time-of-use schedules for electric vehicle loads: A practical perspective. *Power and Energy Technology Systems Journal, IEEE*, 2, 12–20. doi: 10.1109/JPETS.2015.2405069.

60. Bakhshinejad, Alireza; Tavakoli, Abdolreza; and Mirhosseini, Maziar. (2021). Modeling and simultaneous management of electric vehicle penetration and demand response to improve distribution network performance. *Electrical Engineering*, 103. doi: 10.1007/s00202-020-01083-7.

16 Artificial Intelligence for the Operation of Renewable Energy Systems

Ifeanyi Michael Smarte Anekwe
University of the Witwatersrand

Emmanuel Kweinor Tetteh
Durban University of Technology

Edward Kwaku Armah
C. K. Tedam University of Technology and Applied Sciences

Yusuf Makarfi Isa
University of the Witwatersrand

CONTENTS

DOI: 10.1201/9781003331117-16

ABSTRACT

The energy industry is at a critical stage. The growth of digital technologies can significantly alter our energy supply, commerce, and utilisation patterns in the near future. Digitalisation is being fuelled by artificial intelligence (AI) technology, which is becoming increasingly prevalent. It will be operated independently by smart software that optimises decision-making and activities in the incorporation of energy supply, demand, and renewable resources into the energy grid. The deployment of AI will be critical in reaching this goal. This chapter provides a realistic background on the application of AI in the energy sector. In addition, the benefits, challenges, and prospects of AI applications were presented which will enable researchers, readers, and key players in the energy industry to evaluate their AI initiatives and objectives, for effective energy supply.

KEYWORDS

Artificial Intelligence; Bioenergy; Expert System; Energy; Fuzzy Logic; Renewable Energy; Solar Energy; Wind Energy

16.1   INTRODUCTION

Presently, the world economy is inextricably linked to the efficiency with which electrical power is generated, managed, and distributed [1,2], and this is especially true in developing countries. Climate change and global warming are two of the most serious consequences of current energy generation systems. Energy-associated greenhouse gas (GHG) emissions "would result in significant climate damage, with an aggregate global warming of 6°C" [3]. In recent decades, with increased demands for energy use and ecological management, the need for energy technologies in global civilisation has been expanding rapidly [4,5]. As a result, green energy is the most practical approach for ensuring a cleaner and reliable energy supply. It is ecologically beneficial due to the low level of CO_2 emission, which is the primary indicator of the greenhouse effect, which is accountable for ecological damage [6–8]. Stabilising the electricity bills, substituting ageing facilities, enhancing the adaptability and consistency of power systems, decreasing CO_2 discharge, avoiding alteration in the earth's atmosphere, and facilitating dependable power supply to rural areas with incredibly increased power demands have become the key factors to this energy transition [4,9]. Because of the simplicity and reduced cost of upkeep, durability, and the availability of a limitless number of renewable energy (RE) sources, research and development in the field of RE at the national and international levels will result in improved performance and assured compensation in prospective energy demand [10]. RE sources (RES) are also known as alternatives, primarily due to their inability to ensure consistent supply in some particular situations [11,12]. Consequently, improving the efficiency of substitute energy sources is a must if the world is to meet its projected energy demand in the future. This second objective can be accomplished by tackling the restrictions associated with the configuration, efficacy, and performance forecast of the current RE technology, as well as the estimation of weather parameters in the area where the facility is located [13].

The energy storage system (ESS) makes use of renewable forms of energy to achieve the goals of energy conservation and reducing emissions, and as a result, it has undergone significant expansion recently. As a consequence of the unpredictability and irregularity issues that are associated with ESSs, there is variation in both voltage and frequency [14]. These unfavourable effects have slowly grown more obvious owing to the fast-growing use of RES. The synthesis of RE is unstable and erratic, which is not favourable to the uninterrupted activities of the power grid and has an adverse effect on the incorporation of RE including wind and solar power sources [15]. Despite this, the amount of installed potential for RE and distributed energy storage has steadily increased [16]. There have been studies that looked at the integration of RE with ESSs [17], wind-solar hybrid power producing systems, wind-storage accessible power systems [18], and optical storage transmission networks [17]. The advent of innovative techniques has aggravated the challenges associated with the management of RES, as well as the frequency and peak loads of the power grid. In addition, the management of the power grid has become more sophisticated. The ESS can swiftly and dynamically modify the system power and engage several ESS in the power system, thus creating an efficient mechanism for overcoming the aforementioned challenges. Investigations into the dependability of wind, solar, storage, and transmission networks have been carried out [19,20].

The International Agency for Renewable Energy predicts that by the year 2030, the world's total energy storage capacity will have increased from 42% to 68% (based on values from 2017). By the year 2025, India and China will have established themselves as the nations with the rapid rates of expansion for ESS, while Japan and Australia will have established themselves as the nations with the significant proportions of ESS [21]. The adoption of energy storage techniques will aid spread energy peaks and adjusting frequency, smooth variations, and aid generate increased electrical energy. Owing to the economic benefits of the energy storage technique, it can also be used to offer an interim energy source that can be effortlessly shifted off-grid [22–24]. Because of this, it is projected that RES, including but not limited to wind and solar energy, would eventually substitute fossil energy sources, and will steadily grow their share of total energy utilisation [25,26]. Commercial implementation of occasional RE with numerous time scales, on the other hand, has the potential to significantly influence the forecasting and planning precision of the power system as well as implementation security and power quality issues when power is substituted with the power grid, resulting in the demand for conventional transmission [27,28]. As a result, for RE to be employed in operational control and energy management, a control technique that takes into account both the adequacy of the power production capacity and the adaptability of the energy production is considered necessary [29]. A renewable ESS reinforces a RE power source by ensuring its operation over a broader range, maintaining the device's efficiency as well as stability in all grid-connected and island modes, and reducing the interrupted generation of RE that leads to power disruptions [29,30]. It has been proposed to further increase the employment of ESSs to improve the economic and ecological advantages of these techniques [31,32].

Among the most important approaches for the design and methodical research of power systems have been planning, monitoring, and optimisation [31,33]. For the modelling, optimisation, and augmentation of hybrid systems, a range of artificial intelligence (AI) methodologies are employed [34]. A key branch of AI is comprised of neural networks, fuzzy logic control, and concurrent computing algorithms [34,35]. In the study by Maleki and Askarzadeh [34], the efficiency of four heuristic algorithms was evaluated with the aim of reducing the overall yearly expenditure. These algorithms were utilised to optimise the size of a photovoltaic (PV)/wind/FC hybrid system so that it could constantly supply the load requirements at the same time incurring the lowest possible total year expenses. Moreover, an economic evaluation of sustainable energy sources reported by Maleki and Askarzadeh [34] showed an enhanced performance of the hybrid system parts in order to deliver more affordable and reliable electricity while also utilising environmentally sustainable sources. Design and optimisation strategies might be hampered in their conception and implementation by a variety of factors such as resource availability, network infrastructure, system efficiency, and computational models. Improvements in computer technology have made it feasible to address optimisation issues utilising a number of methodologies [36]. Using the concept of intelligent agents systems as a starting point, Russell and Norvig [37] presented a picture of the AI industry that presents, examines while also providing analysis and some light amusement. When it comes to improving machine efficiency and offering economic benefits, intelligent approaches have demonstrated their usefulness [38,39]. They also provide additional advantages that cannot be reached by conventional ways.

This study begins by providing RE storage techniques while presenting a review of AI applications in RE. Furthermore, various difficulties that can be encountered within the context of AI implementation were discussed, as well as prospects for effective AI application in RE systems.

16.2 GLOBAL ENERGY SECTOR

The energy industry is the cornerstone of the world economy; however, it's among the largest emitters of CO_2 [40]. Energy utilisation is increasing annually with China and the United States leading global energy consumption, followed by India, Russia, and other countries [13,41]. The rising rate of energy production and consumption, which is projected to increase as shown in Figure 16.1, is responsible for one-third of the world's CO_2 emissions [42,43]. Consequently, most stakeholders – environmental and social activists, investors, shareholders, and policy makers – are focusing on the sector's negative impact on CO_2 emissions to foster sustainable development [44,45]. Despite calls for environmental sustainability, the energy industry has not made significant headway in its efforts to reduce carbon emissions [40]. For the industry to gain social legitimacy, it needs to develop greater social awareness [46] and take responsibility for the carbon emissions it produces [47,48]. Solar energy, biomass, and thermal energy are examples of non-conventional energy sources that energy companies have integrated into their operating models over the last decade [49]. This change is necessary to achieve the goal of sustainable development [50]. Hence, it is vital to explore and implement sustainability measures in generation and utilisation to establish a solid link between green development and performance [40].

16.2.1 RENEWABLE ENERGY SOURCES

Wind energy, solar energy, hydro energy, geothermal energy, bioenergy, ocean energy, hydrogen energy, and hybrid energy are the most common sources of RE, which are classified as per the

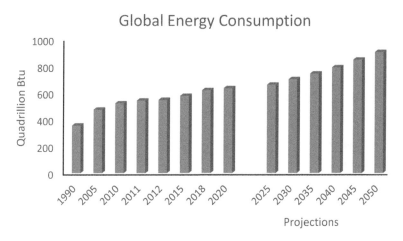

FIGURE 16.1 Global energy consumption from 1990 to 2020 and predictions from 2025 to 2050.

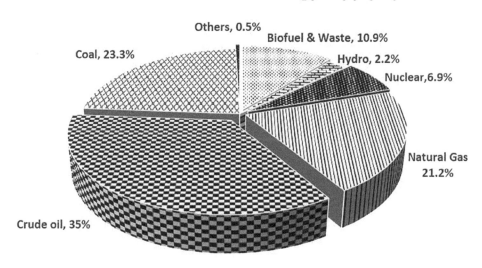

FIGURE 16.2 Global energy supply by sources. (EIA, 2019.)

source of production [1–3,6,51]. Figure 16.2 depicts the various types of RES available. A succinct explanation of some of the most major forms of RE is presented below.

16.2.1.1 Wind Energy

The wind is caused by the motion of the globe and the unbalanced convergence of the sun's radiation on the superficial part of the planet [52,53]. Windmills and wind turbines have been used to transform the kinetic energy of the wind into mechanical energy for many centuries, and the utilisation of wind as a major energy source has continued to this day. Proof of past advancements has been discovered in Asia (Persia and China) (200 BCE) and the United States (1850–1970 CE) [54–56]. The world's earliest wind turbine, with an output of 12 kW, was installed in Ohio, United States, between 1887 and 1888 [57]. Following that, a large number of wind turbines with increased output were constructed in various nations to fulfil the requirement for electric power. China had the highest number of wind power units installed between 2010 and 2015 [13]. In the same period, there has been a significant development in the installed capacity of wind power in the US and Germany. A total development trend of 9.96% is evident when comparing the installed wind power capacity in the past decade in the EU. Aside from that, the yearly operational output has risen from 48.0 to 141.1 GW from 2006 to 2015, which is a compound yearly increase rate of more than 9%. The Vestas V161 wind turbine, which stands 220 m tall and has a diameter of 164 m, is the globe's biggest and most efficient (8.0 MW) wind turbine. It was commissioned at the Danish National Test Centre in Denmark in 2014 [13].

16.2.1.2   Solar Energy

The sun is a critical energy source for all living things on the planet, including humans [58,59]. Humans have been using solar radiation for a variety of reasons for many centuries [58,59]. The earliest known proof incident back to the 7th century BCE, when the sun's rays were utilised to start a fire after being concentrated with a piece of glass [58], provides a comprehensive overview of the historic advancement of solar energy from antiquity (the 7th century BCE to the 1200s CE) through the contemporary age (1767–2001). The invention of PV phenomena in the year 1839 was a watershed moment in the history of science. Solar energy is employed mostly in active systems (PV, thermal, and so on) and passive systems (solar water heating). Photovoltaics is the procedure of converting solar energy into electricity for application in power generation [60], whereas the thermal procedure involves first converting solar energy into some mechanical energy, which is then utilised for electricity generation [59,60]. The passive system collects and transmits solar energy throughout a house without the need for any electrical devices, as opposed to the active systems previously discussed [61]. Innovative and effective components [62–65], international regulations [66], configuration [67], implementation [68], storage [69], and low-energy buildings [70,71] are among the areas in which research is being conducted in solar energy for enhanced performance and quality. Several reviews [72–75] have also been written on the subject of mathematical modelling of solar energy systems.

16.2.1.3   Geothermal Energy

The generation of Lava is caused by the steady disintegration of radioactive materials in the earth. Due to the obvious migration of tectonic plates, the Lava is broken, resulting in the formation of a geothermal reservoir (geothermal energy source) [75–77]. There have been several reviews [78–85] of the research and development in the field of geothermal energy, with the most recent being reported by the authors of Refs. [86,87]. These studies are centred on the identification of accessible resources, the present condition of equipment, and the applications, advantages, and implementation of geothermal energy attributes and impacts [83]. Many research articles [85,86,88,89] have also been written about the modelling and simulation of geothermal energy [13].

16.2.1.4   Hydro Energy

Building the natural (waterfall) or regulated movement of water (utilising a manmade barrage on a river) to create electricity is known as hydroelectricity [90,91]. Hydroelectric power facilities were categorised into three primary groups depending on their capability for generating electric power. Several reviews summarise the research and progress in the area of hydro energy [75,92–99], particularly, storage plants and their constraints, dam maintenance and performance [94], hydrokinetic energy transformation system [95,96], slit erosion strategies in hydro turbines [97], optimal installation of small hydropower systems [98], and socio-technical constraints of hydropower.

16.2.1.5   Bioenergy

In this area of RE, electric power is created from biomass sources such as wood, organic wastes, agricultural byproducts and wastes, algae, microorganisms, vegetable oils, and other natural resources [100–102]. Many review reports [103–110] collate considerable investigations and advancement in the field of bioenergy, especially in the areas of global synthesis and utilisation of bio-ethanol [103], microalgae in biodiesel synthesis and applications [104], microbial fuel cells (MFCs) in bioenergy [111], bio-transformation operations of organic feedstock into bioenergy [105], and pyrolysis of biomass to bio-oil [106].

16.2.1.6 Hydrogen Energy

In fuel products (hydrocarbons), H_2 is an inherent component that can be independently used in a variety of applications. Aside from that, the electrolysis of H_2O and the bioprocess of microbes (bacteria and algae) both release H_2, which, when burned, generates a large amount of energy and can be utilised as a RES for electric power generation [101,112,113]. For the latter technique, fuel cells are frequently employed. Hydrogen energy can meet the need for power regularly, which is a restriction of RE systems depending on wind and solar energy [101,112–114]. The most notable research and advancement findings in the area of H_2 energy are summarised in numerous studies [114–119] primarily on the current status, photo-production of hydrogen [101], affecting parameters in hydrogen synthesis [117], storage technique [118], hydrogen fuel cell [119], current and prospective approaches of hydrogen, technological scenario, and economic aspect [114].

16.2.1.7 Hybrid Renewable Energy System (HRES)

A HRES integrates numerous RES to enhance the performance and reliability of power sources over and above what could be obtained utilising a single RES. PV-diesel, Wind-diesel, PV-hydrogen, Wind–hydrogen, and other HRES are some of the most regularly used HRES [120,121]. Numerous published studies [122–128] provide an outline of research and development findings in the area of HRES, with a particular emphasis on applications [122], design and regulation [123], optimal configuration [124], software tools for incorporation [125], present state and prospect [126], a storage system [127], and mathematical modelling [128].

16.3 ARTIFICIAL INTELLIGENCE – OVERVIEW

Generally, AI is aimed to understand human thought to create intelligent entities that can solve complex problems quickly [129]. Herein, understanding the intricate reasoning of a human brain is a challenging topic to solve; hence, AI researchers are still working on it [129–131]. This development can be traced back to a series of experimental demonstrations, prospects, and expectations [13,129–132]. For instance, the energy industry has been using AI in making prototype robots with a variety of intelligent behaviours [13,132]. AI was initially developed by Alan Turning and was the pioneer in introducing the "imitation game" notion [129]. This concept of AI was coined in 1955 and had a list of processing programs proposed for solving and reading algebraic word problems [131,132]. However, between 1975 and 1980, the progress of AI became very sluggish, as there was low interest due to financing constraints, shortage of processing power, and lack of novelty and logical ideas. Further advancement resulted in the theoretical framework of AI known as an artificial neural network (ANN), which gain widespread acceptance in 1982. Also, a notable AI invention, including logistics planning for US military applications, was investigated between 1990 and 2015 [13,130]. Figure 16.3 shows a graphical trend based on published articles on the development of AI technology adapted from Google Scholar from 2012 to May 2022. It was evident that AI contributes effectively in databases, accounting, information retrieval, product design, production planning and distribution, medical, food quality monitoring, biometrics, forensics, and so on [129,133,134]. AI is built on multiple learning concepts, such as statistical learning, neural learning, and evolutionary learning [133]. Out of these, neural learning has been one of the most used AI techniques in a variety of situations [13,130,132]. The most fundamental neural learning technique is ANN. The AI technologies employed in huge data management, massive computational power, information technology, and better machine learning (ML) and deep learning (DL) algorithms are all handled by the ANN [133,135].

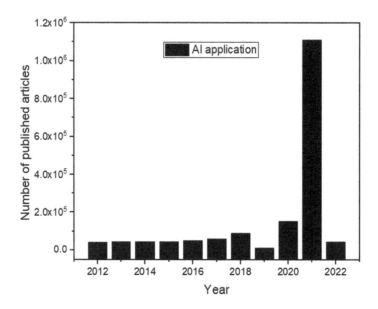

FIGURE 16.3 Number of published publications of AI applications from 2012 to May 2022. (Adapted from Google Scholar.)

16.4 CLASSIFICATION OF AI FOR RENEWABLE ENERGY APPLICATION – REVIEW OF AI TECHNIQUES

Advancing AI has resulted in technologies popularly known as machine learning (ML) and deep learning (DL) [133,135]. The ML infers to the ability of a system to learn automatically based on previous experience expressly programmed without any human interference. ML models are classified into three kinds (supervised, unsupervised, and reinforcement learning) for processing training and model evaluation [133]. There are two types of supervised learning: regression and classification. Diagnostics, identity fraud identification, client retention, and picture categorisation are all examples of applications for classification models. Regression models are frequently utilised in a variety of applications, including market forecasts, RE facility life span estimations, climate forecasting, population growth prediction, and marketing trend predictions. In addition, there are two forms of unsupervised models, for example, aggregation and dimensional reduction. Elicitation, structure identification, massive data visualisation, and compression are all accomplished through the application of dimensional reduction models. However, customer categorisation, tailored advertising, and system considerations are all accomplished through the usage of aggregation models. Among the applications of reinforcement learning include real-time decision-making, AI game theory, skill development, and robot navigation [131]. On the other hand, DL algorithms can directly use raw data and extract automatically processed information to improve the ML system. Some DL techniques include deep convolutional neural networks (CNNs), long short-term memory, deep belief networks, generative adversarial networks, deep convolutional and other hybrid combination systems [136].

16.4.1 Artificial Neural Networks or Neural Network

ANN is made up of several nodes (neurons) structured and interconnected in layers, where each link is a signal that is processed and conveyed from one layer to the next until an output response is produced [13,130]. Feedforward neural networks, recurrent neural networks, and CNNs are all common types of ANN [130,132,133,136]. They are grouped based on the number of layers, the number of neurons in each layer, and their connections. ANNs are widely employed in applications such as false alarm detection and fatigue estimation and classification by state SCADA (supervisory control and data acquisition) data and vibration signals are the most typical sources of information [13]. Combining ML and Big Data techniques with others, like Decision Trees, Particle Swarm Optimisation (PSO), Genetic Algorithms (GAs), Mahalanobis distance, and Bayesian filters, has been the common way to improve system performance [13,132,136]. Faults in blades are frequently discovered or forecasted by analysing signals from ultrasonic and visual sensors, where CNN performs most of the visual analysis. The most prevalent defects are structural damage, where the common monitoring signals for bearings are vibration, temperature, and SCADA data [13,131,136].

16.4.2 Wavelet and Neural Networks (WNNs)

WNNs combine wavelet concepts and neural networks while applying wavelets as the fundamental building blocks of a network. A wavelet function is a local function that affects the output of the network only in a limited range [130,134]. The WNN is surprisingly effective in resolving the common issues of poor convergence or even divergence described by other networks. The wavelet transform is derived from the Fourier transform and has been an innovative technique for signal processing [134,137]. The time–frequency localisation of the wavelet transform defines its trait. This trait can be characterised by its adaptable basis functions, which depend on the type of signal being examined [134]. In recent years, wavelets have received tremendous interest in both theoretical and applied fields. In fact, breakthroughs in the field are occurring so rapidly that the definition of "wavelet analysis" is constantly evolving to embrace new concepts [130,134,137].

16.4.3 Genetic Algorithms and Particle Swarm Optimisation

GA and PSO are similar in how they both think of different factors as classes of individuals interacting with each other to accomplish a common goal [131,134]. GAs are algorithms simulated by evolutionary characteristics including selection, mutation, and hybridisation, responding to the environment by simulating the collective evolution activity of the individuals [137]. The PSO utilises parameters as a collection of animals hunting for food, either cooperating or competing, where each member modifies its search pattern and positions itself against other members [134,137]. Both GA and PSO are commonly used as optimisation mechanisms for microgrid sizing and configuration, as well as for evaluating the design and various components of solar panels, fuel cells, diesel generators, and wind farms [134]. According to Mellit et al. [137], for the construction of wind farms, GA is often used in conjunction with other AI approaches to optimise the design of the system. The foundations, generators, and blades are the most studied components, while wind farm size, layout, power dispatch, operational cost, and environmental indicators are the characteristics that need to be optimised [134,137]. Also, GA can be used for large-scale maintenance planning, crew allocation, and feasibility analysis.

16.4.4 Fuzzy Logic

Fuzzy systems can use a set of fuzzy logic to address vague and partial data, whereby an object's membership can be anywhere between 0 and 1 [130,137]. But in the conventional set concept, an object can become either a member of a set or not [134]. This is used in decision-making agreements on the premise of the degree of efficiency is weighed against the contextualised values associated with the system. Figure 16.4 shows the components that make up a typical fuzzy logical system.

Fuzzy logic is often integrated with other AI technologies to facilitate their operation in different situations [131,137]. This includes reducing costs and improving reliability, making decisions, reducing risks, and doing preventive maintenance [131–133,137]. In terms of optimisation and

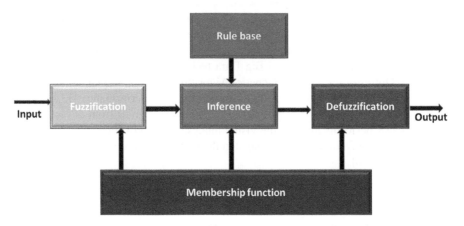

FIGURE 16.4   The main component of the fuzzy logic system. (Adapted from Belu [134].)

control engineering, the most common variables usually considered include the costs and mode of configuration [137]. In ensuring good maintenance practice, fuzzy logic is employed for early defect identification, fault prediction, and false alarm detection [13,131,137]. Moreover, when linguistic parameters are utilised, these degrees may be handled by functions depending on the circumstances. It depends on fuzzy logic reasoning, which uses IF-THEN-ELSE statements to implement linguistic rules. To improve the overall performance of a given application, fuzzy systems are distinguished by two primary criteria [137].

   i. Fuzzy systems are well suited for reasoning that is uncertain or approximative, and they are useful for situations in which the mathematical model of the system is difficult to derive.
   ii. Decisions can be made with estimated values using fuzzy logic even when the available information is inadequate or ambiguous.

A layer neural network receives an input vector from a fuzzy interface block whenever the block is activated in response to linguistic assertions (Figure 16.5). The neural network is capable of being reprogrammed, also known as taught, to produce the desired judgements or outputs (Figure 16.5a). The fuzzy inference process is under the control of a neural network with many layers (Figure 16.5b). The fuzzy neuron is an essential idea that is utilised in a wide variety of methods for incorporating fuzzy and neural techniques. In networks that map fuzzy input to crisp output, nodes in each layer may contain changed neurons. The input comprises a set of fuzzy values, and the weights that connect the node with nodes in the layer above it also have fuzzy values associated with them. Membership functions each represent the input values and the weights independently.

16.4.5   STATISTICAL METHODS

The statistical methodologies rely on a considerable amount of data to evaluate both the statistical features of the methods and their applicability to a system [137]. Bayesian analysis, Markov processes, and Monte Carlo simulations are among the AI techniques commonly used for statistical analysis [130,134,137]. The Bayesian analysis has been the primary method utilised in decision-making and condition monitoring. It also presents knowledge about the system's chance of an event occurrence, which can be detected as a potential failure or success. However, the Bayesian networks can be improved by coupling them with either ANN or ML [134]. This application is very common when working on including the gearbox or the blades or identifying defects in wind

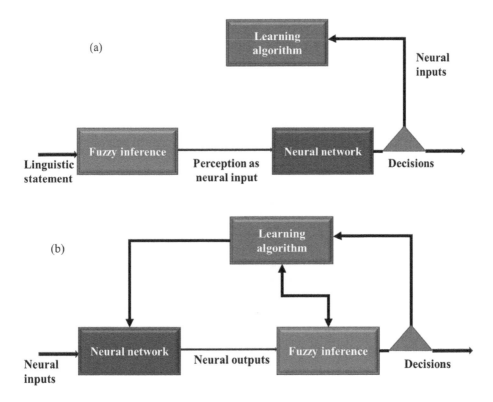

FIGURE 16.5 Fuzzy neural network for (a) first and (b) second model system. (Adapted from Belu [134].)

turbine state transition (Table 16.1). In terms of decision-making, Bayesian analysis emphasises applying techniques that reduce operational, maintenance, and life-cycle costs. Furthermore, it can be used in assessing risk analysis by taking into consideration failure modes, environmental parameters, and assessment and repair expenses, while evaluating the rationality of the information [13,134].

Nevertheless, the primary focus of Markov involves the minimisation of costs and the evaluation of the need for maintenance. In these processes, the likelihood that an event will occur is determined by the state of the system [130,134,137]. Uncertainty and costs, weather conditions, and failure mechanisms are some of the variables that are taken into consideration for this purpose [137]. However, the combinations of Markov with other AI techniques such as utilising clustering algorithms and ANN are used for maintenance evaluation as reported by Mellit et al. [137]. In finding the solution to an existing problem, Monte Carlo simulations are randomly employed to optimise the cost and evaluation for decision-making. Taking into account the economic impact, weather uncertainties and failure mechanisms, wind turbine blade faults can be detected by using the Monte Carlo simulation [130,132]. Table 16.1 presents substantial examples of work centred on statistical approaches with AI application for maintenance, cost analysis, fault detection, risk assessment, and decision-making.

16.4.6 DECISION-MAKING TECHNIQUES

Generally, the primary focus of decision-making involves planning (both short and long term) and evaluating the variables involved in a situation [129,135]. The planning involves the determination of the components suited for long or short term, power forecast, route optimisation, risk assessment, and mitigation of fault, whereas the purpose of variable assessment is to determine which variables

TABLE 16.1

Types of AI Techniques and Their Functionality

Algorithms	Applications	Data/Variables
Bayesian analysis	Fault detection	Study on model residue
Bayesian analysis	Fault detection	Fault data
Bayesian analysis, ANN	Maintenance optimisation	Fault data
Bayesian analysis	Decision-making	Costs, project data
Bayesian analysis	Risk assessment	Weather, costs
Bayesian analysis, ANN	Condition monitoring	Material, geometric features
Markov processes, clustering algorithm	Maintenance assessment	SCADA data, anomaly indexes
Markov process	Decision-making	Economic, wind turbine components
Markov process	Cost analysis	Economic, simulated data
Markov process	Cost optimisation	Mechanical loads, economic weather
Markov process, ANN	Cost optimisation	SCADA data
Monte Carlo methods, GA	Cost optimisation	Economic
Monte Carlo simulation	Decision-making	Economic
Monte Carlo simulation	Risk assessment	Economic, wind turbine component
Monte Carlo simulation	Maintenance planning	Economic weather

Source: Adapted from Jha et al. [13].

ANN, Artificial neural network; GA, Genetic algorithm; SCADA, Supervisory control and data acquisition.

are the most appropriate for carrying out operations and maintenance tasks, to cut down on computational costs, and to forecast the appearance of faults while taking into account signal uncertainties or failure probability. The use of ANNs was found feasible in estimating useful variables for monitoring wind turbine farms [132]. Multi-Criteria Decision Analysis was employed to deduce the condition that should be implemented for offshore wind energy [13,137]. In this case, the effect of three elements (reliability, maintainability, and logistics) on the availability of offshore wind farms was considered to evaluate the upkeep and the process of making decisions [132]. It is vital to optimise both cost and time to guarantee the proper operating system while taking into account factors like the weather, availability, and power output [13,137]. However, long-term scheduling is the process of making strategic decisions to increase the efficiency of wind farm life-cycle services. The maintenance strategy, energy generation policy, risk and investment evaluation, and end-of-life strategies are examples of these strategic decisions [13,134]. The primary uses of various decision-making strategies are presented in Table 16.2.

16.4.7 HYBRID SYSTEM

Hybrid systems involving the combination of either fuzzy logic, neural networks, or GAs systems are proven to be useful in solving several real-world issues [13]. The hybridised technique can combine more than one of the algorithms as mentioned earlier, either as an integrated method in finding a remedy or to execute a specific function that is then accompanied by a second procedure [13,134,137]. For example, analytic computer programs are frequently employed in the modelling, performance prediction, and control of processes including RES [13,136]. To provide accurate predictions, most algorithms general require a great deal of processing power over a significant period. Herein, hybridised AI systems (Table 16.1) can discover the important information patterns inside a multi-dimensional domain, as opposed to using complex rules and mathematical processes. For the design, control, and operation of RES like PV or solar-thermal energy systems, it is often necessary to have a comprehensive long-term set of meteorological data. These systems are utilised in the

TABLE 16.2

Applications of AI Techniques with Various Decision-Making Approaches

Application	Algorithms	Data/Variables
Scheduling	AI decision intervention systems	Economic, topographical data
	Modelling	Economic
	Polynomial regression	Wind turbine operating status
Long-term decision-making	Planning model	Power generation, environmental aspects costs
	Technological readiness level	Literature and reports
	Mathematical cost model	Economic, weather
Maintenance planning and optimisation	Grey fuzzy and internal value	Statistical data
	GA	Costs
	Decision-making tool	Damage economic
	Big data	SCADA
	Stackelberg game	Operator decisions
	Decision support system	Wind speed and wind gust data
	Dynamic relative thresholds	Maintenance and climate data
	Markov decision process	Fault data
	Cloud-based platforms	Economic
Layout planning	Machine learning, decision trees	Ultrasounds
	GA	Location economic
Cost optimisation	Mixed-integer optimisation	Economic
Risk assessment and mitigation	Risk-based decision models	Risks and costs
	Analytical network	Quantitative variables
Route optimisation	M-planning	Uncertainties, maintenance data
Power optimisation and prediction	Non-parametric statistics, mixed integer, non-linear and non-convex	Economic

Source: Adapted from Jha et al. [13].
GA, Genetic algorithm; SCADA, Supervisory control and data acquisition.

production of electricity [131,134]. However, obtaining long series of meteorological parameters for immediate use is a difficulty for designers of such systems. This is a common difficulty for system designers in localities with metrics of poor quality of data and missing data.

16.5 AI ROLE AND APPLICATION IN THE RENEWABLE ENERGY SYSTEM

In the field of energy, AI technologies can be of assistance in seizing the expanding prospects that arise with the implementation of RES and the widespread use of the Internet of Things (IoT) [13,136]. The facilities of the electricity grid have become too dilapidated, ineffective, outdated, inconsistent, and do not offer appropriate security under fault scenarios over several years. However, the generation of energy, the planning of its distribution, and the continued viability of financial systems are of critical importance to the world economy [13,130]. Therefore, the smart energy business can be outfitted with modern components for effective performance [129]. Conventionally, power grids were not built to handle the incorporation of RES. As a result, any change in the features of RES has the potential to alter the variable load efficiency of the power grid [134,137]. The RE industry is currently transforming as a direct result of the developments being made in AI techniques (such as ML, DL, IoT, big data, etc.) (Figure 16.6). Table 16.3 presents the application of AI in the energy industry being used to carry out a variety of activities, including controlling, forecasting, optimisation, control, monitoring, and operating power systems in an efficient manner.

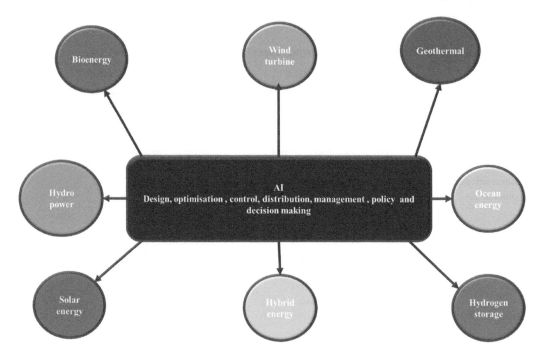

FIGURE 16.6   Application of AI techniques in renewable energy sources.

16.5.1   AI IN WIND ENERGY

In estimating wind speed and its generated power, models commonly used include the physical model, statistical model, correlation model, and neural network. Also, the data mining approach can be used in estimating wind power for short- to long-term decision-making [133,137]. Similarly, probabilistic models for wind power can be estimated using coupled methodologies for short-term wind speed and power prediction. However, the utilisation of AI in wind energy often incorporates neural, statistical, and evolutionary learning or their integration as a hybrid AI technique. The drive for most "wind energy AI" is accurately forecasting wind speed and wind power. It has been reported that the feedforward backpropagation neural network (BPNN) has been the most effective method for estimating the amount of electricity generated by the wind [132]. Also, the combination of diverse ANN, BPNN, and adaptive linear element network can be employed as hybridised technique.

16.5.2   ROLE OF AI IN HYDROGEN ENERGY

In membrane fuel cell systems, the AI-based model is used in the diagnostics of the proton exchange during the hydrogeneration process [13,132]. The ANNs are commonly incorporated to provide the statistical and signal processing information in the area of H_2 energy. Additionally, the BPNN in conjunction with multi-gene genetic programming is applied in the process of predicting the production voltage of MFCs. The analysis of H_2 energy has also applied fuzzy logic techniques. For instance, the fuzzy logic method is employed in the estimation of the time required for the ignition of a hydrogen-powered automobile by utilising three distinct kinds of membership functions. The fuzzy logic controller that is dependent on variable optimisation with the GA is used in fuel cell hybrid vehicles in order to regulate the amount of H_2 that is consumed [13,134].

TABLE 16.3

Applications of AI Techniques in Renewable Energy Industry

Application	Method	Outcome
Wind power prediction	BPNN	RMSE: 0.0065
Wind speed prediction	BPNN and ADALINE	RMSE: 1.254
Wind speed and power prediction	BPNN	20%–40% improved accuracy
Design of wind generation system	Fuzzy method	3.5 kW
Wind speed and power prediction	Hybrid method (Fuzzy-GA)	29.7% improved accuracy
Wind speed	Hybrid (MM5-ANN)	MSE: 0.12%
Risk optimisation in wind energy	Hybrid method (PSO-ANFIS)	Profit estimates for risk level (0–0.1)
Stream flow prediction	BPNN (LM)	MAPE: < 5%
Hydropower plant	BPNN, KNN	BPNN: 0.011% cost-effective
Hydrogeneration	CHGA, SGA, NP	$301 profit
River flow prediction	Hybrid method (HD-PPT-ANN)	Std: 26.48
Hydraulic energy prediction	Hybrid method (ANN-ABC)	MAPE: 4.6%
Solar irradiance prediction	BPNN	Correlation: 94%–99%
Mean temperature prediction	BPNN and batch learning ANN	MPD: 2.13–4.17 for BPNN
Global solar radiation prediction	BPNN and regression methods	RMSE: 0.867 for BPNN
Solar energy and hot water quantity prediction	BPNN	R^2: 0.998 and 0.9973
PV power prediction	Hybrid method (ANN-GA-PSO)	Prediction: 0–35 kW
Solar radiation prediction	Hybrid method (ANN-TDNN)	RMSE: 25–300
Solar power prediction	Hybrid (SVM-FFA)	RMSE: 0.728
Geothermal power prediction	BPNN (LM, PCG, SCG)	R^2: 0.999
Geothermal map prediction	BPNN	Correlation: 0.9253
Pressure prediction in a geothermal plant	BPNN	MAPE: <2.3%
Geothermal temperature prediction	Hybrid method (GMDH-GAS-SVD)	R^2: 0.9899
Sea-level variation prediction	BPNN	RMSE: 10% of tidal range
Sea wave height prediction	BPNN	84% for a lead time of 6 hours
Wave hindcasting	Hybrid method (NWM-BPNN)	Correlation: 0.93
CO_2 flux prediction	Hybrid method (CVR-SVR)	Mean accuracy: 96.3%
Biogas production	Hybrid method (BPNN-GA)	8.64% increase in production
Methane prediction from waste	Hybrid method (BPNN-GA)	R^2: 0.8703
Biomass boiler control	Hybrid method (BPNN-GA)	3.5% increase in turbine output
Power and energy prediction	BPNN, Fuzzy logic	2.4%–14% accuracy
Size optimisation	Markov-GA	Cost: < 0.5 M$
Power flow control	Hybrid method (BPNN-Fuzzy)	SOC: 40%–80%

Source: Adapted from Jha et al. [13].

ABC, Artificial bee colony; BPNN, Backpropagation neural network; ADALINE, Adaptive linear network; KNN, k-nearest neighbour; LM, Levenberg–Marguardt; PSO, Particle swarm optimisation; GA, Genetic algorithm; TDNN, Time delay neural network; SVM, Support vector machine; FFA, Firefly algorithm; PCG, Pola–Ribiere Conjugate gradient; SCG, Scaled conjugate gradient; GMDH, Group method data handling; GAS, Genetic algorithm system; SVD, Singular value decomposition; NWM, Numerical wave model; CVR, Case-based reasoning; SVR, Support vector regression; CHGA, Chaotic hybrid genetic algorithm; PPT, Power point tracking; RMSE, Root mean square error; MAPE, Mean absolute percentage error; SOC, Storage of charge; Std, Standard deviation; MSE, Mean square error.

16.5.3 AI IN HYDROPOWER ENERGY

In the development and control of hydropower facilities, new AI technologies (GA, ANN, Fuzzy) are used in obtaining information for decision-making [137]. Also, the wavelet pre-processor-based hybrid AI techniques have been employed for the estimation of important hydrologic cycle activities.

Table 16.3 provides some of the applications of single and hybrid AI methodologies in the field of hydropower energy. The BPNN method is applied to optimally schedule the activities of Taiwan's hydropower facilities, which are sourced from ten reservoirs [137]. BPNN method simulation was also used in a rainfall-runoff process to estimate its discharge peak time for linear and non-linear reservoirs [129,131,137]. When applied to non-linear reservoirs, BPNN attains greater reliability in forecasting peak discharge, than when applied to linear reservoirs. The GA, called the chaotic hybrid-GA, has also been to tackle the problem of the existing water lag period as a restriction in the brief hydrogenation scheduling [137]. This resulted significantly in a higher yield profit.

16.5.4 AI IN SOLAR ENERGY

Over the years, global solar radiation (GSR) is predicted with the BPNN using temperature and humidity as inputs, of recent, the BPNN has used temperature and humidity as inputs to make predictions for GSR [13,134,137]. Solar energy applications also made use of several evolutionary AI approaches (Table 16.3). The use of GA in solar tracking can also improve the functionality of PV systems. The ideal layout of a solar water heating system can be achieved with the help of GA. To be more specific, a solar PV plate collector area is usually tuned with the GA to reach $63\,m^2$ with about 92% improvement of the solar panels [134]. The maximum power point tracking of a PV array coupled to a battery has been accomplished with the help of GA. Additionally, it was reported that the combination of AI approaches as a hybrid technique can improve the accuracy of the solar energy predictions [13,130,134]. In essence, ANN has been the most used AI technique in the solar process for modelling, design, estimating heating load, and building the PV system. The deployment of agent-based intelligent control systems helps in managing system energy consumption.

16.5.5 AI IN BIOENERGY

Generally, stochastic modelling is utilised for the optimisation of forest biomass in RE generation, more particularly the optimal configuration of the distribution chain [132,137]. This modelling is intended to maximise the efficiency with which RE is produced. Table 16.3 provides a summary of the utilisation of single AI technologies as well as hybrid AI tools for bioenergy analysis. It has been demonstrated that ANN is capable of accurately predicting both the octane rating and the density of diesel fuel [133,134]. The BPNN is employed to estimate the methane content of biomass in bioreactors by utilising several input variables (alkalinity, chemical oxygen demand, chloride, conductivity, pH, sulphate, and temperature) [131]. The biodiesel parameters (density, viscosity, and H_2O and methanol concentration) were estimated by using multiple linear regression and principal component regression [13,135]. The accuracy of this estimation ranges from 69% to 96% [135]. Within the context of the optimisation of the feedstock supply chain, a hybrid AI approach is usually explored [130,132,135]. A similar hybrid strategy is utilised to optimise the synthesis of biogas, which resulted in an increased biogas yield [13,137].

16.5.6 AI IN GEOTHERMAL ENERGY

In geothermal well-drilling, the uses of AI, along with sensors and robots, are for development, control, and optimisation [134,137]. Some computer simulations and modelling of geothermal reservoirs, as well as optimisation of geothermal energy, are shown in Table 16.3. Generally, the numerical models are used to enhance the geothermal systems and reservoir operations [129,137]. Meanwhile, the single and hybrid techniques of AI can be applied to geothermal energy. For example, the BPNN with Levenberg–Marguardt, or Pola–Ribiere conjugate gradient, and scaled conjugate gradient algorithms can be used to forecast the efficiency of a vertical ground-coupled heat pump facility [13,130,134].

16.5.7 AI in Hybrid Renewable Energy

The use of AI technologies in a hybrid RE process can be used for the determination of the ideal layout, capacity, and efficiency of the system [131,132,134,137]. The design techniques of a solar-wind hybrid RE system are anlaysed. The application of AI technologies in hybrid RE is provided in Table 16.3. The few selected single and hybrid AI approaches demonstrated to be effective in optimisation and estimation of the RE efficiency. In terms of a hybrid RE system that is dependent on waterpower supply, BPNN is utilised in the forecast of power consumption as well as the status of the generator (on/off) [137]. The Bee algorithm is also utilised in the process of optimising the performance characteristics of hybrid RE systems [131,137]. These performance parameters include the net present cost, the cost of energy, and the generation cost. A GA has been used in the process of optimising hybrid PV/wind systems for the optimal size of the PV array, wind turbine, and storage capacity [132,137]. More so, to optimise the size and distribution of a hybrid energy system (PV, wind turbine, fuel cell), a multi-objective method is used, resulting in a high voltage stability index. To improve the size of a hybrid wind-PV-diesel system, researchers employ a GA that is based on the Markov model. A hybrid AI technique consisting of an ANN and fuzzy logic-based controller has been created for the purpose of controlling the flow of power between a hybrid RE system and an energy storage unit [131,137].

16.6 BENEFITS OF AI APPLICATION IN RENEWABLE ENERGY SYSTEM

The prospects of AI in RES have been broadly studied by many researchers over the years. With the emergence of AI technology, RE systems and power companies can access better forecasts (energy generation, consumption, trading), better storage options, power grid management, energy efficiency and optimisation, etc. This section explores some benefits of AI application in RES.

16.6.1 Energy Storage

The application of AI in energy storage is seen from many perspectives. Firstly, AI can be used to improve and accelerate the development of batteries for the storage of RE. In this vein, AI systems will show an improved knowledge of the degradation limit of battery storage systems utilising data-driven equipment. This will facilitate the optimisation of the storage systems and cut down the cost [138]. Again, AI-based battery storage systems help in stabilising electrical or energy grids. For example, in 2017, Tesla built the biggest lithium-ion stationary battery ESS worldwide. This AI-based facility operates using auto-bidder, which can intelligently contribute to the stability of the Southern-Australian grid [139]. Another example of an AI-based ESS is seen in the development of AI-based battery systems by a US-based firm known as Advanced Microgrid Solutions (AMS). This was a consequence of the successes chalked by Tesla's lithium-ion stationary battery ESS. AMS noted that smart trading systems for AI-based batteries are more effective than the best human traders [140].

Another perspective on the employment of AI in energy storage is seen in the incorporation of different RES together. The dynamics of the generation and storage of energy from renewable sources differ from source to source. For example, a PV system uses direct normal and diffused radiations to generate electrical energy while temperature, pressure, and wind speed are the main drivers of turbine power. Again, solar and wind energy may not always be available, as the resources that power them are not always predictable. That is, to integrate and store different RES together, many factors and data would be required. This is a complex process that can only be achieved by deploying AI systems. In a very ambitious project by Saudi Arabia in NEOM city [141], efforts are being made to power the city with only RE. Specifically, the project seeks to integrate solar thermal and PV, wind, and battery energy storage. This is to be achieved by employing AI technology to predict energy generation from different sources and forecasting the energy load. This is highly possible, as AI can analyse data from the past to optimise present conditions to accurately predict the future [141].

16.6.2 Fault Prediction

Fault prediction plays a vital function in the overall efficiency of energy systems. Being able to anticipate faults or detect them in advance makes for their timely resolution, which enhances the smooth running of systems. AI has been applied successfully in this area in RES. Various applications of AI in fault detection in RES are presented in Table 16.4.

16.6.3 Energy Efficiency Decision-Making

Efficiency in the use of energy forms an integral part of continuous improvements in energy systems. More especially in recent times when the energy sector faces pressures of increased demands, systems must be put in place to enhance the efficiency of energy that is generated. AI has been shown to affect energy efficiency decisions in many ways. For example, AI technologies such as DL and computer vision can screen out and sort volumes of information, to reorganise and re-create knowledge [151]. This enhances technological innovations with better energy-efficient applications. Researchers have exerted a significant amount of work towards the development of energy-efficient models for power systems and the incorporation of RES into pre-existing grids.

For example, Arévalo and Jurado [152] came up with an AI-based techno-economic model for the evaluation of RE systems. This model analysed storage, power source, and fuel generators of multiple power plants in Southern Ecuador that operate in a hybrid fashion – incorporating RES into the traditional power grid. This model led to the development of energy management protocols with the potential of optimising energy use. In a similar vein, Attonaty et al. [153] developed and suggested a thermodynamic and economic evaluation model for renewable innovative electricity

---

TABLE 16.4

Applications of AI in Fault Detection in RES

Renewable Energy System	AI Application	References
Solar power	Supervised ML and graph signal processing technologies to identify issues in solar arrays from device measurements	[142]
	Flagging of potential anomalies in nearby solar panels using the graphical model approach of power output correlations between the panels	[143]
Wind power	Identification of wind turbine failures through the building of supervised models that predict anomalous gearbox temperatures based on historic data	[144]
	Fault detection in wind turbine converters using improved Octave convolutional neural network structure known as AOC-ResNet50 network	[145]
	Application of deep belief network and signal processing to diagnose faults in turbine gearbox	[146]
	Application of stacked multilevel-denoising autoencoders approach for wind turbine gearbox fault diagnosis	[147]
Nuclear power plants	Detection of anomalies in nuclear reactor data through a method of supervised learning, clustering, and denoising	[148]
	Automatic detection of cracks in nuclear power system facilities by using a CNN-based technology to video data	[149]
	Detection of sensor fault in nuclear plants utilising multivariate-phase prediction technology, supported by vector machine	[150]

storage system. This thermodynamic and economic evaluation is based on a heat energy power–power system where energy is transformed into heat and thermal cycle within a primary heating loop. It must be noted that the idea of the development of all these models is to improve energy performance and consequently affect decision-making within the energy sector. Other similar models were developed and proposed by other researchers [154–156].

16.6.4 Utility Energy Planning and Management

AI provides robust systems for utility planning, control, and distribution. A robust utility management system can control load demand, enable self-healing, negotiate measures, and improve a wide variety of additional services and products. It is also proficient in regulating grid stations [131]. This is the future of effective utility management and control, and AI provides useful trends for energy consumption and supply. In terms of demand-side management (DSM), AI is utilised to automate and optimise systems for load curtailment (reduction of electricity usage) and load relocation (shifting of energy usage to off-peak periods) which have huge potential for matching energy demand and supply [157]. For example, using AI technology, demand response (DR) programs (programs or technologies that focus on shifting energy use in the short term to balance energy use) can be developed to enhance the flexibility of DSM operations [157]. Furthermore, utilising AI technologies in DSM simplifies the complex problem of energy demand optimisation, taking into account shared energy assets such as RE and their associated variable pricing tariffs [157].

16.6.5 Using AI to Identify Theft of Energy

Energy theft cuts across a variety of operations such as bypassing of electric meters, corruption among energy officials, direct hooking to electrical distribution lines, non-payment of bills, political mismanagement, false reading, damage to electrical meters, etc. [158,159]. It is estimated that energy theft, specifically from electricity, cost about $96 billion worldwide. The issue of energy theft affects not only developing but also developed nations. For example, the UK records over £175 million every year, while the US records about $6 billion annually [160,161]. AI has the potential of curbing energy theft. AI has been identified as a better means of curbing energy theft compared to some conventional methods such as game theory and state-based solutions. Largely, AI techniques use smart meter data such as customers' payment history and energy use patterns to identify irregularities [162–164]. Areas of suspicious activities would be identified and physically visited. One major AI method employed in identifying energy theft is the support vector machine (SVM) [165,166]. Other methods employed are the extreme learning machine model, set theory, decision tree and Naïve Bayesian, lower-upper decomposition, etc. [165,167,168]. The efficiency of these methods in determining energy theft has been witnessed to be at least 60% [131].

A proposed structure by Ref. [131] for the identification of energy theft using AI technology shows that a local control centre is connected to the smart meters of the buildings. This control centre, which monitors the activities of the neighbourhood, serves as a link between the whole neighbourhood and the major control centre. This enhances the monitoring of energy consumption of the area in real-time, noting specifically the discrepancy between the energy supplied to the neighbourhood and the actual energy utilised by the neighbourhood. Any irregularities are identified and followed up on.

16.6.6 Predictive Maintenance Monitoring and Energy Trading

Predictive maintenance is a data-dependent and proactive method designed to help predict when the maintenance of a system should be carried out. In RES, predictive maintenance control contributes significantly to minimising the overall operation and maintenance cost. It helps a great deal to identify faults at very early stages. Once faults aggravate to a severe stage, the cost of repair exponentially increases. This is particularly detrimental to the RE industry. For example, for an offshore wind farm, identifying faults at severe stages will cost large amounts of money to fix, judging from the location and transport of the damaged parts [169,170]. AI has revolutionised predictive maintenance control. For example, instead of the human-only inspection, where humans identify faults within a system, AI-based methods including CNNs, region-based CNNs, drones, and big data, IoT, cloud storage and analytics are being used to detect faults, transmit relevant data, detect irregularities, and prevent cyberattacks. This has replaced the risky and time-consuming manual inspections [171,172]. An illustration of this application can be found in the collaboration between the power firm (Con Edison) serves New York City and Columbia University, which resulted in the establishment of a predictive maintenance system for the power grid that serves the city. The power system in the city saw significant improvement as a result of the project, with 1,468 network days out of 4,590 network days being failure-free, which is a significant increase from the previous records of 908 failure-free network days [173].

In energy trading, excess energy from individuals or companies is given back to a nearby grid. Particularly, entities that generate their RE send the residual energy to the grid. This comes with remuneration and requires that the grid be balanced in terms of demand and supply in real-time [174]. Due to the fluctuation in demand and supply, as well as market prices, energy trading is based on complex parameters that require expertise and accuracy. AI systems have been developed in this regard to automatically trade and make a profit. For example, robotic trading is a fully automated process that uses data from live energy markets to make decisions on purchases or sales on the market [174]. Generally, AI applications in the energy market span across robo-advisory, market intelligence, algorithmic trading, smart operations and routing, fraud identification, trade surveillance, and risk evaluation, recourse improvement, and automated trade's historical data evaluation and handling for energy trading, etc. [131]. These processes are complex and challenging to humans.

16.6.7 INFORMING POLICY

Making effective policies to integrate RE into conventional energy systems require adequate data, which will serve as solid bases for implementation and adoption. This is a complex process which needs a proper understanding of the dynamics within the energy sector, backed by a proper market framework and regulations. AI can simplify the process of policy-making. For instance, some previous research works on solar panel deployment aimed at analysing the socio-economic factors that form the bases of installations of solar panels using AI. Again, some efforts have been made using AI systems to monitor GHG emissions in real-time across some facilities within the energy sector and beyond. Information from these studies can serve as guides towards the development of climate change policies, rules, and regulations [175]. The use of AI to study methane and CO_2 emissions from satellite images can be used to inform policy-making and improve already existing ones.

Other areas of application of AI which inform policy-making include analysis of a large body of the document, prediction of market patterns, and filling gaps of missing data. For instance, the authors of Ref. [176] analysed the different technological innovations within a large data set of solar PV patents using topic modelling techniques. In a similar vein, the authors of [177] used topic modelling to analyse research on climate change to expose and understand any available gaps. All these contribute immensely to validating decisions taken towards effective policy-making.

16.6.8 REDUCING FOSSIL FUEL IMPACTS

Reduction in the impact of fossil fuels remains a major challenge in the energy sector. It is, however, noteworthy that many efforts have been made in the right direction to improve the situation. Applications of AI in this area can be noted in the automatic detection of oil spills, methane leaks, and pipeline corrosion, amongst others. Keramitsoglou et al. [178] applied AI in the detection of oil spillage by training a fuzzy logic-based classifier to identify such leakages from satellite images. In terms of methane leakage, Wan et al. [179] applied a SVM model to sensor data to detect methane leakage in natural gas pipelines. In a similar vein, Wang et al. [180] detected methane leakage from infrared videos by training a CNN in a test facility simulating natural gas production. Other notable areas of fossil fuel impact reduction include emission reduction during fuel production energy efficiency in oil and gas production and CO_2 leakage monitoring [181,182].

16.7 LIMITATIONS OF AI APPLICATION IN THE RES

Despite the great benefits that AI brings to the RE industry, many bottlenecks need to be tackled. In the first place, the reliance of AI systems on big data to train machines comes with the challenge of availability of these data and these data come at high costs. Again, building and integrating AI technology into already existing systems pose huge technological issues. Other drawbacks include legal and compliance issues, and safety and security challenges [183,184]. Some of these challenges are discussed further.

16.7.1 Lack of Theoretical Background

Many entities are still unfamiliar with AI and its perks in the RE industry. There appears to be very little information out there to back decision-making. Consequently, stakeholders have little understanding and skills to make informed decisions on how to best deploy AI in the operation of RE systems. As such, there is slow growth and patronage of the technology.

16.7.2 Lack of Practical Expertise

Practical expertise is needed to effectively operate, manage, and optimise AI systems to maximise application. This expertise is, however, difficult to come by. This area of technology remains an emerging area, and as such coming across experts to develop reliable AI-powered systems with maximum practical benefit is a challenge. Besides, many companies are reluctant to take risks in delving into new areas of AI application in the RE industry as a result of the sensitivity and high cost of the risks associated. Entities prefer to stay within their comfort zone, working with the little expertise they have.

16.7.3 Outdated Infrastructure

RE systems consist of different types of infrastructures that are interconnected to produce energy. Consequently, AI operations or systems rely on this infrastructure for data collection, processing, storage, and information dissemination. That is, AI systems exist within other technologies, social arrangements, and existing structures [185,186]. Due to the sophisticated and powerful nature of AI, ultra-modern and hi-tech infrastructures are required. This is, however, problematic for existing conventional power grid infrastructures, which were not originally the use of AI technology; as such the available infrastructure is unable to support the proper application of AI. Again, it must be noted that most conventional power grids were not developed to allow for the incorporation of RE. This makes their adoption and subsequent application of the related technology a challenge to existing plants [187]. The dependence of AI on other infrastructure has been noted by Star [185] to cover certain properties such as embeddedness, transparency, and modular; unfortunately, not all existing infrastructure can support these properties adequately. Periodic updating of these infrastructures is required to continuously support the application of AI technology.

16.7.4 Economic or Financial Pressure

The development of AI systems for use in RE systems involves financial commitments. As mentioned earlier, AI systems rely on data. The acquisition of suitable data storage systems and associated infrastructure requires large financial investments. Furthermore, the development of software that is exclusively suitable for the operation of RE systems is very time-consuming. In most cases, building such an AI system requires hiring an established software provider to create, configure, maintain, and manage it. In addition, there are other peripheral devices. The implementation of AI systems for the operation of RE plants thus requires enormous investments.

16.7.5 VULNERABILITY: TO CYBERATTACKS

One of the risks associated with the application of AI is cyberattacks. Since AI application depends on the integration of data from diverse owners, formats, and structures, the probability of cyberattack associated with this process is equally high [188]. It must be noted that even though cyber risks are minimised through siloed approaches, these methods are not efficient and hence require careful integration to avoid hackers from accessing valuable data [189]. Again, IT systems are being developed to be more powerful, like quantum computing which is more efficient than super computers and can process tons of data within short periods. While this is beneficial for running AI systems, it also makes hacking faster [190]. In the United Kingdom, it is estimated that cyberattacks on the London energy network reach as high as £111 million per day [191]. Hackers are constantly launching attacks on energy systems. It is reported by the US Home Security Department that since 2011, over 400 attempts were made at intruding into US energy installations. In a similar vein, a study in 2016 found that about 75% of energy companies experienced cyberattacks at least once. That is, much work still abounds in securing AI-based RE systems from cyberattacks. Cybersecurity risk management must be backed by policies and stringently implemented.

16.8 PROSPECTS AND ADVANCEMENT IN ARTIFICIAL INTELLIGENCE FOR EFFECTIVE APPLICATION IN RENEWABLE ENERGY SYSTEMS

Recent advancements in AI techniques including big data explosion, advancement in DL and ML models, increased computing power, intelligent robotics, cyberattack protection, rise in RE incorporation, and the position of the IoT in the smart grid, among other advancements in AI technologies are briefly discussed.

16.8.1 THE PROLIFERATION OF DATA AND THE ADVANCEMENT OF ML MODELS

The prospect of AI has been harnessed for increasing computational power and the creation of massive amounts of data. AI technologies derive value from new data and assist in the management of complicated energy systems. It is a regular prerequisite for an intelligent instrument to effectively optimise and evaluate a vast quantity of data sets created by power systems that there be a relationship between big data and AI. Big data and AI techniques are being used to significantly improve organising and judgement, examinations, status surveillance, supply chain optimisation, and authentication to enhance the performance and correctness of current energy systems.

When it comes to policy-making and regulation for the digital energy system, there will be a slew of new considerations. New rival companies, evolving market models, and more proactive customer involvement would necessitate the use of professionals who were also flexible in their approach, and regulatory intervention in challenges associated with customer health, confidentiality, and data security. Because of the rapid advancements in evolving techniques and data analytics, regulatory permissions for innovative products and services can be effectively provided in the current environment. The data collected by the equipment will also be clear and available to the greatest extent possible. Overall, organisations should strive to develop organisational information and expertise in interpreting not only the techniques but also customer perception and behaviour; data acquisition, evaluation, and protection; the related impacts of automation; and cyber threat, in order to guarantee that they reap the greatest possible benefits from the new potentials presented by AI digitisation. As a result, digitalisation in the energy system should not be regarded as a separate issue, but rather as an inherent element of decarbonisation through the application of AI [131].

Enhanced ML models: DL is the emerging approach in AI and ML. The application of DL models has resulted in significant advancements in ML and AI. The absence of the extraction function in DL as compared to ML is a significant advancement. Conventional ML models including SVMs, decision trees, logistic regression, and Naïve Bayes models have all been utilised previously alongside DL. These classical models are not capable of being deployed immediately to raw data including images, .csv files, text, and other types of text [131]. The extraction function, which is part of the processing phase, is responsible for raw data extraction. It is a complicated procedure that necessitates a great deal of data for the fault domain. In recent years, the improvement of DL models has rendered the extraction function component superfluous. The DL now identifies the traits and can make precise forecasts without the need for human intervention or interference. To meet the growing requirement for data size for various energy systems implementations including PV production prediction [192], battery state-of-charge computation [193], planning of household energy [194], the estimation of energy demands for the transmission and personal gadgets, building energy utilisation planning and maintenance [195], DL models have shown their hierarchical interpretation behind data, in conjunction with optimistic DL research, and help to lessen optimisation challenges and build successful scaling approaches [196]. Furthermore, DL models are also widely employed in a variety of fields, including cybersecurity, data science, image identification, data surveillance, and protection.

Sophisticated DL models are frequently described as "non-explainable," which means that they function as a black box in which inputs are used to assist in the transformation of invisible units into target output. In addition, the detection and elucidation of these concealed levels are a significant difficulty [171]. Approaches such as ML are engineered to develop models for certain functions based on a particular quantity of data inputs. These models have been constructed with specimens to educate the algorithm in a sequence of developmental iterations that produce the outcomes that were intended. ML models are classified into three kinds (supervised, unsupervised, and reinforcement learning). There are two types of supervised learning: regression and classification. Diagnostics, identity fraud identification, client retention, and picture categorisation are all examples of applications for classification models. Regression models are frequently utilised in a variety of applications, including market forecasts, RE facility life span estimations, climate forecasting, population growth prediction, and marketing trend predictions. In addition, there are two forms of unsupervised models, for example, aggregation and dimensional reduction. Elicitation, structure identification, massive data visualisation, and compression are all accomplished through the application of dimensional reduction models. However, customer categorisation, tailored advertising, and system considerations are all accomplished through the usage of aggregation models. Among the applications of reinforcement learning include real-time decision-making, AI game theory, skill development, and robot navigation [131].

16.8.2 Increased Computational Ability and Intelligent Robotics

To keep up with the progress of AI technologies and the growing demand for data collection, the increasing potential to execute computations is required. They would not be able to take leverage of this unless there was an expansion in computer processing capability. There are two distinct historical periods in the development of the computational capacity of AI in training systems; for instance, the first era covers the years 1959–2011, while the second era (the contemporary era) covers the years 2012 to the present. This history of AI explains the ups and downs of computing power. Since 2012, there is a significant acceleration in the improvement of computational power technologies [131]. The explosion of big data, cloud computing, mobile internet [197,198], and AI has facilitated both corporate and scientific studies research in these fields [199]. The application of AI research in intelligent robotic industries is helping to improve the performance and intelligence of robots in the fields of design, advancement, and manufacturing [200].

The intelligence of robotics and AI machines has had profound effects on governments, corporations, and the public. It is used to address cognitive challenges involving human intellect. Advances in digital imagery, voice identification, and robotics have sparked fear among businesses, scientists, legislators, and government officials that AI may substitute, automate, and transcend human intelligence. Robots were utilised by energy corporations for a variety of functions, including security, surveillance, and servicing. Distribution lines, cable lines, and wind turbines are among the facilities that are susceptible to this kind of assessment, and robots equipped with powerful AI have performed this task in a technologically savvy manner. Drone technique will be used by electrical utilities for a variety of purposes, including the monitoring of transmission lines and the examination of wind turbine blades. Wind turbine blades are painted, sanded, and polished using robotics in the wind turbine industry. Robots play critical functions in the automation of solar system manufacturing. The deployment of AI robots in the smart energy industry has the potential to boost performance, save time, and enhance the efficiency of RE optimisation [131].

16.8.3 THE USE OF ARTIFICIAL INTELLIGENCE TO GUARD AGAINST AND IDENTIFY CYBER-CRIME

It has lately been popular to conduct substantial research on jammer attacks and false data assaults [201,202] against smart grids. Jamming attacks are typically divided into two classes: reactive and active jamming. Reactive jamming refers to strikes that occur as a result of a network failure. The goal of an active jammer is to maintain the channel dominated regardless of whether or not it is being used. The reactive jammer is only effective when the communication channel is in operation. Among the most significant concerns associated with jamming is the ease with which it can be launched [191]. In recent years, hackers have targeted energy firms and power network suppliers daily. According to the findings of a 2016 survey, hackers launched at least one cyberattack against 75% of energy firms in the previous year [131]. Since 2011, more than 400 invasions into US energy installations have been reported to the US Home Security Department, according to a report published in 2017 [131].

Cyberattacks on energy systems are one of the most pressing concerns facing modern society and mitigating them is a top priority. Smart houses, Internet-connected devices, smart gas and electricity metres, and other innovations are making energy use more efficient. They are essential for energy control of the grid facility and the precise synchronisation of energy requirements and production. They do, however, provide a plethora of options for cyberattackers. Solar and wind energy production, electric vehicles, and ESSs all operate under comparable circumstances because they are all linked to the power grid as well as the Internet. In addition to commercial spies and cybercriminals, adversaries comprise intelligence agencies, state-sponsored hackers, and military cyber commandos with significant ability and resources. Integration of AI models is extremely beneficial for identifying cyber risks depending on the examination of misleading data, and this data assist in exploiting the fault before it damages RE infrastructure. In order to keep accounts safe from fraudsters, passwords must be used, and AI systems can offer biometric authentication for energy processes that are safe and trustworthy. Many investigations have been performed on cyber-attack mitigation, including the advancement of IoT cybersecurity [203,204], security of wide-area surveillance systems [205], cyber-physical mechanisms [206,207], security of the industrial control system [208], and protection of the commercial control system [206,207]. In order to minimise large-scale disruption and limit the likelihood of potentially devastating failures, wide-area surveillance entails using system-wide data to avert commercial interruption and communicate targeted basic information to a distant location. Furthermore, AI harmonises the disparity between security systems and aids in the design of advanced security systems.

16.8.4 ENHANCE RENEWABLE ENERGY INTEGRATION AND ENERGY EFFICIENCY OPTIMISATION

With the incorporation of a substantial quantity of RE and the deployment of novel AI control models, the traditional grid is evolving. The traditional grid does not have sufficient potential to account for the diversification of RES and the intrinsic fluctuation of wind and solar energy, which poses a range of issues in meeting the fluctuating load demand created by these sources. To maintain the reliability of the power system and satisfy their real-time requirement, power providers and utilities are attempting to balance these variable loads. The application of AI can alleviate these types of obstacles and elevate RE to the status of a comparable participant in the nation's energy supply. Using AI technologies and the smart transactive energy architecture, centralised energy control centres will be improved, as well as the incorporation capabilities of microgrids [209]. It is hoped that the AI optimisation technique will improve the output values (for example, the system topology and how it operates) as well as the deployment of distributed RES [210]. According to research findings from different studies [211–214], AI could be used to create smarter equipment including sensors that can measure massive volumes of data and offer prompt judgements. It enables utilities to make intelligent adjustments to the disparity between demand and supply. Additionally, AI assists in checking the health of control systems, identifying energy loss, and understanding energy utilisation patterns.

The accessibility of the facilities of production is critical to the generation of electricity. Optimisation approaches and AI models have been widely adopted by utilities to increase the return on investment by increasing the performance of energy production while making prompt modifications. The incorporation of digital wind farms with AI systems has the potential to boost energy production by approximately 20% while also generating an extra profit stream of ~$100 million throughout the wind turbine's lifetime [131]. It facilitates real-time communication with the numerous wind farm locations, as well as the monitoring of changes in wind pattern and velocity, as well as the condition of the power grid. The employment of AI aids in the analysis of inadequacies and the specification of solutions that optimise energy consumption. The use of AI in utilities can help optimise breakdowns and startup of moving components, reduce unscheduled outages, and forecast potential maintenance demands based on the performance depreciation of their equipment.

16.8.5 THE RELEVANCE OF ARTIFICIAL INTELLIGENCE IN THE SMART GRID AND THE INTERNET OF THINGS

Electricity transmission systems for commercial, industrial, and office buildings have typically been provided via traditional grid stations that are one-way in nature. The smart grid runs in a bidirectional manner and can perform real-time load changes and surveillance functions. When multiple forms of energy sources and IoT infrastructure are combined, the smart grid becomes more efficient and stable. AI can analyse a vast number of data sets to improve the performance and stability of these new information sources. The IoT is a concept that refers to the improved application of devices and things that have integrated sensors and Internet connectivity. The smart grid's power irregularities are controlled by AI technologies. AI can be employed to optimise smart grids [215], numerous types of smart grid techniques [216], smart load control [217], DR [218], smart grid-enabled IoT [219], management of district-level load, and identification of electricity theft in the smart grid environment. By 2025, it is projected that over 75 billion IoT devices will be in operation worldwide [131].

It is expected that the IoT will contribute significantly to infrastructure planning in aspects of energy performance and analytics; asset management; grid control, simulation, and planning; digital twin; smart metres; virtual power plants; and digital substations. The IoT is being implemented in a variety of ways. Consumers will be cognizant of smart home appliances. The IoT techniques are utilised in the industrial domain to handle process control, generation processes, and optimisation. The IoT is already being employed in several areas in the energy industry. For many energy consumers, the smart metre, which allows for real-time measurement of energy usage as well as regular load uploads, is the most frequent type of metre. Despite this, connected sensors are still widely employed in power grids to monitor energy flow and network issues.

16.8.6 Precision Stabilisation and Dependability, and Information Transfer and Communication

AI can assess data from sensor streaming, infer and recognise trends in the data, and create a response for energy network control appliance loops. Visual intelligence is the potential to analyse and interpret the visual content of grid loops and energy networks and to do it in real-time. Physical intelligence offers it the ability to enact events and strategies for cyberattacks, unforeseen events, and fault circumstances in a human-like manner, including independent driving, without relying on AI. Knowledge intelligence develops texts containing data on energy use that is tracked in real-time, and it has even assisted in human communication. These AI features are critical in maintaining a healthy balance between accuracy and dependability in the system. The use of AI enhances the effective utilisation of generators, storage systems, portfolio and workflow management, and the effective deployment of resources and asset management. The application of AI will assist the electricity sector in capitalising on the interruptions caused by decarbonisation, digitisation, and decentralisation. When it comes to energy forecasting utilising big data, ML models are the best choice because they provide the least amount of forecast error. Utilities benefit from the lowest possible forecasting error, since it helps them to stabilise power and supply. In order to optimise and manage highly complicated grid infrastructure, power generation, linked devices, and residences AI approaches are required [131]. This cutting-edge innovation will aid in the intelligent control of the grid and the effective management of emerging decentralised systems by utilities. To increase dependability and prevent restriction in RES, AI will determine which energy source is the most appropriate at any particular period. The AI models can discover a connection between the production of fossil fuels and RE. The AI models will be modified to encourage the usage of RES while simultaneously reducing fossil fuel utilisation. Temperature estimation and electrical theft can be detected using AI technologies. AI technique improves the stability of the electrical system by boosting battery storage and climate predictions.

In the field of AI, knowledge transfer continues to be a concern [220]. The energy sector, on the other hand, is being transformed by existing sophisticated AI approaches and breakthroughs in smart grid techniques. The AI concepts will enable experts and energy strategists to develop novel models for different energy implementations, including RES (e.g. wind and solar), load planning and control, cybersecurity, big data processing, energy theft identification, and other related areas.

16.9 CONCLUSIONS

The energy sector is undergoing significant transformation due to the disruption brought about by decentralisation and digitisation. Advancements in AI can assist the energy business in optimising the power grid while also maintaining stability and dependability. It is possible to split AI potential into three fields: assessment (for example, collecting and recognising data), inferring (for example, understanding and processing from this data), and reaction. Big data explosions, advances in DL and ML, smart robotics for infrastructure development and power grid surveillance, enhanced incorporation of RE, a substantial rise in IoT in the energy sector, cyberattack protection and safety advantages, and improved computing power are all examples of how AI is playing a leading role in solving many issues in the energy industry. Analysis of data necessitates the use of high-accuracy data sets, whereas ML and AI rely on learning data sets.

Variations in AI policies differ from one country to the next. Governments will decide to concentrate on different parts of AI regulation, based on the national limitations and strengths of each country. Governments must provide funding for both basic and applied studies to accomplish significant advancements in AI theory, technology, and implementations. Nations need skilled AI personnel to undertake AI research and development and to apply AI techniques in the business and governmental sectors of the energy industry. AI has the potential to change several energy industries to spur development in the next decades. Government funding in important industries, as well as the establishment of AI clusters and ecosystems, should be made to stimulate private-sector firms. Several studies have demonstrated that both the public and commercial energy sectors are capable of improvement. The application of AI improves the efficiency of the software. The ability of AI to function is dependent on its ability to consume energy and communicate data. Because of this, countries are striving to expose their databases and establish platforms to encourage safe, private data interchange. There have been a lot of ethical debates around confidentiality, algorithmic bias, and safety in recent months. The goal of government-sponsored utilities is to set codes and ethical standards for implementing and improving AI to prevent harm. Utilities and government AI regulations attempt to maximise AI's wide variety of advantages while considerably lowering its drawbacks and hazards. This is a critical feature of society and its economy, and utilities and government AI policies are a crucial part of that. This study will encourage the AI energy industry to improve its processes and actively assist the creation of RE sources.

ACKNOWLEDGEMENT

This work is based on research funded partly by the National Research Foundation of South Africa (NRF, BRIC190321424123).

REFERENCES

1. A. J. Wood, B. F. Wollenberg, and G. B. Sheblé, *Power Generation, Operation, and Control.* John Wiley & Sons: Hoboken, NJ, 2013.
2. T. B. Johansson, H. Kelly, A. K. Reddy, and R. H. Williams, "Renewable fuels and electricity for a growing world economy: Defining and achieving the potential," in: *Renewable Energy: Sources for Fuels and Electricity.* Island Press: Washington, DC, 1992, pp. 1–71.
3. IEA, "Technology roadmap: Solar photovoltaic energy," International Energy Agency: Paris, 2014.
4. T. Hargreaves and L. Middlemiss, "The importance of social relations in shaping energy demand." *Nature Energy,* vol. 5, no. 3, pp. 195–201, 2020.
5. A. A. Kadhem, N. I. A. Wahab, I. Aris, J. Jasni, A. N. Abdalla, and Y. Matsukawa, "Reliability assessment of generating systems with wind power penetration via BPSO," *Journal of International Journal on Advanced Science, Engineering and Information Technology,* vol. 7, pp. 1–7, 2017.
6. S. G. Chalk and J. F. Miller, "Key challenges and recent progress in batteries, fuel cells, and hydrogen storage for clean energy systems," *Journal of Power Sources,* vol. 159, no. 1, pp. 73–80, 2006.
7. P. V. Kamat, "Meeting the clean energy demand: Nanostructure architectures for solar energy conversion," *The Journal of Physical Chemistry C,* vol. 111, no. 7, pp. 2834–2860, 2007, doi: 10.1021/jp066952u.
8. N. L. Panwar, S. C. Kaushik, and S. Kothari, "Role of renewable energy sources in environmental protection: A review," *Renewable and Sustainable Energy Reviews,* vol. 15, no. 3, pp. 1513–1524, 2011, doi: 10.1016/j.rser.2010.11.037.
9. M. S. Nazir et al., "Wind generation forecasting methods and proliferation of artificial neural network: A review of five years research trend," *Sustainability,* vol. 12, no. 9, p. 3778, 2020.
10. D. Elliott, "Renewable energy and sustainable futures," *Futures,* vol. 32, no. 3, pp. 261–274, 2000, doi: 10.1016/S0016-3287(99)00096-8.
11. A. Evans, V. Strezov, and T. J. Evans, "Assessment of sustainability indicators for renewable energy technologies," *Renewable and Sustainable Energy Reviews,* vol. 13, no. 5, pp. 1082–1088, 2009, doi: 10.1016/j.rser.2008.03.008.
12. E. R. Berndt and D. O. Wood, "Technology, prices, and the derived demand for energy," *The Review of Economics and Statistics,* vol. 57, no. 3, pp. 259–268, 1975.
13. S. K. Jha, J. Bilalovic, A. Jha, N. Patel, and H. Zhang, "Renewable energy: Present research and future scope of Artificial Intelligence," *Renewable and Sustainable Energy Reviews,* vol. 77, pp. 297–317, 2017.
14. M. S. Nazir, Y. Wang, A. J. Mahdi, X. Sun, C. Zhang, and A. N. Abdalla, "Improving the performance of doubly fed induction generator using fault tolerant control: A hierarchical approach," *Applied Sciences,* vol. 10, no. 3, p. 924, 2020.
15. D. Infield and L. Freris, *Renewable Energy in Power Systems.* John Wiley & Sons: Hoboken, NJ, 2020.

16. S.-E. Razavi et al., "Impact of distributed generation on protection and voltage regulation of distribution systems: A review," *Renewable and Sustainable Energy Reviews,* vol. 105, pp. 157–167, 2019.

17. L. A. Wong, V. K. Ramachandaramurthy, P. Taylor, J. Ekanayake, S. L. Walker, and S. Padmanaban, "Review on the optimal placement, sizing and control of an energy storage system in the distribution network," *Journal of Energy Storage,* vol. 21, pp. 489–504, 2019.

18. O. Krishan and S. Suhag, "Techno-economic analysis of a hybrid renewable energy system for an energy poor rural community," *Journal of Energy Storage,* vol. 23, pp. 305–319, 2019.

19. A. Abdukhakimov, S. Bhardwaj, G. Gashema, and D.-S. Kim, "Reliability analysis in smart grid networks considering distributed energy resources and storage devices," *International Journal of Electrical and Electronic Engineering & Telecommunications,* vol. 8, no. 5, pp. 233–237, 2019.

20. A. Azizivahed et al., "Risk-oriented multi-area economic dispatch solution with high penetration of wind power generation and compressed air energy storage system," *IEEE Transactions on Sustainable Energy,* vol. 11, no. 3, pp. 1569–1578, 2019.

21. L. Gaudard and K. Madani, "Energy storage race: Has the monopoly of pumped-storage in Europe come to an end?" *Energy Policy,* vol. 126, pp. 22–29, 2019, doi: 10.1016/j.enpol.2018.11.003.

22. P. C. Sekhar and R. R. Tupakula, "Model predictive controller for single-phase distributed generator with seamless transition between grid and off-grid modes," *IET Generation, Transmission & Distribution,* vol. 13, no. 10, pp. 1829–1837, 2019, doi: 10.1049/iet-gtd.2018.6345.

23. M. S. Nazir et al., "Comparison of small-scale wind energy conversion systems: Economic indexes," *Clean Technologies,* vol. 2, no. 2, pp. 144–155, 2020. [Online]. Available: https://www.mdpi.com/2571-8797/2/2/10.

24. A. A. Kadhem, N. I. A. Wahab, I. Aris, J. Jasni, and A. N. Abdalla, "Advanced wind speed prediction model based on a combination of weibull distribution and an artificial neural network," *Energies,* vol. 10, no. 11, p. 1744, 2017. [Online]. Available: https://www.mdpi.com/1996-1073/10/11/1744.

25. M. S. Nazir, M. Bilal, H. M. Sohail, B. Liu, W. Chen, and H. M. Iqbal, "Impacts of renewable energy atlas: Reaping the benefits of renewables and biodiversity threats," *International Journal of Hydrogen Energy,* vol. 45, no. 41, pp. 22113–22124, 2020.

26. A. A. Kadhem, N. I. A. Wahab, I. Aris, J. Jasni, and A. N. Abdalla, "Computational techniques for assessing the reliability and sustainability of electrical power systems: A review," *Renewable and Sustainable Energy Reviews,* vol. 80, pp. 1175–1186, 2017.

27. M. S. Nazir, N. Ali, T. Yongfeng, A. N. Abdalla, and H. M. J. Nazir, "Renewable energy based experimental study of doubly fed induction generator: Fault case analysis," *Journal of Electrical Systems,* vol. 16, no. 2, pp. 235–245, 2020.

28. W. Ahmed, N. Ali, M. S. Nazir, and A. Khan, "Power quality improving based harmonical studies of a single phase step down bridge-cycloconverter," *Journal of Electrical Systems,* vol. 15, no. 1, pp. 109–122, 2019.

29. A. A. Kadhem, N. I. A. Wahab, and A. N. Abdalla, "Differential evolution optimization algorithm based on generation systems reliability assessment integrated with wind energy," in *2019 International Conference on Power Generation Systems and Renewable Energy Technologies (PGSRET),* 2019, IEEE, pp. 1–6.

30. A. Ali Kadhem, N. I. Abdul Wahab, and A. N. Abdalla, "Wind energy generation assessment at specific sites in a Peninsula in Malaysia based on reliability indices," *Processes,* vol. 7, no. 7, p. 399, 2019. [Online]. Available: https://www.mdpi.com/2227-9717/7/7/399.

31. B. Xu, P. Li, and C. Chan, "Application of phase change materials for thermal energy storage in concentrated solar thermal power plants: A review to recent developments," *Applied Energy,* vol. 160, pp. 286–307, 2015, doi: 10.1016/j.apenergy.2015.09.016.

32. R. Sioshansi and P. Denholm, "The value of concentrating solar power and thermal energy storage," *IEEE Transactions on Sustainable Energy,* vol. 1, no. 3, pp. 173–183, 2010.

33. M. Resener, S. Haffner, L. A. Pereira, and P. M. Pardalos, "Optimization techniques applied to planning of electric power distribution systems: A bibliographic survey," *Energy Systems,* vol. 9, no. 3, pp. 473–509, 2018, doi: 10.1007/s12667-018-0276-x.

34. A. Maleki and A. Askarzadeh, "Comparative study of artificial intelligence techniques for sizing of a hydrogen-based stand-alone photovoltaic/wind hybrid system," *International Journal of Hydrogen Energy,* vol. 39, no. 19, pp. 9973–9984, 2014, doi: 10.1016/j.ijhydene.2014.04.147.

35. S. M. Zahraee, M. Khalaji Assadi, and R. Saidur, "Application of artificial intelligence methods for hybrid energy system optimization," *Renewable and Sustainable Energy Reviews,* vol. 66, pp. 617–630, 2016, doi: 10.1016/j.rser.2016.08.028.

36. Y.-K. Liu, "Convergent results about the use of fuzzy simulation in fuzzy optimization problems," *IEEE Transactions on Fuzzy Systems,* vol. 14, no. 2, pp. 295–304, 2006.

37. S. Russell and P. Norvig, *Artificial Intelligence: A Modern Approach.* Prentice Hall: Hoboken, NJ, 2002.

38. A. Youssef, M. El-Telbany, and A. Zekry, "The role of artificial intelligence in photo-voltaic systems design and control: A review," *Renewable and Sustainable Energy Reviews,* vol. 78, pp. 72–79, 2017, doi: 10.1016/j.rser.2017.04.046.

39. J. Yu, L. Xi, and S. Wang, "An improved particle swarm optimization for evolving feedforward artificial neural networks," *Neural Processing Letters,* vol. 26, no. 3, pp. 217–231, 2007, doi: 10.1007/s11063-007-9053-x.

40. A. Kumar, J. Gupta, and N. Das, "Revisiting the influence of corporate sustainability practices on corporate financial performance: An evidence from the global energy sector," *Business Strategy and the Environment,* 2022, doi: 10.1002/bse.3073.

41. Z. Chen and G. Chen, "An overview of energy consumption of the globalized world economy," *Energy Policy,* vol. 39, no. 10, pp. 5920–5928, 2011.

42. D. Morrone, R. Schena, D. Conte, C. Bussoli, and A. Russo, "Between saying and doing, in the end there is the cost of capital: Evidence from the energy sector," *Business Strategy and the Environment,* vol. 31, no. 1, pp. 390–402, 2022.

43. EIA, "Analysis & projections," in US Energy Information Administration, 2020. [Online]. Available: https://www.eia.gov/analysis/.

44. H. Saleem, M. B. Khan, M. S. Shabbir, G. Y. Khan, and M. Usman, "Nexus between non-renewable energy production, CO_2 emissions, and healthcare spending in OECD economies," *Environmental Science and Pollution Research,* vol. 29, pp. 1–12, 2022.

45. A. Florini and S. Saleem, "Information disclosure in global energy governance," *Global Policy,* vol. 2, pp. 144–154, 2011.

46. B. Bui, O. Moses, and M. N. Houqe, "Carbon disclosure, emission intensity and cost of equity capital: Multi-country evidence," *Accounting & Finance,* vol. 60, no. 1, pp. 47–71, 2020.

47. M. Shahbaz, M. A. Nasir, E. Hille, and M. K. Mahalik, "UK's net-zero carbon emissions target: Investigating the potential role of economic growth, financial development, and R&D expenditures based on historical data (1870–2017)," *Technological Forecasting and Social Change,* vol. 161, p. 120255, 2020.

48. M. Shahbaz, C. Raghutla, M. Song, H. Zameer, and Z. Jiao, "Public-private partnerships investment in energy as new determinant of CO_2 emissions: The role of technological innovations in China," *Energy Economics,* vol. 86, p. 104664, 2020.

49. M. Shahbaz, C. Raghutla, K. R. Chittedi, Z. Jiao, and X. V. Vo, "The effect of renewable energy consumption on economic growth: Evidence from the renewable energy country attractive index," *Energy,* vol. 207, p. 118162, 2020.

50. S. B. Banerjee and A. M. Bonnefous, "Stakeholder management and sustainability strategies in the French nuclear industry," *Business Strategy and the Environment,* vol. 20, no. 2, pp. 124–140, 2011.

51. B. Sorensen, *Renewable Energy: Physics, Engineering, Environmental Impacts, Economics and Planning.* Academic Press: Cambridge, MA, 2017.

52. T. Burton, N. Jenkins, D. Sharpe, and E. Bossanyi, *Wind Energy Handbook.* John Wiley & Sons: Hoboken, NJ, 2011.

53. J. F. Manwell, J. G. McGowan, and A. L. Rogers, *Wind Energy Explained: Theory, Design and Application.* John Wiley & Sons: Hoboken, NJ, 2010.

54. M. J. Pasqualetti, P. Gipe and R. W. Righter, eds., *Wind Power in View: Energy Landscapes in a Crowded World.* Academic Press, 2002.

55. P. D. Fleming and S. D. Probert, "The evolution of wind-turbines: An historical review," *Applied Energy,* vol. 18, no. 3, pp. 163–177, 1984, doi: 10.1016/0306-2619(84)90007-2.

56. J. K. Kaldellis and D. Zafirakis, "The wind energy (r)evolution: A short review of a long history," *Renewable Energy,* vol. 36, no. 7, pp. 1887–1901, 2011, doi: 10.1016/j.renene.2011.01.002.

57. K. Prameela, C. Lahari, G. S. Kishore, K. V. Nikhil, and P. Hemanth, "Artificial Intelligence: A New Era in renewable energy systems," in: S. L. Tripathi, M. K. Dubey, V. Rishiwal, and S. K. Padmanaban (Eds.), *Introduction to AI Techniques for Renewable Energy Systems.* CRC Press: Boca Raton, FL, 2021, pp. 1–14.

58. J. F. Kreider and F. Kreith, *Solar Energy Handbook.* McGraw-Hill: New York, 1981.

59. M. A. Green, *Solar Cells: Operating Principles, Technology, and System Applications.* Prentice-Hall: Englewood Cliffs, NJ, 1982.

60. B. Parida, S. Iniyan, and R. Goic, "A review of solar photovoltaic technologies," *Renewable and Sustainable Energy Reviews,* vol. 15, no. 3, pp. 1625–1636, 2011, doi: 10.1016/j.rser.2010.11.032.

61. J. D. Balcomb, *Passive Solar Buildings.* MIT Press: Cambridge, MA, 1992.

62. R. D. Schaller and V. I. Klimov, "High efficiency carrier multiplication in PbSe nanocrystals: Implications for solar energy conversion," *Physical Review Letters,* vol. 92, no. 18, p. 186601, 2004.

63. G. K. Mor, O. K. Varghese, M. Paulose, K. Shankar, and C. A. Grimes, "A review on highly ordered, vertically oriented TiO_2 nanotube arrays: Fabrication, material properties, and solar energy applications," *Solar Energy Materials and Solar Cells,* vol. 90, no. 14, pp. 2011–2075, 2006, doi: 10.1016/j.solmat.2006.04.007.

64. C. G. Granqvist, "Transparent conductors as solar energy materials: A panoramic review," *Solar Energy Materials and Solar Cells,* vol. 91, no. 17, pp. 1529–1598, 2007, doi: 10.1016/j.solmat.2007.04.031.

65. O. Mahian, A. Kianifar, S. A. Kalogirou, I. Pop, and S. Wongwises, "A review of the applications of nanofluids in solar energy," *International Journal of Heat and Mass Transfer,* vol. 57, no. 2, pp. 582–594, 2013, doi: 10.1016/j.ijheatmasstransfer.2012.10.037.

66. K. H. Solangi, M. R. Islam, R. Saidur, N. A. Rahim, and H. Fayaz, "A review on global solar energy policy," *Renewable and Sustainable Energy Reviews,* vol. 15, no. 4, pp. 2149–2163, 2011, doi: 10.1016/j.rser.2011.01.007.

67. O. V. Ekechukwu and B. Norton, "Review of solar-energy drying systems II: An overview of solar drying technology," *Energy Conversion and Management,* vol. 40, no. 6, pp. 615–655, 1999, doi: 10.1016/S0196-8904(98)00093-4.

68. J. Blanco, S. Malato, P. Fernández-Ibañez, D. Alarcón, W. Gernjak, and M. I. Maldonado, "Review of feasible solar energy applications to water processes," *Renewable and Sustainable Energy Reviews,* vol. 13, no. 6, pp. 1437–1445, 2009, doi: 10.1016/j.rser.2008.08.016.

69. S. Mekhilef, R. Saidur, and A. Safari, "A review on solar energy use in industries," *Renewable and Sustainable Energy Reviews,* vol. 15, no. 4, pp. 1777–1790, 2011, doi: 10.1016/j.rser.2010.12.018.

70. K. E. N'Tsoukpoe, H. Liu, N. Le Pierrès, and L. Luo, "A review on long-term sorption solar energy storage," *Renewable and Sustainable Energy Reviews,* vol. 13, no. 9, pp. 2385–2396, 2009, doi: 10.1016/j.rser.2009.05.008.

71. I. Sartori and A. G. Hestnes, "Energy use in the life cycle of conventional and low-energy buildings: A review article," *Energy and Buildings,* vol. 39, no. 3, pp. 249–257, 2007, doi: 10.1016/j.enbuild.2006.07.001.

72. T. Khatib, A. Mohamed, and K. Sopian, "A review of solar energy modeling techniques," *Renewable and Sustainable Energy Reviews,* vol. 16, no. 5, pp. 2864–2869, 2012, doi: 10.1016/j.rser.2012.01.064.

73. F. Besharat, A. A. Dehghan, and A. R. Faghih, "Empirical models for estimating global solar radiation: A review and case study," *Renewable and Sustainable Energy Reviews,* vol. 21, pp. 798–821, 2013.

74. A. K. Yadav and S. S. Chandel, "Tilt angle optimization to maximize incident solar radiation: A review," *Renewable and Sustainable Energy Reviews,* vol. 23, pp. 503–513, 2013, doi: 10.1016/j.rser.2013.02.027.

75. P. Verma, and S. K. Singal, "Review of mathematical modeling on latent heat thermal energy storage systems using phase-change material," *Renewable and Sustainable Energy Reviews,* vol. 12, no. 4, pp. 999–1031, 2008, doi: 10.1016/j.rser.2006.11.002.

76. H. C. H. Armstead, *Geothermal Energy: Its Past, Present and Future Contributions to the Energy Needs of Man.* E. & F.N. Spon: London, 1978.

77. M. H. Dickson and M. Fanelli, *Geothermal Energy: Utilization and Technology.* Routledge: England, 2013.

78. E. Barbier, "Nature and technology of geothermal energy: A review," *Renewable and Sustainable Energy Reviews,* vol. 1, no. 1, pp. 1–69, 1997, doi: 10.1016/S1364-0321(97)00001-4.

79. E. Barbier, "Geothermal energy technology and current status: An overview," *Renewable and Sustainable Energy Reviews,* vol. 6, no. 1, pp. 3–65, 2002, doi: 10.1016/S1364-0321(02)00002-3.

80. J. W. Lund and D. H. Freeston, "World-wide direct uses of geothermal energy 2000," *Geothermics,* vol. 30, no. 1, pp. 29–68, 2001, doi: 10.1016/S0375-6505(00)00044-4.

81. I. B. Fridleifsson, "Geothermal energy for the benefit of the people," *Renewable and Sustainable Energy Reviews,* vol. 5, no. 3, pp. 299–312, 2001, doi: 10.1016/S1364-0321(01)00002-8.

82. J. W. Lund, D. H. Freeston, and T. L. Boyd, "Direct application of geothermal energy: 2005 Worldwide review," *Geothermics,* vol. 34, no. 6, pp. 691–727, 2005, doi: 10.1016/j.geothermics.2005.09.003.

83. J. W. Lund, D. H. Freeston, and T. L. Boyd, "Direct utilization of geothermal energy 2010 worldwide review," *Geothermics,* vol. 40, no. 3, pp. 159–180, 2011, doi: 10.1016/j.geothermics.2011.07.004.

84. S. Haehnlein, P. Bayer, and P. Blum, "International legal status of the use of shallow geothermal energy," *Renewable and Sustainable Energy Reviews,* vol. 14, no. 9, pp. 2611–2625, 2010, doi: 10.1016/j.rser.2010.07.069.

85. M. J. O'Sullivan, K. Pruess, and M. J. Lippmann, "State of the art of geothermal reservoir simulation," *Geothermics,* vol. 30, no. 4, pp. 395–429, 2001, doi: 10.1016/S0375-6505(01)00005-0.

86. E. R. Okoroafor, C. M. Smith, K. I. Ochie, C. J. Nwosu, H. Gudmundsdottir, and M. Aljubran, "Machine learning in subsurface geothermal energy: Two decades in review," *Geothermics,* vol. 102, p. 102401, 2022, doi: 10.1016/j.geothermics.2022.102401.

87. S. M. Pinnoo, N. R. Hart-Wagoner, B. Pollett, R. Pilko, and J. Chen, "Advancing geothermal energy exploration," *Presented at the Offshore Technology Conference,* 2022. [Online]. Available: https://doi.org/10.4043/32109-MS.

88. O. Bamisile et al., "An innovative approach for geothermal-wind hybrid comprehensive energy system and hydrogen production modeling/process analysis," *International Journal of Hydrogen Energy,* vol. 47, no. 27, pp. 13261–13288, 2022, doi: 10.1016/j.ijhydene.2022.02.084.

89. Q. Zhang, S. A. R. Shah, and L. Yang, "Modeling the effect of disaggregated renewable energies on ecological footprint in E5 economies: Do economic growth and R&D matter?" *Applied Energy,* vol. 310, p. 118522, 2022, doi: 10.1016/j.apenergy.2022.118522.

90. S. A. Alnaqbi, S. Alasad, H. Aljaghoub, A. H. Alami, M. A. Abdelkareem, and A. G. Olabi, "Applicability of hydropower generation and pumped hydro energy storage in the middle east and North Africa," *Energies,* vol. 15, no. 7, p. 2412, 2022. [Online]. Available: https://www.mdpi.com/1996-1073/15/7/2412.

91. D. K. Okot, "Review of small hydropower technology," *Renewable and Sustainable Energy Reviews,* vol. 26, pp. 515–520, 2013, doi: 10.1016/j.rser.2013.05.006.

92. J. P. Deane, B. P. Ó Gallachóir, and E. J. McKeogh, "Techno-economic review of existing and new pumped hydro energy storage plant," *Renewable and Sustainable Energy Reviews,* vol. 14, no. 4, pp. 1293–1302, 2010, doi: 10.1016/j.rser.2009.11.015.

93. C.-J. Yang and R. B. Jackson, "Opportunities and barriers to pumped-hydro energy storage in the United States," *Renewable and Sustainable Energy Reviews,* vol. 15, no. 1, pp. 839–844, 2011, doi: 10.1016/j.rser.2010.09.020.

94. J. W. Labadie, "Optimal operation of multireservoir systems: State-of-the-art review," *Journal of Water Resources Planning and Management,* vol. 130, no. 2, pp. 93–111, 2004, doi: 10.1061/(ASCE)0733-9496(2004)130:2(93).

95. M. J. Khan, G. Bhuyan, M. T. Iqbal, and J. E. Quaicoe, "Hydrokinetic energy conversion systems and assessment of horizontal and vertical axis turbines for river and tidal applications: A technology status review," *Applied Energy,* vol. 86, no. 10, pp. 1823–1835, 2009, doi: 10.1016/j.apenergy.2009.02.017.

96. M. J. Khan, M. T. Iqbal, and J. E. Quaicoe, "River current energy conversion systems: Progress, prospects and challenges," *Renewable and Sustainable Energy Reviews,* vol. 12, no. 8, pp. 2177–2193, 2008, doi: 10.1016/j.rser.2007.04.016.

97. M. K. Padhy and R. P. Saini, "A review on silt erosion in hydro turbines," *Renewable and Sustainable Energy Reviews,* vol. 12, no. 7, pp. 1974–1987, 2008, doi: 10.1016/j.rser.2007.01.025.

98. S. Mishra, S. K. Singal, and D. K. Khatod, "Optimal installation of small hydropower plant: A review," *Renewable and Sustainable Energy Reviews,* vol. 15, no. 8, pp. 3862–3869, 2011, doi: 10.1016/j.rser.2011.07.008.

99. B. K. Sovacool, S. Dhakal, O. Gippner, and M. J. Bambawale, "Halting hydro: A review of the socio-technical barriers to hydroelectric power plants in Nepal," *Energy,* vol. 36, no. 5, pp. 3468–3476, 2011, doi: 10.1016/j.energy.2011.03.051.

100. D. Yue, F. You, and S. W. Snyder, "Biomass-to-bioenergy and biofuel supply chain optimization: Overview, key issues and challenges," *Computers & Chemical Engineering,* vol. 66, pp. 36–56, 2014, doi: 10.1016/j.compchemeng.2013.11.016.

101. I. M. S. Anekwe, L. Khotseng, and Y. M. Isa, "The place of biofuel in sustainable living; prospects and challenges," in: F. W. Protozoa (Ed.), *Reference Module in Earth Systems and Environmental Sciences.* Elsevier: Amsterdam, Netherlands, 2021.

102. T. Demirbas and A. H. Demirbas, "Bioenergy, green energy. biomass and biofuels," *Energy Sources, Part A: Recovery, Utilization, and Environmental Effects,* vol. 32, no. 12, pp. 1067–1075, 2010, doi: 10.1080/15567030903058600.

103. M. Balat and H. Balat, "Recent trends in global production and utilization of bio-ethanol fuel," *Applied Energy,* vol. 86, no. 11, pp. 2273–2282, 2009, doi: 10.1016/j.apenergy.2009.03.015.

104. T. M. Mata, A. A. Martins, and N. S. Caetano, "Microalgae for biodiesel production and other applications: A review," *Renewable and Sustainable Energy Reviews,* vol. 14, no. 1, pp. 217–232, 2010, doi: 10.1016/j.rser.2009.07.020.

105. O. P. Karthikeyan and C. Visvanathan, "Bio-energy recovery from high-solid organic substrates by dry anaerobic bio-conversion processes: A review," *Reviews in Environmental Science and Bio/Technology,* vol. 12, no. 3, pp. 257–284, 2013, doi: 10.1007/s11157-012-9304-9.

106. D. Mohan, C. U. Pittman, and P. H. Steele, "Pyrolysis of wood/biomass for bio-oil: A critical review," *Energy & Fuels,* vol. 20, no. 3, pp. 848–889, 2006, doi: 10.1021/ef0502397.

107. P. McKendry, "Energy production from biomass (part 2): Conversion technologies," *Bioresource Technology,* vol. 83, no. 1, pp. 47–54, 2002, doi: 10.1016/S0960-8524(01)00119-5.

108. S. Gold and S. Seuring, "Supply chain and logistics issues of bio-energy production," *Journal of Cleaner Production,* vol. 19, no. 1, pp. 32–42, 2011, doi: 10.1016/j.jclepro.2010.08.009.

109. M. Fatih Demirbas, "Biorefineries for biofuel upgrading: A critical review," *Applied Energy,* vol. 86, pp. S151–S161, 2009, doi: 10.1016/j.apenergy.2009.04.043.

110. A. P. C. Faaij, "Bio-energy in Europe: Changing technology choices," *Energy Policy,* vol. 34, no. 3, pp. 322–342, 2006, doi: 10.1016/j.enpol.2004.03.026.

111. Z. Du, H. Li, and T. Gu, "A state of the art review on microbial fuel cells: A promising technology for wastewater treatment and bioenergy," *Biotechnology Advances,* vol. 25, no. 5, pp. 464–82, 2007, doi: 10.1016/j.biotechadv.2007.05.004.

112. F. Dawood, M. Anda, and G. M. Shafiullah, "Hydrogen production for energy: An overview," *International Journal of Hydrogen Energy,* vol. 45, no. 7, pp. 3847–3869, 2020, doi: 10.1016/j.ijhydene.2019.12.059.

113. S. Sharma and S. K. Ghoshal, "Hydrogen the future transportation fuel: From production to applications," *Renewable and Sustainable Energy Reviews,* vol. 43, pp. 1151–1158, 2015, doi: 10.1016/j.rser.2014.11.093.

114. A. Midilli, M. Ay, I. Dincer, and M. A. Rosen, "On hydrogen and hydrogen energy strategies: I: Current status and needs," *Renewable and Sustainable Energy Reviews,* vol. 9, no. 3, pp. 255–271, 2005, doi: 10.1016/j.rser.2004.05.003.

115. M. Momirlan and T. N. Veziroglu, "Current status of hydrogen energy," *Renewable and Sustainable Energy Reviews,* vol. 6, no. 1, pp. 141–179, 2002, doi: 10.1016/S1364-0321(02)00004-7.

116. J. R. Bolton, "Solar photoproduction of hydrogen: A review," *Solar Energy,* vol. 57, no. 1, pp. 37–50, 1996, doi: 10.1016/0038-092X(96)00032-1.

117. J. Wang and W. Wan, "Factors influencing fermentative hydrogen production: A review," *International Journal of Hydrogen Energy,* vol. 34, no. 2, pp. 799–811, 2009, doi: 10.1016/j.ijhydene.2008.11.015.

118. B. Sakintuna, F. Lamari-Darkrim, and M. Hirscher, "Metal hydride materials for solid hydrogen storage: A review," *International Journal of Hydrogen Energy,* vol. 32, no. 9, pp. 1121–1140, 2007, doi: 10.1016/j.ijhydene.2006.11.022.

119. X. Cheng et al., "A review of PEM hydrogen fuel cell contamination: Impacts, mechanisms, and mitigation," *Journal of Power Sources,* vol. 165, no. 2, pp. 739–756, 2007, doi: 10.1016/j.jpowsour.2006.12.012.

120. S. Al-Hallaj and K. Kiszynski, *Hybrid Hydrogen Systems: Stationary and Transportation Applications.* Springer Science & Business Media: Berlin, Germany, 2011.

121. A. N. Agrawal, R. Wies, and R. A. Johnson, *Hybrid Electric Power Systems: Modeling, Optimization and Control.* VDM Publishing: Saarbrücken, Germany, 2007.

122. P. Bajpai and V. Dash, "Hybrid renewable energy systems for power generation in stand-alone applications: A review," *Renewable and Sustainable Energy Reviews,* vol. 16, no. 5, pp. 2926–2939, 2012.

123. M. Nehrir et al., "A review of hybrid renewable/alternative energy systems for electric power generation: Configurations, control, and applications," *IEEE Transactions on Sustainable Energy,* vol. 2, no. 4, pp. 392–403, 2011.

124. O. Erdinc and M. Uzunoglu, "Optimum design of hybrid renewable energy systems: Overview of different approaches," *Renewable and Sustainable Energy Reviews,* vol. 16, no. 3, pp. 1412–1425, 2012.

125. P. Nema, R. Nema, and S. Rangnekar, "A current and future state of art development of hybrid energy system using wind and PV-solar: A review," *Renewable and Sustainable Energy Reviews,* vol. 13, no. 8, pp. 2096–2103, 2009.

126. D. Connolly, H. Lund, B. V. Mathiesen, and M. Leahy, "A review of computer tools for analysing the integration of renewable energy into various energy systems," *Applied Energy,* vol. 87, no. 4, pp. 1059–1082, 2010.

127. A. Etxeberria, I. Vechiu, H. Camblong, and J.-M. Vinassa, "Hybrid energy storage systems for renewable energy sources integration in microgrids: A review," *in 2010 Conference Proceedings IPEC,* 2010, IEEE, pp. 532–537.

128. B. Bhandari, S. R. Poudel, K.-T. Lee, and S.-H. Ahn, "Mathematical modeling of hybrid renewable energy system: A review on small hydro-solar-wind power generation," *International Journal of Precision Engineering and Manufacturing-Green Technology,* vol. 1, no. 2, pp. 157–173, 2014.

129. V. S. B. Kurukuru, A. Haque, M. A. Khan, S. Sahoo, A. Malik, and F. Blaabjerg, "A review on artificial intelligence applications for grid-connected solar photovoltaic systems," *Energies,* vol. 14, no. 15, p. 4690, 2021.

130. A. Mellit and S. Kalogirou, "Artificial intelligence and internet of things to improve efficacy of diagnosis and remote sensing of solar photovoltaic systems: Challenges, recommendations and future directions," *Renewable and Sustainable Energy Reviews,* vol. 143, p. 110889, 2021, doi: 10.1016/j.rser.2021.110889.

131. T. Ahmad et al., "Artificial intelligence in sustainable energy industry: Status Quo, challenges and opportunities," *Journal of Cleaner Production,* vol. 289, p. 125834, 2021, doi: 10.1016/j.jclepro.2021.125834.

132. F. P. García Márquez and A. Peinado Gonzalo, "A comprehensive review of artificial intelligence and wind energy," *Archives of Computational Methods in Engineering,* pp. 1–24, 2021.

133. M. Shehab et al., "Machine learning in medical applications: A review of state-of-the-art methods," *Computers in Biology and Medicine,* vol. 145, p. 105458, 2022.

134. R. Belu, "Artificial intelligence techniques for solar energy and photovoltaic applications," in: Management Association, Information Resources (Ed.), *Robotics: Concepts, Methodologies, Tools, and Applications.* IGI Global: Hershey, PA, 2014, pp. 1662–1720.

135. S. Gupta and L. Li, "The potential of machine learning for enhancing CO_2 sequestration, storage, transportation, and utilization-based processes: A brief perspective," *JOM,* vol. 74, pp. 1–15, 2022.

136. Y. Guo, Y. Liu, A. Oerlemans, S. Lao, S. Wu, and M. S. Lew, "Deep learning for visual understanding: A review," *Neurocomputing,* vol. 187, pp. 27–48, 2016.

137. A. Mellit, S. A. Kalogirou, L. Hontoria, and S. Shaari, "Artificial intelligence techniques for sizing photovoltaic systems: A review," *Renewable and Sustainable Energy Reviews,* vol. 13, no. 2, pp. 406–419, 2009.

138. K. A. Severson *et al.*, "Data-driven prediction of battery cycle life before capacity degradation," *Nature Energy,* vol. 4, no. 5, pp. 383–391, 2019.

139. T. A. Faunce, J. Prest, D. Su, S. J. Hearne, and F. Iacopi, "On-grid batteries for large-scale energy storage: Challenges and opportunities for policy and technology," *MRS Energy & Sustainability,* vol. 5, p. E11, 2018.

140. P. Boza and T. Evgeniou, "Artificial intelligence to support the integration of variable renewable energy sources to the power system," *Applied Energy,* vol. 290, p. 116754, 2021.

141. A. Boretti, "Integration of solar thermal and photovoltaic, wind, and battery energy storage through AI in NEOM city," *Energy and AI,* vol. 3, p. 100038, 2021.

142. S. Rao et al., "Machine learning for solar array monitoring, optimization, and control," *Synthesis Lectures on Power Electronics,* vol. 7, no. 1, pp. 1–91, 2020.

143. S. Iyengar, S. Lee, D. Sheldon, and P. Shenoy, "SolarClique: Detecting anomalies in residential solar arrays. ," in *ACM SIGCAS Conference on Computing and Sustainable Societies (COMPASS'18),* New York, 2018.

144. R. Orozco, S. Sheng, and C. Phillips, "Diagnostic models for wind turbine gearbox components using SCADA time series data," in *2018 IEEE International Conference on Prognostics and Health Management (ICPHM),* 2018, IEEE, pp. 1–9.

145. C. Xiao, Z. Liu, T. Zhang, and X. Zhang, "Deep learning method for fault detection of wind turbine converter," *Applied Sciences,* vol. 11, no. 3, p. 1280, 2021.

146. L. Xiuli, Z. Xueying, and W. Liyong, "Fault diagnosis method of wind turbine gearbox based on deep belief network and vibration signal," in *2018 57th Annual Conference of the Society of Instrument and Control Engineers of Japan (SICE),* Nara, Japan, 2018, IEEE, pp. 1699–1704.

147. G. Jiang, H. He, P. Xie, and Y. Tang, "Stacked multilevel-denoising autoencoders: A new representation learning approach for wind turbine gearbox fault diagnosis," *IEEE Transactions on Instrumentation and Measurement,* vol. 66, no. 9, pp. 2391–2402, 2017.

148. F. Caliva et al., "A deep learning approach to anomaly detection in nuclear reactors," in *2018 International Joint Conference on Neural Networks (IJCNN),* 2018, IEEE, pp. 1–8.

149. F.-C. Chen and M. R. Jahanshahi, "NB-CNN: Deep learning-based crack detection using convolutional neural network and Naïve Bayes data fusion," *IEEE Transactions on Industrial Electronics,* vol. 65, no. 5, pp. 4392–4400, 2017.

150. Z. Nela and C. G. Kenny, "Sensor fault detection in nuclear power plants using multivariate state estimation technique and support vector machines," *Presented at the Third International Conference of the Yugoslav Nuclear Society YUNSC 2000,* October 2–5, 2000, Belgrade, Yugoslavia, 2000. [Online]. Available: https://www.osti.gov/servlets/purl/766314.

151. L. J. Catania, *Foundations of Artificial Intelligence in Healthcare and Bioscience: A User Friendly Guide for IT Professionals, Healthcare Providers, Researchers, and Clinicians.* Academic Press: Cambridge, MA, 2020.

152. P. Arévalo and F. Jurado, "Performance analysis of a PV/HKT/WT/DG hybrid autonomous grid," *Electrical Engineering,* vol. 103, no. 1, pp. 227–244, 2021.

153. K. Attonaty, J. Pouvreau, A. Deydier, J. Oriol, and P. Stouffs, "Thermodynamic and economic evaluation of an innovative electricity storage system based on thermal energy storage," *Renewable Energy,* vol. 150, pp. 1030–1036, 2020.

154. S. E. Haupt et al., "Combining artificial intelligence with physics-based methods for probabilistic renewable energy forecasting," *Energies,* vol. 13, no. 8, p. 1979, 2020.

155. C. Tomazzoli, S. Scannapieco, and M. Cristani, "Internet of things and artificial intelligence enable energy efficiency," *Journal of Ambient Intelligence and Humanized Computing,* pp. 1–22, 2020.

156. K. P. Tsagarakis, "Shallow geothermal energy under the microscope: Social, economic, and institutional aspects," *Renewable Energy,* vol. 147, pp. 2801–2808, 2020.

157. I. Esnaola-Gonzalez, F. J. Diez, D. Pujic, M. Jelic, and N. Tomasevic, "An artificial intelligent system for demand response in neighbourhoods," in *Proceedings of the Workshop on Artificial Intelligence in Power and Energy Systems (AIPES 2020),* 2020.

158. P. Jokar, N. Arianpoo, and V. C. Leung, "Electricity theft detection in AMI using customers' consumption patterns," *IEEE Transactions on Smart Grid,* vol. 7, no. 1, pp. 216–226, 2015.

159. M. Lydia and S. S. Kumar, "A comprehensive overview on wind power forecasting," *in 2010 Conference Proceedings IPEC,* Singapore, 2010, IEEE, pp. 268–273.

160. S. S. S. R. Depuru, L. Wang, and V. Devabhaktuni, "Electricity theft: Overview, issues, prevention and a smart meter based approach to control theft," *Energy policy,* vol. 39, no. 2, pp. 1007–1015, 2011.

161. P. McDaniel and S. McLaughlin, "Security and privacy challenges in the smart grid," *IEEE Security & Privacy,* vol. 7, no. 3, pp. 75–77, 2009.

162. M.-M. Buzau, J. Tejedor-Aguilera, P. Cruz-Romero, and A. Gómez-Expósito, "Hybrid deep neural networks for detection of non-technical losses in electricity smart meters," *IEEE Transactions on Power Systems,* vol. 35, no. 2, pp. 1254–1263, 2019.

163. M.-M. Buzau, "Machine learning algorithms for the detection of non-technical losses in electrical distribution networks," 2020.

164. N. F. Avila, G. Figueroa, and C.-C. Chu, "NTL detection in electric distribution systems using the maximal overlap discrete wavelet-packet transform and random undersampling boosting," *IEEE Transactions on Power Systems,* vol. 33, no. 6, pp. 7171–7180, 2018.

165. A. Nizar, Z. Dong, and Y. Wang, "Power utility nontechnical loss analysis with extreme learning machine method," *IEEE Transactions on Power Systems,* vol. 23, no. 3, pp. 946–955, 2008.

166. J. Nagi, K. S. Yap, F. Nagi, S. K. Tiong, S. Koh, and S. K. Ahmed, "NTL detection of electricity theft and abnormalities for large power consumers in TNB Malaysia," in *2010 IEEE Student Conference on Research and Development (SCOReD),* Malaysia, 2010, IEEE, pp. 202–206.

167. A. H. Nizar, Z. Y. Dong, and P. Zhang, "Detection rules for non technical losses analysis in power utilities," in *2008 IEEE Power and Energy Society General Meeting-Conversion and Delivery of Electrical Energy in the 21st Century,* Pittsburgh, PA, 2008, IEEE, pp. 1–8.

168. S. Salinas, M. Li, P. Li, and Y. Fu, "Dynamic energy management for the smart grid with distributed energy resources," *IEEE Transactions on Smart Grid,* vol. 4, no. 4, pp. 2139–2151, 2013.

169. W. Shin, J. Han, and W. Rhee, "AI-assistance for predictive maintenance of renewable energy systems," *Energy,* vol. 221, p. 119775, 2021.

170. N. T. Tran, H. T. Trieu, V. T. Tran, H. H. Ngo, and Q. K. Dao, "An overview of the application of machine learning in predictive maintenance," *Petrovietnam Journal,* vol. 10, pp. 47–61, 2021.

171. Y. Merizalde, L. Hernández-Callejo, O. Duque-Perez, and V. Alonso-Gómez, "Maintenance models applied to wind turbines. A comprehensive overview," *Energies,* vol. 12, no. 2, p. 225, 2019.

172. Y. Merizalde, L. Hernández-Callejo, O. Duque-Pérez, and V. Alonso-Gómez, "Diagnosis of wind turbine faults using generator current signature analysis: A review," *Journal of Quality in Maintenance Engineering,* 2019. doi: 10.1108/JQME-02-2019-0020.

173. C. Rudin et al., "Machine learning for the New York City power grid," *IEEE Transactions on Pattern Analysis and Machine Intelligence,* vol. 34, no. 2, pp. 328–345, 2011.

174. GridBeyond, "Artificial intelligence in energy: Everything you need to know," 2021.

175. P. L. Donti and J. Z. Kolter, "Machine learning for sustainable energy systems," *Annual Review of Environment and Resources,* vol. 46, pp. 719–747, 2021.

176. S. Venugopalan and V. Rai, "Topic based classification and pattern identification in patents," *Technological Forecasting and Social Change,* vol. 94, pp. 236–250, 2015.

177. M. W. Callaghan, J. C. Minx, and P. M. Forster, "A topography of climate change research," *Nature Climate Change,* vol. 10, no. 2, pp. 118–123, 2020.

178. I. Keramitsoglou, C. Cartalis, and C. T. Kiranoudis, "Automatic identification of oil spills on satellite images," *Environmental Modelling & Software,* vol. 21, no. 5, pp. 640–652, 2006.

179. J. Wan, Y. Yu, Y. Wu, R. Feng, and N. Yu, "Hierarchical leak detection and localization method in natural gas pipeline monitoring sensor networks," *Sensors,* vol. 12, no. 1, pp. 189–214, 2011.

180. J. Wang et al., "Machine vision for natural gas methane emissions detection using an infrared camera," *Applied Energy,* vol. 257, p. 113998, 2020.

181. B. Chen, D. R. Harp, Y. Lin, E. H. Keating, and R. J. Pawar, "Geologic CO_2 sequestration monitoring design: A machine learning and uncertainty quantification based approach," *Applied Energy,* vol. 225, pp. 332–345, 2018.

182. S. Mo, Y. Zhu, N. Zabaras, X. Shi, and J. Wu, "Deep convolutional encoder-decoder networks for uncertainty quantification of dynamic multiphase flow in heterogeneous media," *Water Resources Research,* vol. 55, no. 1, pp. 703–728, 2019.

183. S. Guo, Y. Lin, N. Feng, C. Song, and H. Wan, "Attention based spatial-temporal graph convolutional networks for traffic flow forecasting," *Proceedings of the AAAI Conference on Artificial Intelligence,* vol. 33, no. 01, pp. 922–929, 2019.

184. L. Zhao, L. Tang, and Y. Yang, "Comparison of modeling methods and parametric study for a piezoelectric wind energy harvester," *Smart Materials and Structures,* vol. 22, no. 12, p. 125003, 2013.

185. S. L. Star, "The ethnography of infrastructure," *American Behavioral Scientist,* vol. 43, no. 3, pp. 377–391, 1999.

186. S. Robbins and A. van Wynsberghe, "Our new artificial intelligence infrastructure: Becoming locked into an unsustainable future," *Sustainability,* vol. 14, no. 8, p. 4829, 2022.

187. K. W. Kow, Y. W. Wong, R. K. Rajkumar, and R. K. Rajkumar, "A review on performance of artificial intelligence and conventional method in mitigating PV grid-tied related power quality events," *Renewable and Sustainable Energy Reviews,* vol. 56, pp. 334–346, 2016.

188. R. Nishant, M. Kennedy, and J. Corbett, "Artificial intelligence for sustainability: Challenges, opportunities, and a research agenda," *International Journal of Information Management,* vol. 53, p. 102104, 2020.

189. J. P. Meltzer, "Governing digital trade," *World Trade Review,* vol. 18, no. S1, pp. S23–S48, 2019, doi: 10.1017/S1474745618000502.

190. P. J. Denning, "Design thinking," *Communications of the ACM,* vol. 56, no. 12, pp. 29–31, 2013.

191. E. J. Oughton et al., "Stochastic counterfactual risk analysis for the vulnerability assessment of cyberphysical attacks on electricity distribution infrastructure networks," *Risk Analysis,* vol. 39, no. 9, pp. 2012–2031, 2019.

192. H. Wang et al., "Deterministic and probabilistic forecasting of photovoltaic power based on deep convolutional neural network," *Energy Conversion and Management,* vol. 153, pp. 409–422, 2017.

193. E. Chemali, P. J. Kollmeyer, M. Preindl, and A. Emadi, "State-of-charge estimation of Li-ion batteries using deep neural networks: A machine learning approach," *Journal of Power Sources,* vol. 400, pp. 242–255, 2018.

194. I. M. Coelho, V. N. Coelho, E. J. D. S. Luz, L. S. Ochi, F. G. Guimarães, and E. Rios, "A GPU deep learning metaheuristic based model for time series forecasting," *Applied Energy,* vol. 201, pp. 412–418, 2017.

195. E. Mocanu, P. H. Nguyen, M. Gibescu, and W. L. Kling, "Deep learning for estimating building energy consumption," *Sustainable Energy, Grids and Networks,* vol. 6, pp. 91–99, 2016.

196. M. R. Minar and J. Naher, "Recent advances in deep learning: An overview," arXiv preprint arXiv:1807.08169, 2018.

197. J. Chen, K. He, Q. Yuan, M. Chen, R. Du, and Y. Xiang, "Blind filtering at third parties: An efficient privacy-preserving framework for location-based services," *IEEE Transactions on Mobile Computing,* vol. 17, no. 11, pp. 2524–2535, 2018.

198. Y. Chen, Y. Sun, and Z. Meng, "An improved explicit double-diode model of solar cells: Fitness verification and parameter extraction," *Energy Conversion and Management,* vol. 169, pp. 345–358, 2018.

199. X. Ge, L. Pan, Q. Li, G. Mao, and S. Tu, "Multipath cooperative communications networks for augmented and virtual reality transmission," *IEEE Transactions on Multimedia,* vol. 19, no. 10, pp. 2345–2358, 2017.

200. L. Hu, Y. Miao, G. Wu, M. M. Hassan, and I. Humar, "iRobot-Factory: An intelligent robot factory based on cognitive manufacturing and edge computing," *Future Generation Computer Systems,* vol. 90, pp. 569–577, 2019.

201. Y. Li, L. Shi, P. Cheng, J. Chen, and D. E. Quevedo, "Jamming attacks on remote state estimation in cyber-physical systems: A game-theoretic approach," *IEEE Transactions on Automatic Control,* vol. 60, no. 10, pp. 2831–2836, 2015.

202. G. Liang, J. Zhao, F. Luo, S. R. Weller, and Z. Y. Dong, "A review of false data injection attacks against modern power systems," *IEEE Transactions on Smart Grid,* vol. 8, no. 4, pp. 1630–1638, 2016.

203. J. Li et al., "Optimal investment of electrolyzers and seasonal storages in hydrogen supply chains incorporated with renewable electric networks," *IEEE Transactions on Sustainable Energy,* vol. 11, no. 3, pp. 1773–1784, 2019.

204. F. Li, Y. Shi, A. Shinde, J. Ye, and W. Song, "Enhanced cyber-physical security in internet of things through energy auditing," *IEEE Internet of Things Journal,* vol. 6, no. 3, pp. 5224–5231, 2019.

205. A. Ashok, M. Govindarasu, and J. Wang, "Cyber-physical attack-resilient wide-area monitoring, protection, and control for the power grid," *Proceedings of the IEEE,* vol. 105, no. 7, pp. 1389–1407, 2017.

206. X. Zhang, X. Yang, J. Lin, G. Xu, and W. Yu, "On data integrity attacks against real-time pricing in energy-based cyber-physical systems," *IEEE Transactions on Parallel and Distributed Systems,* vol. 28, no. 1, pp. 170–187, 2016.

207. Y. Zhang, Y. Xiang, and L. Wang, "Power system reliability assessment incorporating cyber attacks against wind farm energy management systems," *IEEE Transactions on Smart Grid,* vol. 8, no. 5, pp. 2343–2357, 2016.

208. F. Zhang, H. A. D. E. Kodituwakku, J. W. Hines, and J. Coble, "Multilayer data-driven cyber-attack detection system for industrial control systems based on network, system, and process data," *IEEE Transactions on Industrial Informatics,* vol. 15, no. 7, pp. 4362–4369, 2019.

209. M. Marzband, F. Azarinejadian, M. Savaghebi, E. Pouresmaeil, J. M. Guerrero, and G. Lightbody, "Smart transactive energy framework in grid-connected multiple home microgrids under independent and coalition operations," *Renewable Energy,* vol. 126, pp. 95–106, 2018.

210. N. Shaukat et al., "A survey on consumers empowerment, communication technologies, and renewable generation penetration within Smart Grid," *Renewable and Sustainable Energy Reviews,* vol. 81, pp. 1453–1475, 2018.

211. S. Zahraee, M. K. Assadi, and R. Saidur, "Application of artificial intelligence methods for hybrid energy system optimization," *Renewable and Sustainable Energy Reviews,* vol. 66, pp. 617–630, 2016.

212. A. Y. Abdelaziz and E. S. Ali, "Load frequency controller design via artificial cuckoo search algorithm," *Electric Power Components and Systems,* vol. 44, no. 1, pp. 90–98, 2016.

213. S. Mohammadi and A. Mohammadi, "Stochastic scenario-based model and investigating size of battery energy storage and thermal energy storage for micro-grid," *International Journal of Electrical Power & Energy Systems,* vol. 61, pp. 531–546, 2014.

214. A. Mellit and S. A. Kalogirou, "Artificial intelligence techniques for photovoltaic applications: A review," *Progress in Energy and Combustion Science,* vol. 34, no. 5, pp. 574–632, 2008.

215. K. G. Di Santo, S. G. Di Santo, R. M. Monaro, and M. A. Saidel, "Active demand side management for households in smart grids using optimization and artificial intelligence," *Measurement,* vol. 115, pp. 152–161, 2018.

216. V. C. Gungor et al., "Smart grid technologies: Communication technologies and standards," *IEEE Transactions on Industrial Informatics,* vol. 7, no. 4, pp. 529–539, 2011.

217. M. Q. Raza and A. Khosravi, "A review on artificial intelligence based load demand forecasting techniques for smart grid and buildings," *Renewable and Sustainable Energy Reviews,* vol. 50, pp. 1352–1372, 2015.

218. H. Hui, Y. Ding, Q. Shi, F. Li, Y. Song, and J. Yan, "5G network-based Internet of Things for demand response in smart grid: A survey on application potential," *Applied Energy,* vol. 257, p. 113972, 2020.

219. F. Al-Turjman and M. Abujubbeh, "IoT-enabled smart grid via SM: An overview," *Future Generation Computer Systems,* vol. 96, pp. 579–590, 2019.

220. Y. Zhao, T. Li, X. Zhang, and C. Zhang, "Artificial intelligence-based fault detection and diagnosis methods for building energy systems: Advantages, challenges and the future," *Renewable and Sustainable Energy Reviews,* vol. 109, pp. 85–101, 2019.

17 Application of Back Propagation Algorithm for Solar Radiation Forecasting in Photovoltaic System

Kura Ranjeeth Kumar
EPAM Systems India Private Limited

*Poonam Upadhyay, J. Anwesh Kumar, B. Nagaraja Naik,
P.R. Sai Sasidhar, and P. Sathvik Reddy*
VNR Vignana Jyothi Institute of Engineering and Technology

CONTENTS

ABSTRACT

In the present era, non-renewable energy sources are depleting at a faster rate; therefore, it's right time to switch to renewable energy sources. Moreover, the climatic changes occurring due to usage of non-renewable sources are alarming us to move toward the natural sources such as solar, wind, tidal and biogas. Among these, solar energy has become a promising alternative source of energy due to its abundance. The function of the photovoltaic system is to convert the incident solar light into electrical energy. High-fidelity forecasting is required for effective and reliable utilization of solar power. There are lot of ways to forecast, such as genetic algorithm, neuro-fuzzy, Particle Swarm Optimization method and neural network. Multilayer perception takes care of neural network, and error propagation algorithm takes care about the training part. Regression analysis has been done on various training algorithms with different sets of parameters (epochs and Max_fail); variation of regression and performance of the network is observed for these combinations to determine the best training function and the best set of parameters to get satisfactory results. This analysis would help us to predict the solar radiation in a given area.

KEYWORDS

Artificial Neural Network; Solar Radiation Forecasting; Multilayer Perceptron; Back Propagation; Regression

17.1 INTRODUCTION

The demand for the electricity is increasing day by day, and we cannot even imagine a life without electricity, which results in the increase in the demand of power generation. If we use inappropriate methods to generate power, using non-renewable energy sources, it will result in the increase of greenhouse emissions, which has already caused an irreversible damage to our environment.

There are some ways which can be used to increase energy generation without increasing the level of greenhouse gases in the atmosphere; solar energy is the most preferable. Forecasting the solar energy plays a major role, as it will help to have a better photovoltaic (PV) system control and management. It can also be used in facilities like thermal-solar power plants and in power regulation and power scheduling in power grids. This system helps the power system operators in load balancing and for the optimization of energy transfer and also allots the required amount of energy by other resources (conventional generation stations) when there is absence of adequate solar energy. It also helps us plan and maintain functions that take place in production sites, and also takes care about necessary measures from extreme events.

A vast amount of information is required for this on many factors, like long-term old energy data, real-time energy data, site irradiances data, local solar energy resource forecasting, socio-economic conditions, environmental conditions, electricity demand and trading of electric power.

17.2 PROBLEM FORMULATION

Existence of the solar forecast remains unclear. Many methods which include atmospheric physics, machine learning, solar instrumentation, forecasting theory, and remote sensing have been developed and implemented. It has been described that performance may be improvised when the information about cloud covering the island is known 15 minutes before this usual power backup is planned or disconnection for the critical area load demands. Numerical weather estimates the weather conditions by employing current weather conditions as input to the mathematical models, where the model analyses and forecasts solar irradiance for longer duration. But this mathematical model doesn't give good results in small scale and can't predict when the panel is covered with the clouds. But there are some classic approaches, such as time series-based traditional forecasting of solar energy. For long-duration forecasting weather station data and climate time series are also taken into consideration. For the small-scale forecasting, cloud cover must be taken necessary.

Models for very short-time forecasting use of total sky and satellite images are helpful because we can get the inappropriate information about atmosphere by using image processing and cloud tracking techniques. In this chapter, the digital images taken by a ground-placed sun tracking camera are used to analyze the statistics of ramping rates and duration for the cloud covering intermediately. It also tells many image processing techniques are helpful in solar forecasting, speed field calculations, cloud grouping and spatial transformation of images.

17.3   SOLAR ENERGY

There are a lot of ways in which electrical energy can be generated, as energy can neither be generated nor destroyed. For instance, we have hydroelectric power stations, wherein we use the potential energy stored by the water to run the turbine and generate electrical energy. In thermal stations, heat energy is converted into electrical energy; most of the turbines currently employed in the power generation are steam-driven. Wind stations use the kinetic energy of the air to run the turbines and give electrical output. Out of all these, PV systems are widely preferred. There are numerous advantages to support the statement (Figure 17.1).

- They are static systems, which means they have a long life.
- No fuels, liquids, or gases, hence no risk of leakage.
- No pollution; hence are often considered as eco-friendly.
- Very little maintenance and can be unmanned.
- PV cells are made of silicon, which can be found in major portion of earth's crust.
- Solar stations depend on solar energy, which is clean and cheap.

17.3.1   LIMITATIONS OF SOLAR ENERGY

There are a lot of advantages for using PV systems over other generating stations. But there is one drawback while going for solar-based energy, that is, solar cells have poor efficiency (typically 12%–18%). The major reasons for low solar cell efficiency are as follows:

- Part of the light is reflected from the surface due to the fundamental property of material.
- Light may not have sufficient energy to liberate electrons.
- Electron and hole recombination's (either direct or indirect).
- Manufacturing defects.
- Resistance offered to the current flow.
- Degradation in performance after long years in use.
- High temperature and low temperature losses.

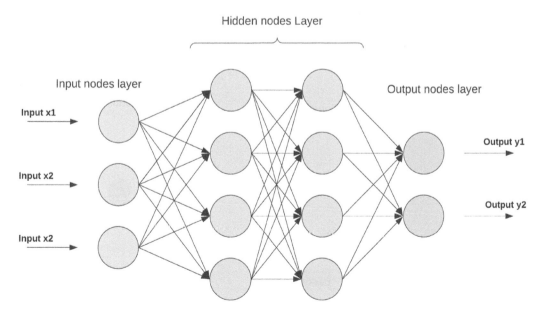

FIGURE 17.1   Block diagram of conversion of solar energy to electrical energy.

17.4 NEURAL NETWORKS

17.4.1 INTRODUCTION

Neural networks are the algorithms that are modeled loosely to mimic human brain. Humans are good with patterns; no matter how random the data, the human mind always tries to find a pattern. Computers, on the other hand, can process huge amounts of data without any problem. But the limitation of computers is its inability to find a pattern and that of humans is to process huge chunks of data. Neural network is the common ground where we can integrate neural activity and algorithms to form a network which can find a pattern by feeding the network with tons of data.

Neural networks are used to find a relation between input and output, if at all there is any, by means of correlation. NNs are used mainly for classification and clustering, for example, in supermarkets, economics, population prediction, marketing and medicine. NNs are preferred when the rules to solve the problem are not quite certain, and are often considered a black box as we cannot really explain how the problem is solved.

17.4.2 NEURAL NETWORKS ARCHITECTURE

Neural networks have three layers: input layer, hidden layer and output layer. Each layer has its own set of nodes and the data processed through these layers. The data are fed to the input layer nodes (number of nodes are equal to the input parameters and the data fed to the input layer are not altered during the learning stage) and every node in the input layer is connected to the hidden layer node (number of nodes depends on the application and requirement). The output from the hidden layer is then fed to the output layer node from which it is passed through an activation function. Connections made between the layers have weights assigned to them; these weights are altered based on the error propagated during the learning stage to achieve the task at hand (Figure 17.2).

17.4.3 BACK PROPAGATION ALGORITHM

This has two phases: one is Forward Phase and the other is Backward Phase. In the Forward Phase, weights are adjusted with respect to input values, and if we get any error, that error will be propagated backward to adjust the weights such that simulated output is almost or approximately equal to the expected output (Figure 17.3).

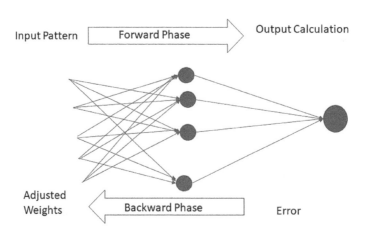

FIGURE 17.2 Architecture of ANN.

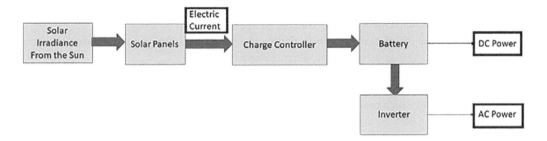

FIGURE 17.3   Back propagation algorithm.

17.4.4   APPLICATION OF NN IN SOLAR FORECASTING

To understand the application of neural networks in the forecasting, we have understood the history on how neural networks came into existence. Humans are wired to recognize a pattern in the given data no matter how random the data are, while computers are good with processing the data in large chunks. But the limitation of computer is that it cannot process creatively; in general terms, it cannot find a pattern in each data no matter how good the processor is or how good the memory is. On the other hand, humans are lousy at handling data; to be precise, we cannot find a pattern when we encounter, say, 600 datasets. Neural networks are the common platform where we can imitate a human brain (which helps us in detecting a pattern) by transferring tons of data to the network, which is an algorithm (which handles data well).

PV output depends on solar irradiance, which in turn depends on geographical location of the site, time of the year, climate and the level of irradiance. While using neural networks, these variable parameters on which the solar irradiance depends on is given as input parameters and the electrical output is taken as the predicted output. Number of nodes needed depends on the collected data; training algorithm depends on the application. The next step will be neural network training based on the available data, following which the validated data is fed and tested on the network. This makes it possible to evaluate performance of the network in all possible scenarios. If the results are satisfactory, then we feed the network with future data, at a location and time, and predict the electrical output of the system. If the output does not meet the requirement, then we can make changes to the proposed plan and run the network again till we get the predicate value.

17.5   SOLAR RADIATION FORECASTING

17.5.1   INPUT PARAMETERS

The choice can be made by observing at the past data. There are certain parameters that are constant or do not change much and have a little effect on the output; we can omit those parameters and save the network some complexity. On the other hand, we have some parameters that are imperative and can lead to false predictions if not taken into account. We have nine input parameters, and these are used to train the network. Of these parameters, daily solar radiation is considered as the most effectual one. It depends on many factors such as Air Temperature, Day Solar Radiation, Wind Speed, Atmospheric Pressure, Temperature of Earth, Relative Humidity Precipitation, Heating Degree-Days (18°C) and Cooling Degree-Days (10°C).

17.5.2   OUTPUT PARAMETER

We are training the neural network in a supervised manner, for which we need to give the predicted output, more precisely a target, while training the network. We take the electrical output of the PV system as the target for the training procedure, and its unit is KWh. RET screen software gives the PV output for a given capacity when fed with the annual data.

We have collected 268 datasets; in every dataset, we are taking annual data to train neural network. Whole information is classified into three variables.

   i. Input (9×248, which means 9 input variables and 248 datasets)
  ii. Target (1×248, which means 1 output variable and 248 datasets proportional to input)
 iii. Sample (9×20, which means 9 input variables and 20 datasets).

As we have three sets of data, we are going to use the input and target datasets to train the network, and then the network is used to simulate the sample datasets to validate our results.

17.5.3 MATLAB® TRAINING

There are seven basic steps for designing a neural network:

1. Data Gathering
2. Network Creation
3. Network Configuration
4. Initialization of Biases and Branch Weights
5. Network Training
6. Validation of Network
7. Network Usage

17.5.3.1 Training Functions

Feed-Forward Back propagation has 13 different types of training functions; they are TRAINBFG, TRAINBR, TRAINCGF, TRAINCGP, TRAINGD, TRAINGDM, TRAINGDA, TRAINGDX, TRAINLM, TRAINOSS, TRAINR, TRAINRP and TRAINSCG.

Each training function has unique functionality and algorithm. To train neural network we used 8 out of 13 training functions as they are preferred for pattern recognition and trained and validated the network to get accurate output values. Some training functions are good with handling weight updates, while the others have good accuracy. Here, we used TRAINRP function.

TRAINRP:
It is Resilient Propagation. It is a function used for training network which updates the values of weight and bias as per Resilient Back Propagation Algorithm (R-prop) (Figure 17.4).

17.5.4 ADAPTATION LEARNING FUNCTIONS

- LEARNGD function is used for learning gradient descent weight and bias.
- LEARNGDM function is used for learning gradient descent along with momentum weight and bias.

Performance Functions:

- MSE: Mean squared normalized error
- SSE: Sum squared error performance function

Transfer Function:

- TANSIG: Hyperbolic tangent sigmoid

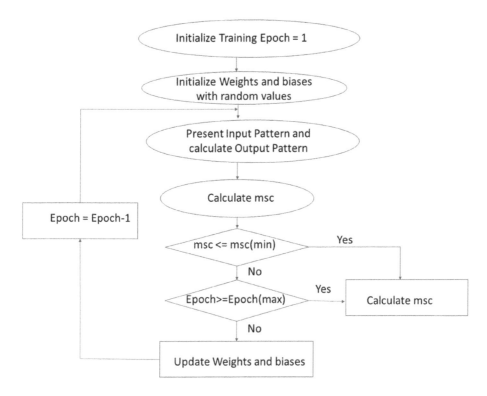

FIGURE 17.4 Flow chart of training process.

- LOGSIG: Log-sigmoid transfer function
- PURELIN: Linear transfer function

17.5.5 Steps to Be Followed to Simulate and Train the Neural Network

1. Collect the datasets (solar data) from any website or software.
2. Collected data should be in Excel format.
3. Now open the MATLAB tool (Artificial Neural Network (ANN) Tool).
4. Import the data from excel sheet to MATLAB Tool in terms of Input, Sample and Target datasets.
5. Select the New option to create the neural network.
6. Assign the specific functions like Training Functions, Adaptation Learning Functions and Performance Functions for the different types of neural networks.
7. Create and open the network.
8. Select Train option and assign the values for training info and training parameters like epochs and Max_fail (different combinations).
9. After assigning the training parameters train the network, a window will open which displays the Performance, Transition state and Regression plots along with gradient values.
10. Number of iterations depends on the epochs value that we give I intent-based networking (IBN) training parameters. More epochs values and more Max_fail values lead to effective training.
11. Now, open the regression plot, in that we have four plots of Training, Validation, Testing and All. The entire dataset that we are giving as input is divided into training, validation and test sets, and output for these values is obtained.
12. Regression value should be 0.75–1.0. Train the network until we get good regression values.

13. As soon as we get good regression values, simulate the network with sample as input.
14. We get output values which are approximately equal to desired output. Train and simulate the network until we get minimum error.
15. Export the network for users after completion of training and simulation.

17.6 RESULTS

The analysis is done with different training functions. We chose epochs and Max_fail as our performance parameters, as they have a powerful effect on the output. Epochs influence the training time, while Max_fail will affect the accuracy of the network. Epochs and Max_fail are varied in sets considering all the possibilities. For the training function regression of the training, test and validation set is tabulated, and the regression plot of the best combination is provided. The network with the combination of best pair is used to simulate a sample test data, and the simulated values are then compared with the expected values and the error is calculated to demonstrate the use of network in real time (Table 17.1 and Figure 17.5).

TABLE 17.1
TRAINRP Regression Table

Epochs	Max_fail	Training Regression	Validation Regression	Test Regression
200	1,000	0.7640	0.8465	0.7691
1,000	200	0.7526	0.8768	0.780
1,000	1,000	0.75525	0.78881	0.8449
200	200	0.7328	0.8087	0.93212
500	500	0.79182	0.85339	0.92073

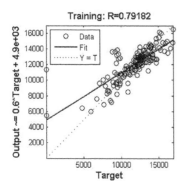

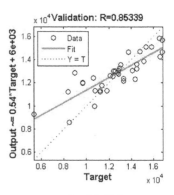

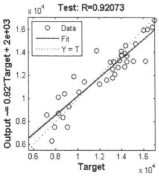

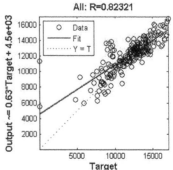

FIGURE 17.5   Regression plot.

TABLE 17.2
Simulation Results

Expected (KWh)	Simulated (KWh)	Error (%)
13,359	13,525.8203	−1.248
14,271	13,414.568	6
12,985	13,199.6415	−1.6
16,011	14,752.128	7.8
13,352	12,655.2754	5.2
12,962	14,014.7353	−8.1
13,075	14,219.3462	−8.7
13,628	14,903.9947	−9.36
13,257	14,400.0719	12.4
12,866	12,865.2643	0.0057

We initialize epochs and Max_fail as 200 each and train the network. We continue training the network until we get a good regression, by checking the regression plot time to time. If we get a good regression, we use the network and simulate the sample data and compare it with expected output. It is seen that the maximum error is 12.4% and the minimum is 0.0057% respectively (Table 17.2).

17.7 CONCLUSION

Fluctuating nature of solar energy is what makes it unreliable and poses a major problem in the grid integration. ANN has proved to be the best solution for the issue. Artificial Neural Network-based solar radiation forecasting is done, so as to evaluate the performance of a PV system before installing the solar plant. The correlation between the nine input parameters and the output is found to be accurate in most of the cases. Comparison with the actual results shows that the predicted results are very close, up to an accuracy of 85%. This method is employed to predict solar radiation for any location in the world.

BIBLIOGRAPHY

1. E.H. Camm and S.E. Williams, "Solar power plant design and interconnection," *IEEE Power and Energy Society General Meeting*, Detroit, MI, 2011.
2. A. Haque and Zaheeruddin, "Research on solar photovoltaic (PV) energy conversion system: An overview", *Third International Conference on Computational Intelligence and Information Technology (CIIT 2013)*, IET, Mumbai, 2013.
3. M. Mishra and M. Srivastava, "A view of artificial neural network", *IEEE International Conference on Advances in Engineering & Technology Research (ICAETR-2014)*, Unnao, India, 2014.
4. P. Kumar and P. Sharma, "Artificial neural network: A study", *International Journal of Emerging Engineering Research and Technology*, vol. 2, issue 2, 2014, pp. 143–148.
5. X. Yan, D. Abbes, and B. Francois, "Solar radiation forecasting using artificial neural network for local power reserve", *International Conference on Electrical Sciences and Technologies in Maghreb (CISTEM)*, Tunis, Tunisia, 2014.
6. A. Ghanbarzadeha, A.R. Noghrehabadia, E. Assareh, and M.A. Behrang, "Solar radiation forecasting based on meteorological data using artificial neural networks", *7th IEEE International Conference on Industrial Informatics*, 2009.
7. J. Ida Jensona and E. Praynlin, "Solar radiation forecasting using artificial neural network", *International Conference on Innovations in Power and Advanced Computing Technologies i-PACT2017*, Vellore, India, 2017.

8. IEEE Technology Navigator for Solar Power, http://technav.ieee.org/tag/1582/solar-power.
9. A Report Submitted to IIT Kanpur by Solar Energy Research Enclave, http://www.iitk.ac.in/olddord/solar%20energy%20proposal%20from%20IIT%20Kanpur.pdf
10. IEEE Technology Navigator for Artificial Neural Networks, http://technav.ieee.org/tag/4791/artificial-neural-networks.

18 Technical and Feasibility Analysis of Interconnected Renewable Energy Sources in Three Separate Regions
A Comparative Study

Rupan Das
Tripura Sundari H.S. School

Somudeep Bhattacharjee and Uttara Das
Tripura University

CONTENTS

DOI: 10.1201/9781003331117-18

ABSTRACT

With the increase in utilization of electricity in household, industrial and commercial sectors, the rate of pollution is also increasing rapidly. The electricity sector is one of the major sources of greenhouse gases as they use non-renewable energies. These greenhouse gases have a direct impact on human health and environment. Therefore, utilizing of clean energy must be maximized to reduce the unnecessary change in climatic conditions. In this study, a hybrid system (HBS) connected to grid is proposed to utilize fresh energies like solar, wind, and biogas for generating electricity. On the basis of actual-time information of climatic components, this HBS has been skillfully examined in three separate sites in Southern Asia. This case study is performed with the aim to minimize carbon emissions in terms of the cheapest electricity production. At last, this study concludes that the proposed total hybrid framework is pollution-free and lucrative as it is using renewable resources.

KEYWORDS

Hybrid Energy System; Sustainable Energy; Economic Analysis; Solar and Wind Energy; Grid

18.1 INTRODUCTION

Pollution is the word for a famously destructive concept in our modern world. The main reason for pollution is the unruly activity of humans. Almost 8 billion metric tons of carbon is emitted every year in the atmosphere due to human activities. Undesirable changes in environmental constraints like air, water, soil stand for environmental pollution. The most important reason for environmental pollution is the use of fossil fuel and deforestation. Fossil fuel combustion is the cause of the inanition of non-conventional energy in near future. It emits 6.5 billion tons of carbon per year. In the world scenario, deforestation is the cause of 1.5 billion tons of carbon discharge every year [1,2].

Now, the focusable factor behind the term climate change is environmental pollution. Discontinuity in rainfall, improper wind speed, unpredicted storm drought, etc. occur due to climate change. Moreover, this climate change has the ability to spread various diseases in the agricultural field as well. Climate change brings monsoon unpredictably in the southern part of Asia [3]. As per Intergovernmental Panel on Climate Change, the most affected region in the world is Asia due to climate change. If we focused on Asia only, Bangladesh is more affected than other countries in Southern Asia. Due to rapid urbanization, most of the countries uses petroleum product which is the cause of diminished the underground layers [4]. Previous research reported that most of the Asian countries are the front-line countries in the area of greenhouse gas production.

India, Bangladesh, and Bhutan fall under the southern part of the Asian continent. Most of the South Asian countries are developing nations. In developing nations like India, the installed electricity generation capacity increased from 1362 MW (1947) to 271 GW (2015) [1]. In Bangladesh, the installed capacity of electricity generation was 667 MW in 1974; now it reached 12,780 MW (2016). Moreover, Bangladesh imported 600 MW of power from India to fulfil its demand [5]. On the other side, 1.6 GW is the total installed electricity generation capacity of Bhutan [6]. Thus, an increase in electricity demand means an increase in using fossil fuels. In most cases, people of developing countries use electricity produced by the burning of fossil fuels. These fossil fuels are collected from mining, which is directly related to deforestation. Further, this deforestation has a relation with land erosion [7]. Overall, the whole process of electricity generation by using fossil fuels is directly related to global warming. Fossil fuels like oil, natural gases, coal, etc. are the main sources of greenhouse gases. So, it can be said that electricity production by burning fossil fuels is a hazardous in the decades to follow [8–13].

Therefore, production of electricity using renewable energy sources can help reduce the risks of global warming and climate change. Electricity production systems using renewable resources can be preferred over other types of electricity generation systems due to their abundance and environment-friendly characteristics [14–16]. Renewable energy can help a nation for its sustainable development. Renewable energy is used as electricity, heat, vehicles fuel, etc. During electricity generation, renewable energy-based system helps to decrease greenhouse gases emission. Thus, renewable energy-based electricity generation is a great option to protect our diversity [17]. If we focus on large-scale load, hybridization of renewable resources is the only solution to curb global warming.

Renewable energies are mostly dependent on climatic conditions. Hybridization of renewable energies is required due to the non-linear behaviour of renewable components. Photovoltaic (PV) and wind energy are the most remarkable alternative sources of energy, among the renewable sources of energy [18]. A study on renewable energy technologies reports that basically PV and wind-based technologies are mostly used in rural areas [19]. Wind speed becomes comparatively low when solar works it best. Renewable energy-based hybrid systems (HBS) give a reliable, cheaper, and environmentally friendly output. Electricity generation by renewable resources is best suited for all local loads. Depending on the location, HBS using renewable energy can operate as grid off or connected to grid systems [20].

South Asian countries like India, Bangladesh, and Bhutan use rice or wheat-based diets to feed their nation. This rice and wheat both leave too much waste products during processing from field to dining table. Thus, managing waste materials is becoming a big headache to every authority throughout these nations. Nowadays, people of different cities use biodegradable waste to generate electricity. A study reported that 550 kWh of electricity can be generated from one tonne of solid waste [21]. Except for electricity, biomass technology is used in the application of heat, chemicals, transport fuels, etc. So, a solar-wind-biomass-based hybrid renewable energy system (HRES) can work as a huge source of energy for rural electrification. In addition, this system is a great option for the location where a solar resource may be moderated and biodegradable waste is available and vice-versa [22].

Various works have been done by authors based on optimization problems. In Ref. [23], a solar-wind-biomass-based HRES was proposed. To minimize the Net Present Cost (NPC) of Total System, unmet load, carbon emission, and cost of energy (COE) are the main focus of the study. To study the change in annual fuel price and wind speed, a sensitivity analysis was performed. Thus, a comparative study was performed to get the proper optimal solution. The authors of Ref. [24] designed an electric vehicle charging station based on biomass resources. To find out the economic parameters, the authors proposed a HOMER-based model. Therefore, this system can reduce carbon emissions more than a grid-operated charging station. The authors Ref. [25] presented a grid-connected HBS to minimize the impact of climate change. Analysis of this HBS was conducted in four separate regions. They also explained the role of renewable energies in combating against climate change. To investigate the application of PV, biomass, and diesel in the field of electricity generation, the authors of Ref. [26] proposed a HBS. In the study, a comparison between solar installation techniques was done to select the proper location. Thus, the result showed that the COE of this optimal system falls under the acceptable range. The authors of Ref. [27] developed a HBS using renewable energy to reduce the contribution of thermal power plants. An algorithm based on energy management is proposed here for storing more energy to charge vehicles. Therefore, four different technologies based on renewable energies were explained (Table 18.1).

In this study, a HRES connected to grid is developed to utilize solar-wind-biomass power. By thinking about the waste materials spread over society, biomass energy has been extensively used here. This HBS is structured in a manner that it will emit less amount of carbon and can produce

TABLE 18.1

Some Optimal Studies Based on a Hybrid System Reported in the Literature

Reference	Year	Description	Result
[23]	2020	A HRES utilizing solar-wind-biomass was proposed to reduce the overall price of the system	The system can meet the peak demand. Carbon emission is also less
[24]	2020	A biogas/biomass based off-grid vehicle charging station was developed to reduce carbon emission	65.61% carbon emission was reduced due to the use of renewable resources
[25]	2020	A solar, wind, and gas generator-based hybrid system was presented to cut down the impact of climate change on the environment	This proposed system is more profitable, and 0.688 $/kWh is their levelized cost of energy
[26]	2019	A grid-connected solar, wind, and diesel-based hybrid system was designed to investigate technical, economical, and environmental features	In case of high diesel price, a biogas generator was used as a backup system. Thus, the energy prices will range from 0.085 to 0.238 $/kWh
[27]	2020	Solar, hydro, wind, and thermal-based HBS was proposed to manage the energy for vehicle charging	This system is profitable and environment friendly. It can fulfil all the demands by storing the required amount of energy

low-cost electricity and thus to minimize the energy bill. This HBS is skilfully analysed for the case study in three different locations of India, Bangladesh, and Bhutan. All these locations receive a sufficient amount of solar irradiance, wind speed, and biodegradable waste; hence, generating electricity from these sources will not be any difficult. Here, a fixed amount of load is presumed for a particular location, so that excess energy is sold to the grid. This case study is completed utilizing actual-time information for a particular area's solar irradiance and wind speed. In the result of this case study, we found that the whole HBS is lucrative and pollution-free, as it is using renewable resources. Therefore, the main aim of this case study is to minimize carbon emissions resulting from electricity production and sell the surplus energy to the grid by satisfying all the load demand of the selected locations.

18.2   PROFILE OF RENEWABLE ENERGY RESOURCES

On the basis of renewable resource availability and potential, the hybrid solar/wind/bio-generator system is to be characterized for installing in most suitable sites. In this work, the monthly standard variation of solar irradiance and wind speed is estimated with the help of the NASA Power Website Database [28], based on satellite observation. The annually mean value of biomass data is assumed to be constant for all the three areas. The geographical location of the three study regions is shown in Figure 18.1. The latitude of three separate investigation areas are 91°50.0′E, 89°38.3′E, and 90°22.0′E, their longitude are 25°34.8′N, 27°28.9′N, 23°42.2′N, and elevation of those regions are 200, 1,000, and 200 m. The NASA GEOS-5 model elevation is utilized for acquiring these investigation regions.

18.2.1   Solar Irradiation and Temperature Parameter

The solar radiation, clearness index, and temperature variation on monthly basis for three different locations (Shillong, India/Desi Lam, Bhutan/Keraniganj, Bangladesh) are shown in Table 18.2. The monthly averaged data have been taken from NASA Power Website Database for three different selected areas [28]. For Shillong, the annual average solar radiation is 4.41 kWh/m²/day with its

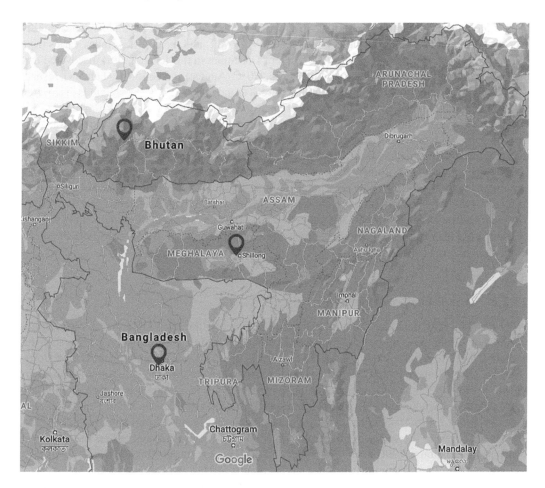

FIGURE 18.1 Location of three investigation regions.

TABLE 18.2
Solar Radiation (kWh/m²), Clearness Index, and Temperature Values (Degrees Celsius) of the Investigation Locations

Month	Shillong, India			Desi Lam, Bhutan			Keraniganj, Bangladesh		
	R	I	T	R	I	T	R	I	T
January	4.290	0.647	16.150	4.030	0.637	0.980	4.360	0.630	19.740
February	4.970	0.640	19.000	4.540	0.605	2.630	4.920	0.614	23.000
March	5.610	0.611	22.730	5.060	0.562	6.440	5.590	0.598	26.450
April	5.680	0.547	24.020	5.620	0.544	9.240	5.760	0.551	27.150
May	4.930	0.446	25.090	5.730	0.517	11.320	5.300	0.481	27.650
June	3.850	0.342	25.940	5.420	0.478	13.760	4.530	0.405	27.950
July	3.580	0.322	25.910	5.100	0.456	14.550	4.230	0.382	27.660
August	3.680	0.347	25.970	4.830	0.457	14.420	4.290	0.404	27.610
September	3.560	0.372	25.050	4.510	0.478	12.910	4.020	0.415	27.010
October	4.270	0.523	23.100	4.710	0.594	9.320	4.320	0.516	25.480
November	4.380	0.638	19.860	4.290	0.653	5.580	4.280	0.599	22.500
December	4.090	0.653	16.990	3.910	0.658	2.850	4.210	0.641	20.200

highest monthly radiation in the month of April and lowest in the month of September. The monthly average solar radiation ranges from 3.560 to 5.680 kWh/m²/day and the clearness index varies from 0.322 in the month of July to 0.653 for the month of December in this location. It is discovered that the highest and lowest values of temperature are 25.970°C (August) and 16.150°C (January), respectively, while the average annual value of temperature is found to be 22.48°C. Desi Lam is discovered to have the maximum and minimum values of monthly solar radiation of 5.730 kWh/m²/day in May and 3.910 kWh/m²/day in December, with an annual mean value of 4.81 kWh/m²/day. The monthly average value of clearness index is observed to be between 0.658 (December) and 0.456 (July), and the annual standard temperature is 8.67°C, where the highest value is recorded in the month of July (14.550°C) and the lowest in the month of January (0.980°C). In the case of Keraniganj, a similar variation is recorded in April and September for highest and lowest values of solar radiation being 5.760 and 4.020 kWh/m²/day, respectively, with an annual standard value equal to 4.65 kWh/m²/day. The clearness index ranges from 0.382 (July) to 0.641 (December); 27.950°C in June and 19.740°C in January are the recorded highest and lowest values of temperature, respectively, where the yearly mean temperature is discovered to be 25.20°C.

18.2.2 SPECIFICATION OF WIND SPEED

Wind resource variations are essential information to depict the wind state that a wind turbine would face during the whole year. Wind speed is the most significant wind source parameter that is recorded on month-to-month average basis, details of which are obtained from NASA Power Database [28]. All the monthly averaged data regarding wind speed of three different locations are shown in Table 18.3. It is noticed that Shillong experiences the maximum wind speed of 3.090 m/s in the month of March and the minimum wind speed of 2.250 m/s in the month of September, with a mean value of 2.64 m/s annually. Desi Lam has the best wind speed data as compared to two other regions, discovering the yearly average value of wind speed being 5.73 m/s, where the maximum wind speed variation is 7.390 m/s in November and the lowest wind speed is 4 m/s in September. In Keraniganj, the same phenomenon is noted in the months of May and October for the highest wind speed data to be 3.2 m/s and lowest wind speed to be 1.850 m/s, respectively, with 2.56 m/s being the annual standard value.

TABLE 18.3
Monthly Average Wind Speed Data of the Investigation Locations

Month	Wind Speed (m/s)		
	Shillong, India	Desi Lam, Bhutan	Keraniganj, Bangladesh
January	2.810	6.520	2.360
February	2.900	6.880	2.660
March	3.090	7.120	2.830
April	2.860	6.350	3.190
May	2.650	5.190	3.200
June	2.540	4.570	3.070
July	2.440	4.260	2.740
August	2.340	4.100	2.410
September	2.250	4.000	2.170
October	2.410	5.460	1.850
November	2.640	7.390	2.080
December	2.700	6.900	2.200

18.2.3 Details of Biomass Resource

The annual average amount of biomass for all the three locations is considered to be 45.75 tonnes/-day, and the monthly mean merit of biomass ranges from 35 tonnes/day (December) to 60 tonnes/day (May). Classically, biogas has a low heating variation as compared to fossil fuels, and 5.50 MJ/kg is seen to be the Lower Heating Value (LHV) (or also called as Net Calorific Value) of the biogas. Here, the percentage (%) of carbon content (C) in biomass is noted to be 5% and the ratio of biomass gasification (kg/kg) is defined as the proportion of the amount of biogas induced to the feedstock of biomass utilized in the gasifier is 0.70. The data of biomass resources are considered using Refs. [29–31].

18.3 EXPLANATION OF HRES

The HBS is made up of sets of solar PV panels, wind energy system, bio-generator, converter (DC to AC), electrical load, and grid connection as shown in Figure 18.2. Table 18.4 shows the component's rating that is utilized in the integrated energy system. In a solar PV system, the generated DC power experiences a Maximum Power Point Tracking (MPPT) so that the highest power can be measured, and MPPT is also utilized in the PV system to handle the controlling of a DC-DC boost converter. Boost converter raises this DC power, and then the inverter converts this DC power into AC and sends to the grid. The electrical load is associated with a converter along the AC bus. The wind energy system and bio-generator are also connected to the AC bus. At a specific state, while wind or solar system isn't able to generate an adequate amount of electricity, the generator operates as a backup power component. The load data are considered to be similar in the case of all the three separate regions due to the simplicity of analysis. 130 kWh/day is considered to be scaled annual standard electric load having the yearly peak load of 14.24 kW. The load factor is seen to be 0.38 kW with the average load being 5.42 kW.

18.3.1 Mathematical Modelling

The proposed HRES system component's mathematical modelling for the optimization of sizing is an important step in ensuring its performance in various conditions. A mathematical model of the individual system component of the HBS is proposed here.

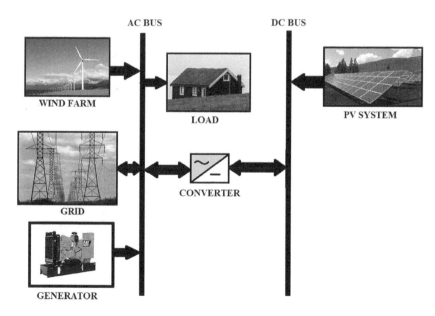

FIGURE 18.2 PV-wind-generator hybrid system configuration.

TABLE 18.4
Rating of System Components

Component	Name and Size	
Generator	Generic 100 kW genset	100 kW
PV panel	Generic flat-plate PV	5,000 kW
Wind turbine	Generic 10 kW	1,000 kW
Electric load	Electric load	130 kW
System converter	System converter	5,000 kW

18.3.1.1 Solar/PV System

In this study, a single-diode mathematical model of solar PV module is used for examination. Equation (18.1) [32,33] is utilized to obtain the voltage of the solar PV module ($V_{\text{solar PV}}$):

$$V_{\text{solar PV}}(t) = V_{\text{mt}}\left[1 + 0.0539\log\left\{G_{\text{mt}}(t)/G_{\text{st}}(t)\right\}\right] + \beta\left\{T_{\text{at}}(t)\right\} + 0.02G_{\text{mt}}(t). \tag{18.1}$$

Here, $V_{\text{mt.}}$ = Module voltage in maximum power (Volts)
  G_{mt} = Measured solar radiation value (kW/m^2)
  G_{st} = Standard solar radiation value (1 kW/m^2)
  β = Temperature Coefficient
  T_{at} = Changeable temperature (K).

Equation (18.2) is used to calculate the solar PV module current output ($I_{\text{solar PV}}$):

$$I_{\text{solar PV}}(t) = I_{\text{phc}}(t) - I_{\text{sc}}(t)\left[e\left\{qV_{\text{solar PV}}/N_s KT_{\text{at}}(t)\right\}A_i\right]. \tag{18.2}$$

Here, I_{phc} = Photo current of solar PV module (A)
  I_{sc} = Saturation current of solar PV module (A)
  q = Charge of electron (1.602×10^{-19}C)
  N_s = Cell's number in the solar PV module
  K = Boltzmann constant (1.381×10^{-23} J/K)
  A_i = Solar PV module ideality factor.

Equation (18.3) is the expression of production of total energy from solar PV module:

$$E_{\text{solar PV}}(t) = \left[N_{\text{solar PV}} \times V_{\text{solar PV}}(t) \times I_{\text{solar PV}}(t) \times \Delta t\right]/1,000, \tag{18.3}$$

Here, $N_{\text{solar PV}}$ = Solar PV module's number
  Δt = Step time.

18.3.1.2 Wind Farm

On the basis of the real power curve provided by the manufacturer, wind turbine modelling is suggested. Wind turbine's characteristic equations were derived by fitting its real power curve utilizing the least-squares method [33,34]. Wind turbine's output power (P_{WT}) in kW is obtained by utilizing the following equations:

$$P_{\text{WT}}(t) = 0, \qquad \text{for } V < V_{\text{sci}} \text{ and } V > V_{\text{sco}},$$

$$P_{\text{WT}}(t) = x_1 V^2 + y_1 V + z_1, \quad \text{for } V_{\text{sci}} \leq V < V_1, \tag{18.4}$$

$$P_{\text{WT}}(t) = x_2 V^2 + y_2 V + z_2, \qquad \text{for } V_1 \leq V < V_2,$$

$$P_{WT}(t) = x_3 V^2 + y_3 V + z_3, \quad \text{for} \quad V_2 \leq V < V_{sco}.$$

Here, V=Real speed of wind turbine
$\quad V_{sci}$=Cut-in wind speed
$\quad V_{sco}$=Cut-out wind speed
$\quad x, y, z$=Quadratic equation coefficients.

Equation (18.5) is used to calculate the wind turbine rated power output (P_{rp}):

$$P_{rwt} = 1/2 C_{pwt} \rho_a \eta_g A_{rwt} v_r^3. \tag{18.5}$$

Here, C_{pwt}=Wind turbine power coefficient
$\quad \rho_a$=Air density
$\quad \eta_g$=Efficiency of the generator
$\quad A_{rwt}$=Wind turbine rotor swept area
$\quad v_r$=Wind turbine rated speed.

The total output energy generation from wind turbine system (E_{WTS}) is obtained by equation (18.6):

$$E_{WTS} = \left[N_{WTS} \times P_{WT}(t) \times \Delta t \right] / 1,000. \tag{18.6}$$

Here, N_{WTS}=Wind turbines number.

18.3.1.3   Generator

Cofired generators use a combination of fossil fuels and biogas to generate electricity. HOMER computes the generator's needed output as well as the appropriate fossil fuel and biogas mass flow percentages for each time step [35]. Several fundamental assumptions are utilized in this calculation:

  i. Regardless of the independent power output of the engine or fuel combination, the biogas substitution ratio (Z_g) remains constant.
 ii. The system strives to use biogas as much as possible while reducing the consumption of fossil fuel.
iii. The fossil fraction must not go down a specific level.
 iv. Even though the de-rating factor for dual-fuel operation is <100%, the generator can generate rated power up to 100% in case of maximum enough fossil fraction.

A cofired generator's fuel curve shows how much fuel the generator uses in pure fossil mode. Therefore, the total amount of fossil fuel consumed in pure fossil mode can be obtained from the following equation [35]:

$$\dot{M}_0 = \rho_{fo} \left(F_{ic} Y_g + F_s P_g \right). \tag{18.7}$$

Here, $\dot{M}_0$=Flow percentage of fossil fuel in pure fossil mode (in kg/hour)
$\quad \rho_{fo}$=Fossil fuel density (in kg/L)
$\quad F_{ic}$=Intercept coefficient of fuel curve of a generator (in L/hour/kW)
$\quad Y_g$=Generator rated capacity (in kW)
$\quad F_s$=Slope of fuel curve of generator (in L/hour/kW)
$\quad P_g$=Generator power output (in kW).

From equation (18.7):

$$\dot{M}_0 = \dot{M}_{fo} + \left(\dot{M}_g / Z_g \right),$$

$$\therefore \dot{M}_g = Z_g \left(\dot{M}_0 - \dot{M}_{fo} \right). \tag{18.8}$$

Here, $\dot{M}_g$ =Flow percentage of biogas in dual-fuel mode (in kg/hour)
$\dot{M}_{fo}$ =Flow percentage of fossil fuel in dual-fuel mode (in kg/hour)
Zg=Biogas substitution ratio.

Now fossil fraction's definition gives:

$$X_{fo} = \left(\dot{M}_{fo} / \dot{M}_0 \right).$$ (18.9)

Here, X_{fo} =Fossil fraction (in %).
From equations (18.8) and (18.9),

$$\dot{M}_g = Z_g \left(\dot{M}_0 - X_{fo} \dot{M}_0 \right),$$

$$\dot{M}_g = Z_g \dot{M}_0 \left(1 - X_{fo} \right).$$ (18.10)

However, because the value of X_{fo} for a specified value of P_{bio} is uncertain, the preceding equation is insufficient to solve the flow rate of the biogas. From equation (18.8), we aim to increase $\dot{M}_g$, which means we aim to reduce X_{fo}. However, based on equation (18.9):

$$X_{fo}^* \leq X_{fo} \leq 1.$$

Here, X_{fo}^* =Minimum fossil fraction (in %) needed for ignition. Therefore, the desired value for $\dot{M}_g$ proportional to $\dot{M}_g^t$. Utilizing equation (18.10):

$$\dot{M}_g^t = Z_g \dot{M}_0 \left(1 - X_{fo}^* \right)$$ (18.11)

Here, $\dot{M}_g^t$ =Target value of flow percentage of biogas (in kg/hour).
However, the real value of $\dot{M}_g$ has two independent higher limits. The generator's output is limited to Y_g^* at the lowest fossil fraction which is expressed as follows:

$$Y_g^* = \tau Y_g.$$

Here, Y_g^* =Generator highest output at lowest fossil fraction (in kW).
τ=De-rating factor, $\tau \leq 1$.
The higher limit on $\dot{M}_g$ proportional to $\dot{M}_g^*$ can be used to accomplish this limitation. The highest value can be calculated by utilizing equations (18.7) and (18.10).

$$\dot{M}_g^* = Z_g \rho_{fo} \left(F_{ic} Y_g + F_s Y_g^* \right) \cdot \left(1 - X_{fo}^* \right)$$ (18.12)

Here, $\dot{M}_g^*$ =Highest value of flow percentage of biogas (in kg/hour).
This higher limit can be viewed as a physical constraint – the highest percentage at which biogas may be consumed in the engine. The remaining higher limit on $\dot{M}_g$ is A_g, the obtainable biomass resource. Therefore, the lowest of $\dot{M}_g^t$, $\dot{M}_g^*$, A_g is the real value of $\dot{M}_g$:

$$\dot{M}_g = \text{Min} \left(\dot{M}_g^t, \dot{M}_g^*, A_g \right).$$ (18.13)

Here, A_g =Obtainable flow percentage of biogas (in kg/hour).
As we know $\dot{M}_g$ value, X_{fo} can be obtained by solving equation (18.10):

$$X_{fo} = 1 - \dot{M}_g / Z_g \dot{M}_0.$$ (18.14)

From equation (18.9),

$$\dot{M}_{\mathrm{fo}} = X_{\mathrm{fo}}\dot{M}_0. \tag{18.15}$$

Therefore, given specific values of P_{bio} and A_g, equations (18.13) and (18.15) can be utilized to compute the flow percentage of the biogas and the fossil fuel, respectively.

18.3.2 Component Parameter Utilized for Simulation

18.3.2.1 Solar/PV System

The PV array is the linked collection of PV modules that produces electricity in DC form, when solar insolation falls on it. Here, a generic flat-plate PV system is used, whose manufacturer is Generic. The maximum rated capacity of this flat-plate-type panel is 5,000 kW. Thirteen percent is the efficiency of the panel, while it operates at 47°C temperature. The Operation & Maintenance (O&M) cost of each panel is 50 $/year with a capital cost of 500 $/kW. Twenty-five years are taken as a lifetime of solar panels. The de-rating factor of the PV panel (scaling factor, i.e. power output of PV array for decreased output in actual-field working states compared to working states at which the panel is rated) is supposed to be 80%. The ground reflectance (albedo that is solar insolation's fragment experiences on the ground which is reflected) is supposed as 20%. The parameters of the solar/PV system are considered using Refs. [36–38].

18.3.2.2 Wind Farm

A total of 100 wind turbines are used in this wind farm, and each wind turbine has a rated capacity of 10 kW, and Generic is the manufacturer. Cost of O&M of each wind turbine is 400$/year with a capital cost of 24,000$. Twenty years is the lifetime of a wind turbine, and 24 m is the hub height. The parameters of a wind farms are considered using Refs. [32,33,37].

18.3.2.3 System Converter

A converter is needed in the case of a system which is contained both DC ad AC elements. The converter is utilized in a different system such as a rectifier that changes AC into DC, an inverter which changes DC into AC or both the system. In this proposed hybrid model, a converter is used as an inverter and the manufacturer is Generic. The system converter has a size of 5,000 kW and the O&M price of this converter is 10.0 $/year with a capital cost of 300 $/kW. Fifteen years is the lifetime of this converter and 95% is the efficiency. The parameters of the converter are considered using Refs. [37–39].

18.3.2.4 Generator

The size of this generator is 100 kW, which is manufactured by Generic. 2$/op is the O&M price of the generator with a capital cost of 40,000 $/kW. This generator has a minimum load ratio of 25% with a LHV of 43.2 MJ/kg. This generator has a lifetime of 15,000 hours, and both biogas and diesel are used as fuel in this generator with a 20% minimum fossil fraction. The de-rating factor is 70%, and 8.5% is the biogas cofired substitution ratio (the ratio of biogas/fossil). The parameters of a generator are considered using Refs. [29,32,40].

18.3.2.5 Grid

Basically, grid connection is needed in case of giving extra electricity to the grid or consuming electricity from the grid. Extra electricity can sell to the grid if there is an additional power supply after gratifying the load demand of the region and electricity can consume from the grid during the lack of power supply from renewable sources. The purchase COE from the grid is 0.1500 $ per kWh with a sellback cost of 0.1200 $ per kWh to the grid. The parameters of the grid are considered using Refs. [13,15].

18.3.3 Problem Formulation

For an economic analysis of any hybrid power station, till now HOMER is the best platform. It shows the annual interest rate, initial capital cost, COE, NPC, replacement or maintenance of each component used in the HBS.

NPC or life cycle cost (LCC) includes all project costs and revenues, which include original system component capital costs, cost of replacement parts in the life of a project, maintenance costs, fuel costs, and grid operational expenses. Economic analysis is carried out in order to allow a sensible comparison of structures according to their NPC. HOMER computes NPC for every component used in the HBS and for the whole system. The following equation is utilized to obtain the NPC [29]:

$$NPC = C_{tac}/CRF(i_r, R_p).$$

Here, C_{tac}=Total annualized cost in \$/year
i_r=Annual real interest rate (discount rate) in %
R_p=Lifetime of project in a year
CRF=Capital recovery factor.

COE or levelized COE is another essential parameter in economic analysis. The mean cost per kWh of a system generating useful electrical energy is defined as the levelized COE. For calculating COE, HOMER divide the yearly expenditure of generating electricity by overall generated useful electrical energy. It is computed as follows [41]:

$$COE = C_{tac}/(M_{PAC} + M_{PDC} + M_{gs}).$$

Here, C_{tac}=Total annualized cost (\$/year)
M_{PAC}=Primary AC load served (kWh/year)
M_{PDC}=Primary DC load served (kWh/year)
M_{gs}=Total grid sale.

The factor by which the recent variation of series of equivalent cash flow is calculated yearly basis in a ratio is defined as Capital Recovery Factor (CRF). CRF formula is as follows [41]:

$$CRF = \left\{i_r^*(1+i_r)^n\right\}/\left\{(1+i_r)^{n-1}\right\}.$$

Here, n=No. of years
i_r=Real interest rate, which is calculated on a yearly basis.

Now, the nominal interest rate (function) is the annual real interest rate. The formula for annual real interest rate calculation is as follows [41]:

$$i_r = (i_n - F_r)/(1 + F_r).$$

Here, i_r=Real interest rate is.
i_n=Nominal interest rate
F_r=Annual inflation rate.

The total cost of the system and the cost of a component to component with the separate specifications can be found through HOMER software. The summary of different components used in three different locations is shown in Table 18.5. This summary is available by introducing the parameter of interest into the HOMER software.

18.3.4 Economic Parameters Introduction

Economic analysis is required before the installation of a hybrid power generation system. HOMER simulation needed the economic input parameters, which are discussed in the following sections.

18.3.4.1 Total Investment Cost

It included in setting up the hybrid model that comprises the cost of materials and components, transportation and delivery cost, the costs related to installation and building, the cost related to consultation service, and power distribution. These costs are described during the starting of a project [42].

18.3.4.2 Initial Capital Cost

The cost of installation and equipment purchased is the initial capital cost [29].

18.3.4.3 Replacement Cost

It is the expenditure of those types of components that may be needed to replace after a certain amount of its lifetime. This cost is different as compared to initial capital cost as only a portion of the element may require replacing [29].

18.3.4.4 Operation & Maintenance Cost

This cost defines as the summation of total programmed expenditure that is utilized for lifetime maintenance and operation of the plant [36].

18.3.4.5 Salvage Value

After the project's lifespan, it is the existing value in a power system component. In order to compute each component's salvage value after the project's lifespan, the equation below [36] is utilized by HOMER:

$$S = C_{rc} R_{rl} / R_{lc}.$$

Here, S = Salvage value
  C_{rc} = Replacement cost of component
  R_{rl} = Component's remaining life
  R_{lc} = Component's lifetime.

18.3.4.6 Life Cycle Cost

It includes complete installation and operation costs for a component or process, over a defined period, usually many years. The analysis of life cycle of HBS is assessed as a financial evaluation process [29]:

LCC = Initial capital cost + Replacement cost + Fuel cost + O&M cost − Salvage value

18.3.4.7 Annualized Cost

To find the annualized cost of the component, HOMER combines replacement, service, capital, and fuel costs together with salvage and any other costs or revenue of each component. This is the assumed annual cost that would generate a net current cost equivalent to that of every single cost and revenue associated with that component during the project's lifespan if it occurred every year. In order to calculate total annualized system costs, HOMER will add up all annualized costs for each portion along with any miscellaneous costs. The value is significant because it is used by HOMER to measure two key economic figures for the system: the levelized COE and the total NPC [36].

18.3.4.8 Operating Cost

All costs (fuel, replacement, operating, and maintaining) and revenues other than initial capital costs, measured on an annual basis, are the operating cost. The operating cost [29] is computed by utilizing the following equation:

$$C_{oc} = C_{tac} - C_{tacc}.$$

Here, C_{oc} = Operating cost
  C_{tac} = Total annualized cost
  C_{tacc} = Total annualized capital cost [29].

18.4 OPTIMIZATION RESULTS

The simulation was executed by considering the project lifetime for 25 years' duration. The NPC summary of hybrid energy scheme for three various locations (Shillong, India/Desi Lam, Bhutan/Keraniganj, Bangladesh) are shown in Table 18.5. After optimization, the total NPC for the three different investigation areas are found to be $8.32M, $8.06M, and $8.25M, respectively, and the levelized COE for the three separate locations are $0.167, $0.155, and $0.164, respectively. Table 18.6 presents the annual outlay of integrated energy configuration based total annualized costs for various areas obtained are as $650,659, $630,251, $645,728, respectively. Table 18.7 presents the yearly production history of three investigation regions. The total power generated is 4,046,284 kWh/year in Shillong, 4,225,233 kWh/year in Desi Lam, and 4,089,315 kWh/year in Keraniganj. The power consumption summary of an electric load is indicated in Table 18.8. From the table, it

TABLE 18.5

NPC Summary of the Three Investigation Locations (Shillong, India; Desi Lam, Bhutan; Keraniganj, Bangladesh)

Name	Capital	Operating	Replacement	Salvage	Resource	Total
Generic 10 kW	$2.40M	$511,334	$623,609	−$349,498	$0.00	$3.19M
	$2.40M	$511,334	$623,609	−$349,498	$0.00	$3.19M
	$2.40M	$511,334	$623,609	−$349,498	$0.00	$3.19M
Generic 100 kW	$40,000	$223,964	$286,926	−$3,728	$1.51M	$2.05M
genset	$40,000	$223,964	$286,926	−$3,728	$1.51M	$2.05M
	$40,000	$223,964	$286,926	−$3,728	$1.51M	$2.05M
Generic flat-	$2.50M	$3.20M	$0.00	$0.00	$0.00	$5.70M
plate PV	$2.50M	$3.20M	$0.00	$0.00	$0.00	$5.70M
	$2.50M	$3.20M	$0.00	$0.00	$0.00	$5.70M
Grid	$0.700	−$5.98M	$0.00	$0.00	$0.00	−$5.90M
	$0.700	−$6.16M	$0.00	$0.00	$0.00	−$6.16M
	$0.700	−$5.96M	$0.00	$0.00	$0.00	−$5.96M
PV dedicated	$68,000	$543,293	$28,374	−$5,281	$0.00	$634,385
converter	$68,000	$543,293	$28,374	−$5,281	$0.00	$634,385
	$68,000	$543,293	$28,374	−$5,281	$0.00	$634,385
System	$1.50M	$639,168	$625,898	−$116,499	$0.00	$2.65M
converter	$1.50M	$639,168	$625,898	−$116,499	$0.00	$2.65M
	$1.50M	$639,168	$625,898	−$116,499	$0.00	$2.65M
System	$6.51M	−$785,892	$1.56M	−$475,007	$1.51M	$8.32M
	$6.51M	−$1.05M	$1.56M	−$475,007	$1.51M	$8.06M
	$6.51M	−$848,920	$1.56M	−$475,007	$1.51M	$8.25M

TABLE 18.6

Annualized Costs of the Three Investigation Areas (Shillong, India; Desi Lam, Bhutan; Keraniganj, Bangladesh)

Name	Capital	Operating	Replacement	Salvage	Resource	Total
Generic 10 kW	$187,744	$40,000	$48,783	−$27,340	$0.00	$249,187
	$187,744	$40,000	$48,783	−$27,340	$0.00	$249,187
	$187,744	$40,000	$48,783	−$27,340	$0.00	$249,187
Generic 100 kW	$3,129	$17,520	$22,445	−$291.63	$117,785	$160,588
genset	$3,129	$17,520	$22,445	−$291.63	$117,785	$160,588
	$3,129	$17,520	$22,445	−$291.63	$117,785	$160,588
Generic	$195,567	$250,000	$0.00	$0.00	$0.00	$445,567
flat-plate PV	$195,567	$250,000	$0.00	$0.00	$0.00	$445,567
	$195,567	$250,000	$0.00	$0.00	$0.00	$445,567
Grid	$0.0548	−$461,498	$0.00	$0.00	$0.00	−$461,498
	$0.0548	−$481,905	$0.00	$0.00	$0.00	−$481,905
	$0.0548	−$4,68,093	$0.00	$0.00	$0.00	−$4,68,093
PV dedicated	$5,319	$42,500	$2,220	−$413.14	$0.00	$49,626
converter	$5,319	$42,500	$2,220	−$413.14	$0.00	$49,626
	$5,319	$42,500	$2,220	−$413.14	$0.00	$49,626
System	$117,340	$50,000	$48,962	−$9,113	$0.00	$207,189
converter	$117,340	$50,000	$48,962	−$9,113	$0.00	$207,189
	$117,340	$50,000	$48,962	−$9,113	$0.00	$207,189
System	$509,100	−$61,478	$122,410	−$37,158	$117,785	$650,659
	$509,100	−$81,885	$122,410	−$37,158	$117,785	$630,251
	$509,100	−$66,408	$122,410	−$37,158	$117,785	$645,728

TABLE 18.7

Production Summary of the Three Investigation Locations

Component	Shillong, India		Desi Lam, Bhutan		Keraniganj, Bangladesh	
	Production (kWh/year)	Percent (%)	Production (kWh/year)	Percent (%)	Production (kWh/year)	Percent (%)
Generic flat-plate PV	3,060,363	75.6	3,238,155	76.6	3,099,244	75.8
Generic 100 kW genset	876,000	21.6	876,000	20.7	876,000	21.4
Generic 10 kW	109,921	2.72	111,078	2.63	114,071	2.79
Total	4,046,284	100	4,225,233	100	4,089,315	100

TABLE 18.8

Power Consumption Summary of the Three Different Locations

Component	Shillong, India		Desi Lam, Bhutan		Keraniganj, Bangladesh	
	Consumption (kWh/year)	Percent (%)	Consumption (kWh/year)	Percent (%)	Consumption (kWh/year)	Percent (%)
AC primary load	47,450	1.22	47,450	1.17	47,450	1.21
DC primary load	0	0	0	0	0	0
Grid sales	3,845,816	98.8	4,015,875	98.8	3,886,903	98.8
Total	3,893,266	100	4,063,325	100	3,934,353	100

TABLE 18.9

Emission Summary of the Three Investigation Regions

	Quantity (kg/year)		
Pollutant	Shillong, India	Desi Lam, Bhutan	Keraniganj, Bangladesh
Carbon dioxide	447,381	447,381	447,381
Carbon monoxide	4,380	4,380	4,380
Unburned hydrocarbons	177	177	177
Particulate matter	17.5	17.5	17.5
Sulfur dioxide	316	316	316
Nitrogen oxides	350	350	350

can be seen that the consumption by AC primary load is the same (47,450 kWh/year) for the three investigation areas. Thus, it is observed that the hybrid energy scheme satisfied the load demand while producing more than essential demand. Table 18.9 depicts summary of contaminants released from hybrid power plant of the three investigation locations. It can be noticed from the table that the emission of contaminants is less due to renewable dependency. The parameter related to purchase from the grid (kWh) each month and sales to the grid (kWh) on a month-to-month basis for the three investigation regions are shown in Table 18.10. The table indicates that the proposed integrated energy scheme infused a huge amount of electricity into the grid as compared to the amount of consumption.

18.4.1 COMPARATIVE ANALYSIS OF OPTIMIZATION RESULTS OF THREE DIFFERENT REGIONS

The comparison of optimization results for three investigation regions are illustrated in Table 8.11. It is noticed that the integrated energy scheme is most profitable in Desi Lam and it is also cost-effective in the other two locations. This table summarizes the variation in parameters: initial cost, Total NPC, operating cost, levelized COE, total production, levelized cost of PV, levelized cost of wind, PV plant production, wind farm production, generator production, grid purchases, grid sales, PV hours of operation, wind turbine hours of operation, generator hours of operation, total fuel, renewable fraction, maximum PV output, maximum wind output, mean output of PV, mean output of wind, PV penetration, wind penetration, capacity factor of PV, capacity factor of wind, generator maximum & minimum output, generator mean output, generator marginal generation cost, and generator capacity factor, between the three different investigation areas.

Figure 18.3 represents the PV plant power generation on an hourly basis in kW for the regions of Keraniganj, Desi Lam, and Shillong. Figure 18.4 illustrates wind farm power generation on an hourly basis in kW for the regions of Shillong, Desi Lam, and Keraniganj. Figure 18.5 represents grid sale power in kW on an hour-to-hour basis for the regions of Keraniganj, Desi Lam, and Shillong. Grid purchases in kW are almost zero. These figures are implemented on the basis of 24-hour real-time data by considering a particular day for the month of April. This analysis is executed by utilizing the optimization results of the proposed integrated energy scheme and it is on the basis of a 24-hour period. The aim of this analysis is for understanding what the integrated energy system can experience in various circumstances. The presence of wind and solar power are usually available in daytime but not during the night. In this condition, when one or both generation systems fails to generate an adequate amount of electricity, then a generator is utilized as a backup supplier of power for gratifying the load requirement. Thus, the consistency supply of power is balanced in each circumstance. By selecting the cheapest possible configuration, this integrated energy system is developed to balance the constant supply of power.

TABLE 18.10

Grid Purchase (kWh) and Grid Sales (kWh) on a Month-to-Month Basis for the Regions of Shillong (India), Desi Lam (Bhutan), and Keraniganj (Bangladesh)

Month	Energy Purchased (kWh)	Energy Sold (kWh)	Net Energy Purchased (kWh)	Peak Demand (kW)	Energy Charge	Demand Charge
January	0	328,592	−328,592	0	−$39,431	$0.00
	0	317,784	−317,784	0	−$38,134	$0.00
	0	311,092	−311,092	0	−$37,331	$0.00
February	0	311,092	−311,092	0	−$37,331	$0.00
	0	306,828	−306,828	0	−$36,819	$0.00
	0	303,146	−303,146	0	−$36,378	$0.00
March	0	350,810	−350,810	0	−$42,097	$0.00
	0	351,397	−351,397	0	−$42,168	$0.00
	0	351,591	−351,591	0	−$42,191	$0.00
April	0	359,664	−359,664	0	−$43,160	$0.00
	0	347,963	−347,963	0	−$41,756	$0.00
	0	355,918	−355,918	0	−$42,710	$0.00
May	0	346,288	−346,288	0	−$41,555	$0.00
	0	365,060	−365,060	0	−$43,807	$0.00
	0	362,510	−362,510	0	−$43,501	$0.00
June	0	303,608	−303,608	0	−$36,433	$0.00
	0	349,069	−349,069	0	−$41,888	$0.00
	0	335,148	−335,148	0	−$40,218	$0.00
July	0	302,263	−302,263	0	−$36,272	$0.00
	0	348,852	−348,852	0	−$41,862	$0.00
	0	328,170	−328,170	0	−$39,380	$0.00
August	0	303,755	−303,755	0	−$36,451	$0.00
	0	333,950	−333,950	0	−$40,074	$0.00
	0	315,416	−315,416	0	−$37,850	$0.00
September	0	282,663	−282,663	0	−$33,920	$0.00
	0	307,920	−307,920	0	−$36,950	$0.00
	0	291,224	−291,224	0	−$34,947	$0.00
October	0	321,787	−321,787	0	−$38,614	$0.00
	0	334,877	−334,877	0	−$40,185	$0.00
	0	315,557	−315,557	0	−$37,867	$0.00
November	0	314,284	−314,284	0	−$37,714	$0.00
	0	328,903	−328,903	0	−$39,468	$0.00
	0	307,053	−307,053	0	−$36,846	$0.00
December	0	321,009	−321,009	0	−$38,521	$0.00
	0	323,272	−323,272	0	−$38,793	$0.00
	0	310,079	−310,079	0	−$37,209	$0.00
Annual	0	3,845,816	−3,845,816	0	−$461,498	$0.00
	0	4,015,875	−4,015,875	0	−$481,905	$0.00
	0	3,886,903	−3,886,903	0	−$466,428	$0.00

TABLE 18.11
Comparison of Optimization Results for the Three Investigations Regions

Cost Type	Shillong, India	Desi Lam, Bhutan	Keraniganj, Bangladesh
Initial cost ($)	$6.51M	$6.51M	$6.51M
Total NPC ($)	$8.32M	$8.06M	$8.25M
Operating cost ($/year)	$141,559	$121,152	$136,629
COE ($/kWh)	$0.167	$0.155	$0.164
Total production (kWh)	4,046,284	4,225,233	4,089,315
Levelized cost of PV ($/kWh)	0.162	0.153	0.160
Levelized cost of wind ($/kWh)	2.27	2.24	2.18
PV plant production (kWh/year)	3,060,363	3,238,155	3,099,244
Wind farm production (kWh/year)	109,921	111,078	114,071
Generator production (kWh/year)	876,000	876,000	876,000
Grid sales (kWh/year)	3,845,816	4,015,875	3,886,903
Grid purchases (kWh/year)	−3,845,816	−4,015,875	−3,886,903
Total fuel (L/year)	102,422	102,422	102,422
Ren. Frac. (%)	90.6	91.0	90.7
Hours of operation of PV (hours/year)	4,366	4,380	4,372
Hours of operation of wind turbine (hours/year)	3,400	3,295	3,332
Generator hours of operation (hours/year)	8,760	8,760	8,760
Maximum PV output (kW)	850	850	850
PV mean output (kW)	349	370	354
PV penetration (%)	6,450	6,824	6,532
PV capacity factor (%)	6.99	7.39	7.08
Maximum wind output (kW)	750	623	794
Wind mean output (kW)	12.5	12.7	13.0
Wind penetration (%)	232	234	240
Wind capacity factor (%)	1.25	1.27	1.30
Generator max. & min. output (kW)	100	100	100
Generator mean output (kW)	100	100	100
Generator marginal generation cost ($/kWh)	0.0582	0.0582	0.0582
Generator capacity factor (%)	100	100	100

In the time period from 7:00 AM to 17:00 PM, both PV plant power and wind farm power are obtainable on a huge scale in all regions; therefore, grid purchase power becomes zero at that duration and grid sale power is immensely high. In the time period from 17:00 PM to 0:00 AM, the generation of PV plant power minimizes to zero; the generation of wind farm power is obtainable during 17:00 PM to 20:00 PM, and after this duration, wind farm power becomes zero. Grid purchase power is also zero in that condition due to the backup power supply of generators and grid sale power is reducing. During the time of 0:00 AM to 4:00 AM, PV plant power is zero for all locations, and wind farm power is only obtainable for the Desi Lam region and zero for other areas. Therefore, the integrated energy system takes power supply from a generator, and thus grid purchasing power is zero, and grid sale power is reducing. In the last condition, during the time period of 4:00 AM to 7:00 AM, PV plant power is improving, and wind farm power is only increasing in the Keraniganj region. In other locations, wind farm power becomes zero, and grid purchasing power is also zero but grid sale power is improving at this duration. One thing that is common in all conditions that is grid purchase is zero due to the presence of a generator that works as a backup supplier of power. All these conditions are considered on the basis of the proposed integrated energy system. This analysis

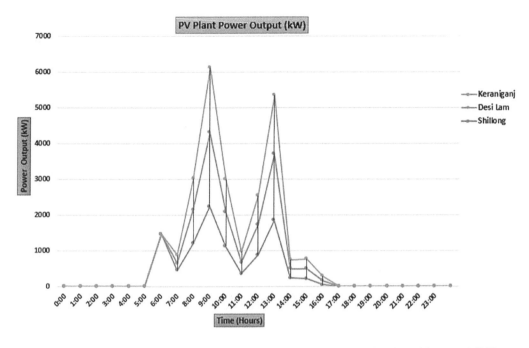

FIGURE 18.3 PV plant power generation (kW) on an hourly basis in Keraniganj, Desi Lam, and Shillong.

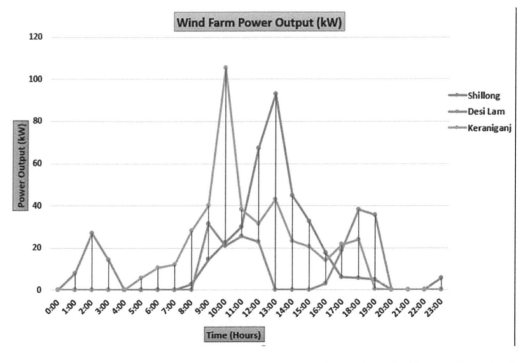

FIGURE 18.4 Wind farm power generation (kW) on an hourly basis in Shillong, Desi Lam, and Keraniganj.

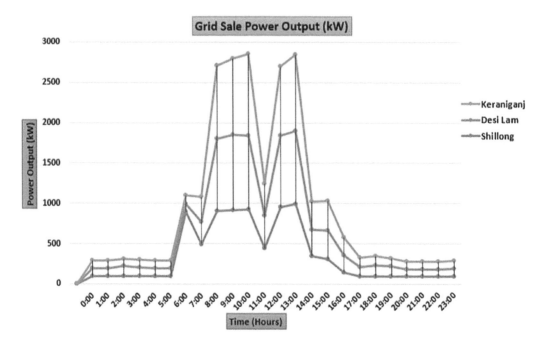

FIGURE 18.5   Grid sale power (kW) on hour-to-hour basis of Keraniganj, Desi Lam, and Shillong.

shows and proves that all three proposed regions have a good opportunity for power production using solar and wind on a huge scale, which is needed to utilize.

18.5   CONCLUSION

To minimize the increasing air pollution in the world, this study tried to explore a hybrid mode of electricity generation with renewable energy sources. Keeping in the mind the availability of renewable energy resources in the selected locations of India, Bhutan, and Bangladesh, an optimized HBS connected to grid using solar, wind, and biomass is designed. From the research results, it is discovered that the rate of major generation in each country is almost 76% which comes from the solar system. Moreover, 22% of power comes from biomass-gasifier, and 3% comes from a wind farm of the HBS. It is also found from the study that these HBS try to utilize the least expensive energy sources to gratify the load requirement and to reduce electricity bills. Here, three locations are taken to examine the feasibility of the HBS, and it is found to be profitable in all locations. Thus, it is possible to say that the recommended HBS for different locations is eco-friendly, economical, and efficient in terms of power production. Hence, we can conclude here with a statement that this proposed HBS can be used to replace a thermal power plant to fight against climate change. In future research, this HBS should be used in different locations based on the availability of renewable resources. Biogas plants can also be replaced by hydroelectric plants. This hybrid hydroelectric plant incorporating a solar system and the wind farm is more suitable and profitable than that with other combinations.

REFERENCES

1. Chandel S S, Shrivastva R, Sharma V, Ramasamy P. (2015) 'Overview of the initiatives in renewable energy sector under the national action plan on climate change in India', *Renewable and Sustainable Energy Reviews*, Vol. 54, pp. 866–873.

2. Bhattacharjee S, Chakraborty S, Jena B B, Deb S, Das R. (2018) 'An optimization study of both on-grid and off-grid solar-wind-biomass hybrid power plant in Nakalawaka', *Fiji. International Journal for Research in Applied Science and Engineering Technology*, Vol. 6, pp. 3822–3834.

3. De Jong P, Barreto Tarssio B, Tanajura Clemente A S, Kouloukoui D, Oliveira-Esquerre K P, Kiperstok A, Torres E A. (2019) 'Estimating the impact of climate change on wind and solar energy in Brazil using a South American regional climate model', *Renewable Energy,* Vol. 141, pp. 390–401.

4. Byravan S, Rajan S C. (2009) 'The social impacts of climate change in South Asia', *JMRI*, Vol. 5, no. 3, pp. 1–16.

5. Islam S, Khan M Z R. (2017) 'A review of energy sector of Bangladesh', *Energy Procedia*, Vol. 110, pp. 611–618.

6. Energy in Bhutan [online] https://en.wikipedia.org/wiki/Energy_in_Bhutan (accessed 31 May 2020).

7. Shushant Parashar V C, Saxena S. (2019) 'Climate change challenges in South Asia: A case of Bhutan', *International Journal of Innovative Technology and Exploring Engineering,* Vol. 8, pp. 469–473.

8. Nandi C, Bhattacharjee S, Reang S. (2018) 'An optimization case study of hybrid energy system based charging station for electric vehicle on Mettur, Tamil Nadu', *International Journal of Advanced Scientific Research and Management,* Vol. 3, pp. 225–231.

9. Bhattacharjee S, Nandi C. (2020) 'Design of an industrial internet of things-enabled energy management system of a grid-connected solar–wind hybrid system-based battery swapping charging station for electric vehicle', In: Mandal J, Mukhopadhyay S, Roy A. (Eds), *Applications of Internet of Things. Lecture Notes in Networks and Systems*, vol. 137. Springer, Singapore, pp. 1–14.

10. Bhattacharjee S, Das U, Chowdhury M, Nandi C. (2020) 'Role of hybrid energy system in reducing effects of climate change', In: Qudrat-Ullah H, Asif M. (Eds), *Dynamics of Energy, Environment and Economy.* Springer International Publishing, New York, pp. 115–138.

11. Bhattacharjee S, Nandi C, Reang S. (2018) 'Intelligent energy management controller for hybrid system', *3rd International Conference for Convergence in Technology (I2CT),* Pune, India, pp. 1–7.

12. Bhattacharjee S, Nandi C. (2019) 'Technical, economic, feasibility and comparative analysis of three different configurations of energy system to control intermittency of renewable energy', *Proceedings of the 2nd International Conference on Information Systems & Management Science (ISMS),* Tripura University, Agartala, Tripura, India, pp. 117–124.

13. Bhattacharjee S, Nandi C. (2020) 'Design of a voting based smart energy management system of the renewable energy based hybrid energy system for a small community', *Energy,* Vol. 214, pp. 1–15.

14. Bouraiou A, Neçaibia A, Boutasseta N, Mekhilef S, Dabou R, Ziane A, Sahouane N, Attoui I, Mostefaoui M, Touaba O. (2019) 'Status of renewable energy potential and utilization in Algeria', *Journal of Cleaner Production*, Vol. 246, pp. 1–59.

15. Bhattacharjee S, Das R, Deb G, Thakur B N. (2019) 'Techno-economic analysis of a grid-connected hybrid system in Portugal Island', *International Journal of Computer Sciences and Engineering*, Vol. 7, pp. 1–14.

16. Herrera I, Rodríguez-Serrano I, Lechon Y, Oliveira A, Krüger D, Bouden C. (2020) 'Sustainability assessment of a hybrid CSP/biomass. Results of a prototype plant in Tunisia', *Sustainable Energy Technologies and Assessments*, Vol. 42, p. 100862.

17. Elum Z A, Momodu A S. (2017) 'Climate change mitigation and renewable energy for sustainable development in Nigeria: A discourse approach', *Renewable and Sustainable Energy Reviews*, Vol. 76, pp. 72–80.

18. Nandi C, Bhattacharjee S, Chakraborty S. (2019) 'Climate change and energy dynamics with solutions: A case study in Egypt', In: Qudrat-Ullah H, Kayal A A. (Eds), *Climate Change and Energy Dynamics in the Middle East.* Springer, Berlin, Germany, pp. 225–257.

19. Ma T, Javed M S. (2018) 'Integrated sizing of hybrid PV-wind-battery system for remote island considering the saturation of each renewable energy resource', *Energy Conversion and Management,* Vol. 182, pp. 178–190.

20. Sinha S, Chandel S S. (2015) 'Review of recent trends in optimization techniques for solar photovoltaic–wind based hybrid energy systems', *Renewable and Sustainable Energy Reviews,* Vol. 50, pp. 755–769.

21. Tiwary A, Spasova S, Williams I D. (2019) 'A community-scale hybrid energy system integrating biomass for localised solid waste and renewable energy solution: Evaluations in UK and Bulgaria', *Renewable Energy,* Vol. 139, pp. 960–967.

22. Hussain C M Iftekhar, N B, Duffy A. (2016) 'Technological assessment of different solar-biomass systems for hybrid power generation in Europe', *Renewable and Sustainable Energy Reviews,* Vol. 68, pp. 1115–1129.

23. Suresh V, Muralidhar M, Kiranmayi R. (2020) 'Modelling and optimization of an off-grid hybrid renewable energy system for electrification in a rural areas', *Energy Reports,* Vol. 6, pp. 594–604.

24. Karmaker A K, Hossain M A, Nallapaneni M K, Jagadeesan V, Jayakumar A, Ray B. (2020) 'Analysis of using biogas resources for electric vehicle charging in Bangladesh: A techno-economic-environmental perspective', *Sustainability,* Vol. 12, pp. 1–19.

25. Bhattacharjee S, Chakraborty S, Nandi C. (2020) 'An optimization case study of hybrid energy system in four different regions of India', In: Bhoi A K, Sherpa K S, Kalam A, Chae G-S. (Eds), *Advances in Greener Energy Technologies.* Springer, Singapore, pp. 399–437.

26. Kasaeiana A, Rahdan P, Vaziri Rad M A, Yan W M. (2019) 'Optimal design and technical analysis of a grid-connected hybrid photovoltaic/diesel/biogas under different economic conditions: A case study', *Energy Conversion and Management,* Vol. 198, pp. 1–15.

27. Bhattacharjee S, Nandi C. (2020) 'Design of a smart energy management controller for hybrid energy system to promote clean energy', In: Bhoi A K, Sherpa K S, Kalam A, Chae G-S. (Eds), *Advances in Greener Energy Technologies.* Springer, Singapore, pp. 527–563.

28. NASA Power Website [online] https://power.larc.nasa.gov/ (accessed 22 January, 2020).

29. Bhattacharjee S, Dey A. (2014) 'Techno-economic performance evaluation of grid integrated PV-biomass hybrid power generation for rice mill', *Sustainable Energy Technologies and Assessments,* Vol. 7, pp. 6–16.

30. Gautam J, Ahmed M I, Kumar P. (2018) 'Optimization and comparative analysis of solar-biomass hybrid power generation system using homer', *International Conference on Intelligent Circuits and Systems (ICICS),* pp. 397–400.

31. Chambona C L, Kariaa T, Sandwell P, Halletta J P. (2020) 'Techno-economic assessment of biomass gasification-based mini-grids for productive energy applications: The case of rural India', *Renewable Energy,* Vol. 154, pp. 432–444.

32. Chauhan A, Saini R P. (2017) 'Size optimization and demand response of a stand-alone integrated renewable energy system', *Energy,* Vol. 124, pp. 59–73.

33. Vendoti S, Muralidhar M, Kiranmayi R. (2021) 'Techno-economic analysis of off-grid solar/wind/biogas/biomass/fuel cell/battery system for electrification in a cluster of villages by HOMER software', *Environment, Development and Sustainability,* Vol. 23, pp. 351–372.

34. Thapar V, Agnihotri G, Sethi, V K. (2011) 'Critical analysis of methods for mathematical modeling of wind turbines', *Renewable Energy,* Vol. 36, pp. 3166–3177.

35. HOMER Energy Website, https://www.homerenergy.com/products/grid/docs/1.6/operation_of_a_cofired_generator.html.

36. Bhattacharjee S, Acharya S. (2015) 'PV–wind hybrid power option for a low wind topography', *Energy Conversion and Management,* Vol. 89, pp. 942–954.

37. Bhakta S, Mukherjee V, Shaw B. (2015) 'Techno-economic analysis of standalone photovoltaic/wind hybrid system for application in isolated hamlets of North-East India', *Journal of Renewable and Sustainable Energy,* Vol. 7, 023126.

38. Saharia B J, Manas M. (2017) 'Viability analysis of photovoltaic/wind hybrid distributed generation in an isolated community of Northeastern India', *Distributed Generation & Alternative Energy Journal,* Vol. 32, pp. 49–80.

39. Rohani G, Nour M. (2014) 'Techno-economical analysis of stand-alone hybrid renewable power system for Ras Musherib in United Arab Emirates', *Energy,* Vol. 64, pp. 828–841.

40. Sharma R, Goel S. (2016) 'Stand-alone hybrid energy system for sustainable development in rural India', *Environment, Development and Sustainability,* Vol. 18, pp. 1601–1614.

41. Nurunnabi M, Roy N K. (2015) 'Grid connected hybrid power system design using HOMER', *3rd International Conference on Advances in Electrical Engineering (ICAEE),* pp. 18–21.

42. Olatomiwa L, Mekhilef S, Huda A S N, Ohunakin O S. (2015) 'Economic evaluation of hybrid energy systems for rural electrification in six geo-political zones of Nigeria', *Renewable Energy,* Vol. 83, pp. 435–446.

19 IoT-Based Prioritized Load Management Technique for PV Battery-Powered Building
Mini Review

Mohit Sharan, Prantika Das, and Apurv Malhotra
Vellore Institute of Technology

Aayush Karthikeyan
University of Calgary

Ramani Kannan
Universiti Teknologi Petronas

O.V. Gnana Swathika
Vellore Institute of Technology

CONTENTS

ABSTRACT

The need of the hour is to come up with sustainable power supply systems. A demand-side load management strategy is necessary for regulating loads within a residential structure in such a way that user satisfaction is maximized while costs are kept to a minimum. We aim to provide an energy-oriented production control method that treats electrical energy as a restricted production capacity with specified load profiles for manufacturing. This energy-oriented production control tries to match manufacturing energy demand with a limited energy supply, lowering energy expenditures. It looks into the significance of adaptive power utilization coupled with IoT technology and data analytics. The adaptive power system paves way for an intelligent real-time interaction platform of power utilization. This improves the sustainability, stability and reliability of power supply. Internet of Things (IoT) technology, with its reliable communication and strong data processing, can effectively improve the adaptive power utilization. Photovoltaics (PVs) is playing a growing role in electricity industry around the world, while Battery Energy Storage Systems are falling in cost and starting to be deployed by energy consumers with PVs. Apartment buildings offer an opportunity to apply central battery storage and shared solar generation to aggregated apartment and common loads through an embedded network or microgrid. We aim to have a hybrid consumption setup which smartly couples the existing power consumption setup with an in-demand and reliable PV energy-based power consumption setup to ensure power for all.

DOI: 10.1201/9781003331117-19

KEYWORDS

Load Management; IoT; Sustainability; Energy Demand; PV

19.1 INTRODUCTION

The need of the hour is to come up with sustainable power supply systems. One of the ways to achieve this is to follow the incentive-based pricing system and the method can be based on "Model Predictive Control" which works in line with the thermal load management of the buildings [1]. Demand-Side Management (DSM) programmes, which are implemented through residential energy management systems (EMSs) for smart cities, have a number of advantages: consumers save money on electricity, and the utility works at lower peak demand. The DSM model for scheduling residential users' appliances is provided in this study, which is based on evolutionary techniques (binary particle swarm optimization [PSO], genetic algorithm, and cuckoo search) [2].

In hybrid renewable energy systems (HRES) sizing, splitting the load into major and minor importance sections is done, and based on those categorizations, the load is managed. The sizing is determined using a smart PSO algorithm implemented in MATLAB®. The results reveal that the cost of generated energy is inversely proportional to the load division percentage [3,4]. Using a demand-side load management strategy, we are managing loads within a residential complex by means of maximum user satisfaction by keeping costs at the minimum possible value [5]. A smart grid setting has Energy Management Units. It is basically by the aid of Artificial Neural Network applied alongside DSM method that helps in peak cutting and load prioritization to incorporate both primitive and modern-day (renewable) energy sources [6]. A production control method is used to produce energy as per the capacity mentioned in the load profiles. This "energy-oriented" production control tries to match energy production with a set energy supply, lowering energy costs [7]. Demand response (DR) is employed in this work for economic and system dependability reasons. The concentration is on a reward-based programme for the consumers with the goal of matching a set consumption unit to the total requirement to provide peak control and lower market costs [8].

We can rely on sensor technologies, communications and control methods for proper energy usage. The use of reinforcement learning in building energy management and control has been an aid in proper energy utilization [9]. Using a DR programme we can have a dual targeted optimization system for the appropriate siting and scaling of a storage system in a microgrid (MG) [10]. This study outlines a broad framework for evaluating prospects for effective load management in houses where renewable energy sources and energy storage facilities are provided in response to changing utility incentives [11–15]. Electrical distribution systems can have distributed energy resources. Control methods are based on primitive operational setup that allows overlapping control layers to be integrated smoothly with the local control methods [16–18]. An EMS with a latest criterion named as criticality level, which is the threshold value integrated in apartments on their energy consumption, is considered. The model here maximizes the power consumed simultaneously with the cost minimization [19].

19.2 INTERNET OF THINGS (IoT) IN SMART BUILDINGS

The paper looks at the importance of flexible power consumption associated with IoT technology. The flexible power system paves the way for a real-time interactive energy field. This improves the viability, stability and reliability of the power supply. IoT technology, and reliable communication and robust data processing can effectively improve flexible power consumption [4]. This paper discusses developments in the energy sector. With the growing adoption of smart housing technologies and smart construction, there is an urgent need to establish an energy-efficient ecosystem. The solution has been recognized as environmentally friendly and as an important global aim to upgrade planetary stability [14]. The sudden urge for availability of electricity and the development of smart grids is a convenience for the emergence of local hybrid energy management systems (HEMSs) to improve energy efficiency. It includes a DR tool that changes the need to improve home energy use considering power bills, environmental botherings, load profile and consumer reassurance. This paper also provides a brief on clever home technologies and load scheduling controllers, which are inculcated with artificial intelligence [20–22]. The goal of the paper is to create a home power control and monitoring system. It consists of esp8266, Arduino, current sensor acs712 and cloud platform as a data storage and analysis service. The solution focuses on reducing energy consumption by repeatedly informing electrical appliances and providing optimized information to users, which will assist the user to reduce unnecessary energy use and thus save more resources and money [23,24]. This paper provides a theoretical and practical review that provides a solution to the problem of the need for electricity supply in Zimbabwe during peak hours. The concept model was made using C++ directed by the load management algorithm. It will continue to monitor the feeder, open the automatic controller and perform the load sharing process based on the important settings that ensure the availability of electricity to be used, which are critical in times of increasing demand. The equipment will be attached to the power grid with the help of smart plugs [25]. The paper introduces an intelligent IoT-focused system for power regulation in buildings. The combination of power and the construction industry will ultimately be an important part of building a smart city. The proposed system extends to building data, power generation, energy bills, climate data and consumer behaviour, in order to produce day-to-day consumer applications [26]. This paper conducts a detailed review of the literature on strategies for efficient use of energy and planning in smart households. Intricate description is made of various materials that contribute to thermal relief, visual comfort and good air quality. Computer systems used in smart homes have also been updated [27]. The goal of this project is to build an Arduino and IoT-focused Hierarchical Multi-Agent System for small-grid loading using a promotional strategy. The effectiveness of the proposed algorithm on the small grid was confirmed in this study [28–32]. This project proposes the construction of a virtual web (IoT) based on HRES, along with a wind turbine, a photovoltaic (PV) setup, a battery storehouse equipment and a diesel generator. The parameters to check are labelled into various categories, which include electrical values, status and environment [33].

19.3 PHOTOVOLTAIC POWER SYSTEMS INTEGRATED TO SMART BUILDINGS

This study provides an optimal EMS for reducing residential building electricity bills. The goal is to acquire peak shaving along with lowering the price of electrical energy for the consumer using a dynamic energy pricing model [12]. This paper also demonstrates how to utilize a device having stored energy such as a battery on a MG with PV along with a load generating energy in the most efficient way possible. The method forecasts daily loads and generation power, and then looks for the best storage action strategy for the energy storage structure based on those forecasts [13]. A system is more difficult to govern because of the fluctuation and uncertainty of output and consumption. In reality, weather conditions have an impact on renewable energy output, whereas occupancy has an impact on power consumption. As a result, reliable temporary projections are required for the perfect unification of renewable energy system (RES) such as solar systems and wind turbines alongside the conventional power grid in MG systems. The research proposes a method anticipating power generation and usage in MG systems, along with the state of charge of the batteries [14,15]. This study focuses on boosting the usage of solar energy, which is often pushed as a viable alternative to fossil fuels due to its ease of use and environmental benefits. To compensate for its erratic and unpredictable qualities, electrical energy storage solutions are being developed to synchronize generated power with building demand. The paper presents an overview of current research, scope of study and optimization of the construction of hybrid PV-electrical energy storage systems (ESS) required for power supply in buildings which can be utilized like a roadmap for future studies in the subject [16,17]. Distributed PVs are becoming more common in the world's electrical businesses, whereas Battery Energy Storage Systems (BESS) are becoming more affordable and being used by users of energy with PVs. By an embedded network or MG, apartment complexes can integrate main battery storage along with PV generation to apartments and communal loads [18–20]. Traditional electricity users are being transformed into electricity prosumers as a result of the smart grid. This research examines how a current household bunch in Sweden was converted into electricity prosumers. For the coupled PV-heat pump-thermal storage-electric vehicle (EV) system, an objective function built on genetic algorithms is formed to optimize the capabilities and stances of PV elements at the cluster level, with the goal of increasing self-consumed electricity under a non-negative net present value during the economic life cycle. As the utilization of renewable energy along with DC-based loads has increased dramatically in recent years, DC MGs are being examined as a probable solution for India's electrification. This paper explains how to manage energy in DC MGs and how to overcome common design issues when putting off-grid power systems in place. The rule of conservation of energy serves as the basis for the system's operation. The algorithm's scalability was demonstrated using a prototype hardware configuration at VIT University's Solar Energy Research Center [21–23]. The influence of PV systems in residential and commercial settings is examined in this research. Its significance in these places has grown as a result of daily electricity usage during the solar hours. It presents a controller-based management solution for power circulation in the electrical grid and PV system and EV batteries. It demonstrates the advantages of deploying EVs as a vital component in balancing energy, resulting in lower energy expenditures [24–27]. PV systems do not produce electricity continuously and are mostly dependent on weather conditions, resulting in lesser energy production than other renewable alternatives. As a result, the ESS may prove to be a viable answer to the aforementioned issue. The practicality in terms of money of lead-acid ESS paired with PV panels in apartments is proposed in this study [28,29]. The efficiency of a PV plant placed on the roof of an educational structure is evaluated, and a strategy for enhancing it using only feeder connections is given. The demand for the electrical system of the structure is then reduced by substituting obsolete lights and ventilation fans with more energy-efficient types [30–34]. The authors of Refs. [35–38] investigate the influence of four distinct strength management techniques (EMSs) on the dependability of PV-powered MGs, which are projected to play a significant part in future distribution networks. In Refs. [39–40], control of the EMS for MGs using PV-based distribution generation systems is discussed. During cloudy situations, the Maximum Power Point Tracking algorithm, in this case Hill Climbing technique, is utilized to harvest the maximum power provided by the PV source, which is assisted by a battery-based ESS.

19.4   CONCLUSION

This paper discusses the importance of realizing sustainable buildings. It also illustrates the importance of IoT technology and data analytics in smart load management in PV-integrated smart buildings. The adaptive power system paves way for an intelligent real-time interaction platform of power utilization. The importance of PV integrated with BESS in MG is analysed as well.

REFERENCES

1. Goy, S. and Sancho-Tomás, A., 2019. Load management in buildings. In: U. Eicker (Ed.), *Urban Energy Systems for Low-Carbon Cities* (pp. 137–179). Academic Press: Cambridge, MA.
2. Javaid, N., Ullah, I., Akbar, M., Iqbal, Z., Khan, F.A., Alrajeh, N. and Alabed, M.S., 2017. An intelligent load management system with renewable energy integration for smart homes. *IEEE Access*, 5, pp. 13587–13600.
3. Eltamaly, A.M., Mohamed, M.A., Al-Saud, M.S. and Alolah, A.I., 2017. Load management as a smart grid concept for sizing and designing of hybrid renewable energy systems. *Engineering Optimization*, 49(10), pp. 1813–1828.

4. Abou Emira, S.S., Youssef, K.Y. and Abouelatta, M., 2019. Adaptive power system for IoT-based smart agriculture applications. In *2019 15th International Computer Engineering Conference (ICENCO)* (pp. 126–131). IEEE.

5. Ogunjuyigbe, A.S.O., Ayodele, T.R. and Akinola, O.A., 2017. User satisfaction-induced demand side load management in residential buildings with user budget constraint. *Applied Energy*, 187, pp. 352–366.

6. Raju, M.P. and Laxmi, A.J., 2017. A novel load management algorithm for EMU by implementing demand side management techniques using ANN. *In 2017 International Conference on Electrical and Computing Technologies and Applications (ICECTA)* (pp. 1–6). IEEE.

7. Schultz, C., Braun, S., Braunreuther, S. and Reinhart, G., 2017. Integration of load management into an energy-oriented production control. *Procedia Manufacturing*, 8, pp. 144–151.

8. Ahmed, S. and Bouffard, F., 2017. An online framework for integration of demand response in residential load management. In *2017 IEEE Electrical Power and Energy Conference (EPEC)* (pp. 1–6). IEEE.

9. Mason, K. and Grijalva, S., 2019. A review of reinforcement learning for autonomous building energy management. *Computers & Electrical Engineering*, 78, pp. 300–312.

10. Nojavan, S., Majidi, M. and Esfetanaj, N.N., 2017. An efficient cost-reliability optimization model for optimal siting and sizing of energy storage systems in a microgrid in the presence of responsible load management. *Energy*, 139, pp. 89–97.

11. Georges, E., Braun, J.E. and Lemort, V., 2017. A general methodology for optimal load management with distributed renewable energy generation and storage in residential housing. *Journal of Building Performance Simulation*, 10(2), pp. 224–241.

12. Bonthu, R.K., Pham, H., Aguilera, R.P. and Ha, Q.P., 2017. Minimization of building energy cost by optimally managing PV and battery energy storage systems. In *2017 20th International Conference on Electrical Machines and Systems (ICEMS)* (pp. 1–6). IEEE.

13. Galván, L., Navarro, J.M., Galván, E., Carrasco, J.M. and Alcántara, A., 2019. Optimal scheduling of energy storage using a new priority-based smart grid control method. *Energies*, 12(4), p. 579.

14. Mangla, M., Akhare, R. and Ambarkar, S., 2019. Context-aware automation based energy conservation techniques for the IoT ecosystem. In: M. Mittal, S. Tanwar, B. Agarwal and L. M Goyal (Eds.), *Energy Conservation for IoT Devices* (pp. 129–153). Springer: Singapore.

15. Elmouatamid, A., Ouladsine, R., Bakhouya, M., Zine-Dine, K. and Khaidar, M., 2019. A control strategy based on power forecasting for micro-grid systems. In *2019 IEEE International Smart Cities Conference (ISC2)* (pp. 735–740). IEEE.

16. Ghaffar, M., Naseer, N., Sheikh, S.R., Naved, M., Aziz, U. and Koreshi, Z.U., 2019. Electrical energy management of buildings using fuzzy control. In *2019 International Conference on Robotics and Automation in Industry (ICRAI)* (pp. 1–8). IEEE.

17. Liu, J., Chen, X., Cao, S. and Yang, H., 2019. Overview on hybrid solar photovoltaic-electrical energy storage technologies for power supply to buildings. *Energy Conversion and Management*, 187, pp. 103–121.

18. Roberts, M.B., Bruce, A. and MacGill, I., 2019. Impact of shared battery energy storage systems on photovoltaic self-consumption and electricity bills in apartment buildings. *Applied Energy*, 245, pp. 78–95.

19. Paul, S. and Padhy, N.P., 2019. Real-time bilevel energy management of smart residential apartment building. *IEEE Transactions on Industrial Informatics*, 16(6), pp. 3708–3720.

20. Huang, P., Lovati, M., Zhang, X., Bales, C., Hallbeck, S., Becker, A., Bergqvist, H., Hedberg, J. and Maturi, L., 2019. Transforming a residential building cluster into electricity prosumers in Sweden: Optimal design of a coupled PV-heat pump-thermal storage-electric vehicle system. *Applied Energy*, 255, p. 113864.

21. Shamshuddin, M. A., Babu, T. S., Dragicevic, T., Miyatake, M. and Rajasekar, N., 2018. Priority-based energy management technique for integration of solar PV, battery, and fuel cell systems in an autonomous DC microgrid. *Electric Power Components and Systems*. DOI: 10.1080/15325008.2017.1378949.

22. Shareef, H., Ahmed, M. S., Mohamed, A. and Al Hassan, E., 2019. Review on home energy management system considering demand responses, smart technologies, and intelligent controllers. *Information*, 10, p. 108. doi: 10.3390/info10030108.

23. Revathi, R, Sathya, M., Sujaini, R.S., Suganya, T. and Thenmozhi, N., 2018. IoT based energy auditing in girls hostel. *International Journal of Engineering Research & Technology (IJERT)*, 6(7). ISSN: 2278-0181.

24. Torres-Moreno, J. L., Gimenez-Fernandez, A., and Perez-Garcia, M. and Rodriguez, F., 2018. Energy management strategy for micro-grids with PV-battery systems and electric vehicles. *Energies*, 11, p. 522. doi: 10.3390/en11030522.

25. Mahlangu, G., Musungwini, S. and Nleya, S., 2018. An algorithm for using Internet of Things (IoTs) to improve load management in electric power grid. *International Journal of Computer Applications (0975–8887),* 179(35).

26. Marinakis, V. and Doukas, H., 2018. An advanced IoT-based system for intelligent energy management in buildings. *Sensors,* 18, p. 610. doi: 10.3390/s18020610.

27. Shah, A. S., Nasir, H., Fayaz, M., Lajis, A. and Shah, A., 2019. A review on energy consumption optimization techniques in IoT based smart building environments. *Information,* 10, p. 108.

28. Cucchiella, F., D'Adamo, I., Gastaldi, M. and Stornelli, V., 2018. Solar photovoltaic panels combined with energy storage in a residential building: An economic analysis. *Sustainability,* 10, p. 3117. doi: 10.3390/su10093117.

29. Mozaffari Legha, M. and Farjah, E., 2020. IOT based load management of a micro-grid using Arduino and HMAS. *Iranian Journal of Electrical and Electronic Engineering,* 16(2), pp. 228–234.

30. Manu, D., Shorabh, S.G., Swathika, O.G., Umashankar, S. and Tejaswi, P., 2022. Design and realization of smart energy management system for Standalone PV system. *In IOP Conference Series: Earth and Environmental Science* (Vol. 1026, No. 1, p. 012027). IOP Publishing.

31. Swathika, O.G., Karthikeyan, K., Subramaniam, U., Hemapala, K.U. and Bhaskar, S.M., 2022. Energy efficient outdoor lighting system design: Case study of IT campus. In *IOP Conference Series: Earth and Environmental Science* (Vol. 1026, No. 1, p. 012029). IOP Publishing.

32. Sujeeth, S. and Swathika, O.G., 2018. IoT based automated protection and control of DC microgrids. In *2018 2nd International Conference on Inventive Systems and Control (ICISC)* (pp. 1422–1426). IEEE.

33. Patel, A., Swathika, O.V., Subramaniam, U., Babu, T.S., Tripathi, A., Nag, S., Karthick, A. and Muhibbullah, M., 2022. A practical approach for predicting power in a small-scale off-grid photovoltaic system using machine learning algorithms. *International Journal of Photoenergy,* 2022.

34. Odiyur Vathanam, G.S., Kalyanasundaram, K., Elavarasan, R.M., Hussain Khahro, S., Subramaniam, U., Pugazhendhi, R., Ramesh, M. and Gopalakrishnan, R.M., 2021. A review on effective use of daylight harvesting using intelligent lighting control systems for sustainable office buildings in India. *Sustainability,* 13(9), p. 4973.

35. Swathika, O.V. and Hemapala, K.T.M.U., 2019. IOT based energy management system for standalone PV systems. *Journal of Electrical Engineering & Technology,* 14(5), pp. 1811–1821.

36. Swathika, O.V. and Hemapala, K.T.M.U., 2019. IOT-based adaptive protection of microgrid. In *International Conference on Artificial Intelligence, Smart Grid and Smart City Applications* (pp. 123–130). Springer, Cham.

37. Kumar, G.N. and Swathika, O.G., 2022. Chapter 19: AI applications to renewable energy an analysis. In: O.V. Gnana Swathika, K. Karthikeyan and S.K. Padmanaban (Eds.), *Smart Buildings Digitalization: IoT and Energy Efficient Smart Buildings Architecture and Applications* (p. 283). CRC Press: Boca Raton, FL.

38. Swathika, O.G., 2022. Chapter 5: IoT-based smart. In: O.V. Gnana Swathika, K. Karthikeyan and S.K. Padmanaban (Eds.), *Smart Buildings Digitalization: IoT and Energy Efficient Smart Buildings Architecture and Applications* (p. 57). CRC Press: Boca Raton, FL.

39. Lal, P., Ananthakrishnan, V., Swathika, O.G., Gutha, N.K. and Hency, V.B., 2022. Chapter 14: IoT-based smart health. In: O.V. Gnana Swathika, K. Karthikeyan and S.K. Padmanaban (Eds.), *Smart Buildings Digitalization: IoT and Energy Efficient Smart Buildings Architecture and Applications* (p. 149). CRC Press: Boca Raton, FL.

40. Chowdhury, S., Saha, K.D., Sarkar, C.M. and Swathika, O.G., 2022. IoT-based data collection platform for smart buildings. In: O.V. Gnana Swathika, K. Karthikeyan and S.K. Padmanaban (Eds.), *Smart Buildings Digitalization: IoT and Energy Efficient Smart Buildings Architecture and Applications* (pp. 71–79). CRC Press: Boca Raton, FL.

20 Application of Artificial Intelligence Techniques in Grid-Tied Photovoltaic System – An Overview

Y. Rekha and V. Jamuna
Jerusalem College of Engineering

I. William Christopher
Loyola-ICAM College of Engineering and Technology

T.V. Narmadha
St. Joseph's College of Engineering

CONTENTS

ABSTRACT

Currently, Artificial Intelligence (AI) usage in the grid-tied photovoltaic systems (GTPVS) emerge as a new surrogate approach compared with the traditional methods owing to their presence in solving complicated problems with accurate prediction and conception. AI learns on its own and solves non-linear problems with high speed, and these qualities open its role in various sectors like medicine, engineering, economics, etc. In the last two decades, the grid-tied PV conquers more installation capacity of PV systems compared with the stand-alone systems because of its merits like non-existent battery, low cost, and low maintenance. Power generated by the grid-tied PV lowers the presence of other sources and feeds the grid safely with newly developed technologies. In this chapter, the applications of several

DOI: 10.1201/9781003331117-20

AI techniques in grid-tied PV systems like PV panel reconfiguration, maximum power point tracking techniques under partial shading conditions, harmonics reduction, islanding, meteorological data, and optimal PV sizing are summarized. Additionally, a comparative evaluation is made considering the various aspects of AI like speed, complexity, tuning, real-time monitoring, and implementation. From this overview, it is envisaged that this study provides a vital reference and convenience for the researchers shortly for their research in the promising area of GTPVS.

20.1 INTRODUCTION

At present with the rising energy contingency, researchers have to investigate and focus their research on promising areas like renewable energy, controls, battery storage, and advanced computing techniques. Among these, the combination of grid-tied photovoltaic systems (GTPVS) from renewable energy and Artificial Intelligence (AI) from advanced computing techniques contributes more to the literature. According to the report given by Global Status of Renewables in 2015, the overall PV-installed capacity around the globe in 2004 is 3.7 GW, and the capacity reached 177 GW at the end of 2014 [1,2]. At present solar energy, conversion sector reached enormous growth with a total collective installed capacity of around 942 GW_{DC} in 2021, out of which 175 GW_{DC} achieved in 2021 alone. The graphical representation of annual PV-installed capacity around the globe in the last decade is given in Figure 20.1. In 2021, China leads the global PV market at 31.37% with an installed capacity of 54.9 GW. Beyond China, with 26.9 GW capacity and 15.37%, the United States of America ranked in the second position following the European countries with 15.31%. India grabs the fourth position with 7.42% and Japan with 3.71% occupies the fifth place [3,4]. Figure 20.2 represents the top ten countries in the global PV market for the year 2021. This rapid growth is possible because of several phenomena like increased efficiency, low cost, wide awareness, and appreciative government policies; this makes the photovoltaic system reach a permanent and constant position in the emerging energy sector.

Compared to the stand-alone system, the grid-connected system occupies 90%–99% of the total installed capacity of the PV system, which creates an alternative path to the conventional system. So, the installation of the PV system on the rooftop is more visible on the roadside. Due to the absence of batteries in the grid-tied system, the photovoltaic system directly feeds the generated power to the electric grid for further applications in the field of electric power systems. This makes the grid system a more economical and convenient one [5]. In general, the production of electricity

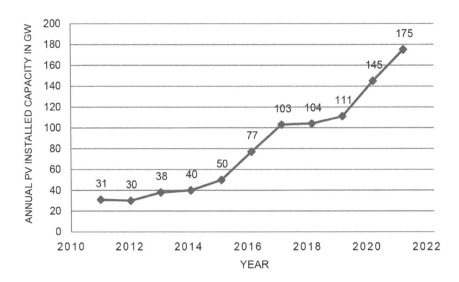

FIGURE 20.1   Annual PV-installed capacity. (Data from IEA PVPS [3].)

Top 10 countries in PV market -2021

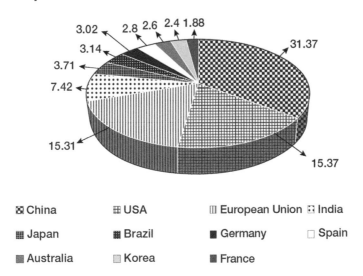

China USA European Union India

Japan Brazil Germany Spain

Australia Korea France

FIGURE 20.2 Top ten countries in 2021 PV market and their shares (in percentage). (Data from IEA PVPS [3].)

from a solar system is a limited one, and it always depends on various factors like temperature, irradiation, and the presence of the sun. So, energy demand arises whenever a deviation occurs in the supply. This could be rectified when a storage battery unit is introduced [6]. The inclusion of storage units increases the total cost and leads to a burden on the PV operators. The grid consumes its energy by the feed-in or self-consumption method. Subsidies are provided in the rural and housing areas to encourage the storage system depending on the battery and PV panel cost [7].

The battery energy concept is best applicable for interim storage only as it involves the process of energy charging, storing, and discharging. In addition, it suffers self-discharge, high losses, and a short life span. So, it is recommended to utilize the output energy from the PV system immediately instead of stowing it [8]. Compared with the stand-alone system, engender of energy for the grid-tied system is always from the photovoltaic array. An off-grid system with a battery backup comes into the picture when the generated energy is inadequate [9]. In general, power electronics play an efficient role in energy conversion with converters and inverters. Usually, the PV stations are amalgamated with the electrical grid system, and the generated energy is insufficient; hence, conventional converters with boosting capability are used with Maximum Power Point Tracking (MPPT) to catch the maximum peak point. Similarly, the inclusion of inverters and rectifiers creates power quality (PQ) issues and devours reactive power [10]. So, attention should be given while selecting an inverter for a grid as it is in charge of the quality of the power which is to be generated. The most important drawbacks of inverters are high switching loss, less efficiency, and harmonics these three components affect the system performance badly. Countries like the USA, Germany, and India frame some standards to handle the regulatory issues particularly flickers, anti-islanding, voltage, and current caused by the PV system, and care must be given to protect the grid system from these regulatory problems [11].

The artificial intelligent technique is one of the smart computing technologies, which arises as an alternate option for the traditional techniques. AI that dispenses complicated and indistinct problems very well. They are self-learners and execute the projection when trained. They acquire the knowledge from the information gained and the system comprehends artificial neural network (ANN), genetic algorithm (GA), fuzzy, data mining, expert systems, etc. Mostly, AI algorithms are complex and prediction is accurate within the required time frame, with colossal advantages and high

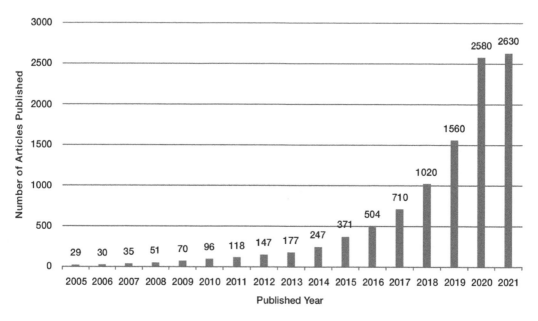

FIGURE 20.3   A graph showing the number of articles published under the topic "AI techniques in GTPVS", from the year 2005 to June 2022 (Data from Google Scholar.)

intelligence nature, AI is applied in various fields like autonomous vehicles, image classifications, AI vision, speech, and voice assistants like Siri, Alexa, Cortana, etc. In addition, the electrical field also benefited particularly in the area of power systems and power electronics [12], including design optimization of heat sink [13], load demand forecasting [14], controller for LED [15], cybersecurity control [16], solar radiation forecasting [17], reinforcement approach for microgrid (MG) [18], etc. Meantime, the evolution of data science and machine learning brings forth enormous information in the form of data which increases the probability of use of AI in the renewable sector. Compared with conventional methods, AI predicts problems smarter and executes the outputs accurately, particularly in the renewable energy system. Based on the literature review, this chapter is divided into two main aspects as role and application of AI techniques in grid-tied PV system and its comparative evaluation. During the time of online exploration in Google Scholar on the topic "Application of AI techniques in grid-tied PV systems", more reviews, overviews, and survey papers appeared. From 2005 to 2010, AI technology starts to boom steadily in areas like algorithms, design, control, and maintenance, particularly in power systems and power electronics engineering. Figure 20.3 shows the number of articles published under the topic "Application of AI techniques in grid-tied photovoltaic systems" from the year 2005 to June 2022 in the Google Scholar search engine. The importance of AI in GTPVS arises expeditiously in 2007, and this encourages the research community to show interest in this area of research.

This chapter is organized as follows. Section 20.1 is the introduction of this review. Section 20.2 provides a summary of AI and grid-tied PV systems. Section 20.3 discusses the applications of AI in grid-tied PV, and each subsection briefly discusses its pros and cons. A comparative evaluation of AI is discussed in Section 20.4. Lastly, Section 20.5 concludes the findings of this review.

20.2   SUMMARY OF AI AND GRID-TIED PV SYSTEM

AI methods play a significant role in the investigation, modelling, and forecasting of the performance of a PV system with grid connection because of its benefits like coherence with precise decision making, the ability to be used in perilous situations, and the computational power with data generation. The PV system's whole architecture for the grid connection which includes the PV system, MPPT, dc-to-dc converter, inverter, and the control techniques with the algorithms have to be functioned in synchrony to augment the whole efficiency of this entire system. The grid-tied PV system has two vital chores to be accomplished; the foremost one is chasing the peak power from the PV system with the support of an appropriate MPPT converter and the subsequent one is that the common coupling voltage must be sinusoidal and at the identical level of the grid voltage [19]. There are five classes of AI-based electrical systems available: optimization, data investigation, cataloging, regression, and gathering. The authors of Ref. [20] proposed linear programming-based optimization for sizing the PV arrays and the energy storage systems (ESS) with a model predictive control for intelligent PV systems. Optimization-based approaches are employed [21] for power system operation and control. The authors of Ref. [22] implemented a long short-term memory (LSTM) network for forecasting the PV system irradiance. For a secured power system, a data-driven method [23] is used to identify the faults. The condition monitoring of a PV module is realized [24] through accruing unlike failure circumstances to form a database and execute an instantaneous calculation with the accomplished database. The authors of Ref. [25] predicted the PV power using a hybrid technique that comprises the GA, particle swarm optimization (PSO), and an ANN. A multi-cluster algorithm [26] is proposed for the ESS sizing via seeing the ambiguity in the power system for optimization. An ordered spectral gathering procedure [27] is introduced and investigated for various grid interconnections. Predictably, circuits with numerical simulations with the PV array corresponding models are deliberated [28,29]. The authors of Ref. [30] presented a GA approach to realize the parameter identification for a duple-diode PV model. A flexible PSO technique is applied for the sole and duple-diode PV modules that result in a lower root mean square error (RMSE) than other techniques. An artificial bee colony (ABC) method is applied [31] for the sole and duple-diode models of the PV cell, which happens quickly and with more accuracy (lesser RMSE) compared with other techniques. Besides, the neural networks (NNs) [32,33] and the adaptive neuro-fuzzy inference system (ANFIS) [34] are extensively applied. A review of solar irradiance prediction presents [35] four diverse machine learning (ML) methods such as ANN, support vector machine (SVM), k-nearest neighbour (k-NN), and deep learning (DL). The DL methods and ramp-enhanced trees are applied [36] to predict solar irradiance directly. The authors of Ref. [37] improved the ability and the accuracy of solar irradiance prediction through the wavelet decomposition-based convolution LSTM networks. The Gaussian process (GP) regression [38,39] method is also applied to achieve the prediction of solar irradiance. The multilayer perceptron NN [40], Naïve Bayes method [41], k-NN [42], and multigene genetic programming [43] are also adopted for solar irradiance prediction.

The control of the inverter is used to regulate the power and frequency at the output side and lessen the harmonics in the grid-connected PV system. Compared with the Person Intelligence (PI) and Public Relation (PR)-based conventional procedures, the AI-based controller accurateness is higher. The anti-islanding protection scheme is categorized into active [44], passive [45], and hybrid [46] islanding detection is widely used in inverter control. The authors of Ref. [47] proposed a dual current controller, under faulty conditions and injecting the reactive power into the grid. The fuzzy logic control (FLC) [48] and computation-based methods [49] are also applied to increase the low-voltage ride-through ability of the inverter controller.

It is essential to achieve maximum power from the PV array at the input side of the inverter. To track the maximum operating point through FLC, the mission profile [50] and a GA [51] are used to optimize the NN controller.

The PV panels may be affected by faults, such as short circuits due to bypass diode, delamination, discoloration, cell crack, snail trail and glass crack, etc. For PV system fault analysis, the overseen learning-based random forest methods [52] are applied. Based on the probabilistic NN (PNN) [53], a swarm intelligence-based ABC method [54] is applied for the PV system fault investigation. The authors of Ref. [55] discussed a DL-based fault cataloging method for PV arrays applying the convolutional neural network. ANN-based fault identification for multilevel H-bridge inverters

[56] is executed. A supervised learning-based PNN is proposed [57] for fault investigation in the diode-clamped multilevel inverters (MLIs). As a result of the ride-through operation, an AI-based lifetime estimation [58] can combine the abrupt fault in the analysis or rapid change in the stress of the components. The authors of Refs. [59,60] proposed that the lifetime estimation is executed using an ANN to analyse the diverse functioning conditions. However, in Ref. [61] a function-based relationship is established between the reliability and design parameters. Henceforth, there exist ample research articles using various AI methods for the grid-connected PV system for better operation with an amended efficiency.

20.3 APPLICATION AND ROLE OF AI TECHNIQUES IN GRID-TIED PV SYSTEMS

In grid-tied PV systems, AI algorithms played a major facet in designing, modelling, sizing, and reconfiguring the photovoltaic parameters. In PV systems, the AI application is subdivided into six categories, which are described in the subsections in detail.

20.3.1 PV PANEL ARRAY RECONFIGURATION

In PV systems, the panel array reconfiguration is an efficient solution to the partial shading issue which reduces the negative power skilfully. Currently, the following array configurations are utilized in all shading schemes: series, parallel, a combination of series-parallel, Honey Comb (HC), Total-Cross-Tied (TCT), and bridge-linked. In Ref. [62], array reconfiguration techniques, such as static and dynamic, are discussed briefly with various control algorithms. Both techniques enhance the array output with their advantage and disadvantage. In static reconfiguration, a novel 8 Queen's (8-Q's) technique is reconfigured in a rectangular-shaped PV array with a TCT configuration. This technique is established in a chessboard and the position of the eight queens is arranged so that they don't charge each other. This method has an $m \times m$ arrangement on the chessboard and obtained solutions for natural numbers only [63]. Current Injection-Dynamic Array Reconfiguration is one of the recent dynamic reconfiguration methods which neglects the occurrence of numerous peak values in the array of the P–V characteristic curl [64]. Further, in Ref. [65], the article explores the different array configurations for a PV model and approaches to diminish the shade effects. The configurations are grouped based on their features like efficiency and dispersion. In Ref. [66], a GA-based procedure is proposed to rid the PV array connection problems, and to improve the processing speed of reconfiguration to obtain maximum power for various mismatching. Later, the algorithm is compared with the Brute force technique, to rid the long-time execution problems. In Ref. [67], precise estimation of irradiance is obtained by the combination of FLC with Recursive Least Square (RLS)-based estimator for different shading. Experimental implementation of the setup is done for residential purposes with a TCT configuration. Compared with the PV voltage and current estimator, the RLS estimator decreases the error by 10% and 4.28%, respectively. In Ref. [68], the Optimal Mileage Array Reconfiguration is implemented with upper and lower-layer optimization reconfiguration where the upper layer is coped with Swarm Reinforcement Learning (SRL) and the lower with the interior point approach. In Ref. [69], the optimal power of the Photovoltaic Generator (PVG) is extracted by using a Fuzzy Logic (FL) algorithm, and this reconfiguration combines irradiance equalization and categorizes the procedure to produce a vital curve with low computing time. Here, PVG is reconfigured with one of the mitigation methods, and a switchover is made in the connections between the PV panels. Similarly, in Ref. [70], the PSO technique is applied to relocate the panels and reduce the difference in row current values. An electrical array reconfiguration method is explored, and for the optimization, a large memory with computing capability is needed. In addition, intense tuning of parameters value will result in an optimum output. In Ref. [71], a novel neuro-fuzzy (N-F) algorithm is implemented to neglect the effects produced due to partial shading and to improve the array performance with low-cost initialization. With a dynamic PV array and global measuring process, this method decreases the possibilities of having more sensors, more processing

TABLE 20.1

AI Techniques for Solar PV Array Reconfiguration

Algorithm & Ref. No.	Year	Author	Purpose	Advantage	Disadvantage
GA [66]	2022	Mariana Durango et al.	Obtain a maximum power output	Low power losses, improved array lifetime	Loss of energy, load side fault
FL [67]	2021	Loubna Bouselham et al.	Dynamic array installation cost is reduced	Improve the array output power	Applicable for small-scale PV applications
SRL [68]	2021	Xiaoshun Zhang et al.	Minimize the mileage payment. Maximize the PV output	Less computational time. Reduced optimization difficulty	Increase the peak(max) value in V–P curve
FL [69]	2019	Loubna Bouselham et al.	Achieve uniformity in shade dispersion	Mismatch loss is low. Fast response	Difficult to implement the controller logic (rules)in large-scale arrays
PSO [70]	2018	Babu et al.	Obtain high power by uniform shading	Switch at once. The complex is less for a large-size array	Require fine-tuning
GA [83]	2015	Deshkar et al.	The increased power is achieved by equalizing the row current values	Accurate selection of switching for reconfiguration	Need a long search and more steps to compute
ANN [84]	2014	Mehmet Karakose et al.	Achieve maximum power output from wide search space	Fastly gives efficient output during hardware implementation	Need trained data

equipment, and high computational time. The N-F algorithm proved its effectiveness in finding the main reason behind the power reduction and also in the placement of shaded modules. Further, in Ref. [72], an Ant Colony Optimization (ACO) approach is discussed for various scenarios such as short, wide, narrow, and long shading. In all shadings, the V–P and I–V curves are drawn to validate output, which confirms that the reconfiguration based on the ACO approach produced maximum power compared to the TCT configuration. In Table 20.1, a brief detail of various AI techniques for solar PV array reconfiguration is presented.

20.3.2 ISLANDING DETECTION

On the distribution side, the distributed generators (DGs) are interconnected to the system, where there is a possibility of islanding issues. Islanding indicates a condition that when a DG power an area, the grid electrical power is faded [73]. So, the distribution system needs a protection setup through which the abnormal conditions are identified with utmost concern and protects the system by disconnecting the utility. Islanding protection setup is recommended and they are categorized as local and remote. Local methods are further divided into active, passive, and hybrid [74,75]. In local and passive methods, detection of islanding is done by measuring and monitoring the parameters such as frequency, voltage, phase, and current. But, in the active method, the system observes an external disturbance to detect the parameter abnormality. Due to the drawbacks of the above methods, the hybrid detection method is proposed where the abnormality is identified by the threshold

TABLE 20.2

AI Techniques for Islanding Detection

Algorithm & Ref. No.	Year	Author	Advantage	Disadvantage
NN [76]	2021	Khan et al	Applicable to multi-inverter grid-connected DG	Output is dependable
ANN [78]	2020	Ananda Kumar et al.	98% of accuracy is achieved Unaware of grid disturbance	Need fine-tuning
GWO-ANN [79]	2020	Samuel Admaise et al.	Perfect for multiple Distributed Energy Resource's (DER's) island detections	Hardware-in-the-loop exploration is needed
ANFIS [80]	2018	Mlakic et al.	Handle multiple DG systems Less detection time	Need large database
ANN [81]	2016	Merlin et al.	Efficiency is high accurate	Processing time is high
PNN [82]	2015	Khamis et al.	Reliable layers are free from learning	

values; meanwhile, the existence of abnormality is verified by adding a perturbation to the system. In the conventional methods, the process of identification of the abnormality is gradual and imprecise, and an alternate option of optimization methods is recommended to overcome the drawbacks [76]. So, an AI technique is preferred, which analyzes the input data and builds its database with all possibilities. Here, the abnormality and accuracy are improved by pre-processing the signal input, which drastically improves the recognition proficiency [77]. In Table 20.2, a detailed outlook of AI techniques for islanding detection is discussed. In Ref. [76], a new technique is proposed by modifying the signal processing unit, where it trains the possible islanding events by extracting, classifying, and processing them.

The new classifier detects islanding in a short period (<0.2 seconds) and achieves its output for 98% of the training data and 97.8% efficiency. Also, in Ref. [78], a Tunable q-factor Wavelet Transform (TqWT) with the ANN technique is proposed, to disconnect the distributed power generation system away from the power grid. During the initial process, parameters are detected, and computing is carried out for various possible events at the grid. The TqWT decreases the signals and the classification of islanding is achieved by the ANN classifier. Through this method, 98% of accuracy is attained; hence, this method is able to provide a good insight into detecting the island issue. In Ref. [79], a Grey Wolf Optimization (GWO) algorithm based on ANN (GWO-ANN) is discussed with an Intelligent-based Islanding Detection Method. The pre-processing is done for each Intrinsic Mode Function (IMF) using the Hilbert transform. The IMF parameters are trained to test the ANN model for islanding events. The testing and training datasets have islanding conditions like sag, various faults, non-linear, loads, etc. Further in Ref. [80], an ANFIS-based islanding detection technique is suggested for the low-voltage inverter-interfaced grids. ANFIS exploits its recognition ability and the mapping relation with the inputs. In Ref. [81], distributed synchronous generators identify their island events by ANN classifier. Here, ANN identifies and samples the waveform measured at DGs. In Ref. [82], an effective passive islanding method called PNN is proposed, which perturbs and extracts the phase voltage in the DG terminal. The revealing features of the Phase Space (PS) technique obtain the initial characteristics and later examine the dimensional space. The PNN with PS technique identifies the islanding and disturbances condition accurately.

20.3.3 Harmonics Reduction

In general, power system networks experience harmonics owing to the presence of unavoidable non-linear loads which distorted the sine waveform nature drastically. Similarly, in a grid-tied PV, the inverter stage next to the converter regulates the ac outputs like the voltage, current, power, and frequency with less harmonic content. So, conventional methods like pulse width modulations, space vector modulation, etc., are used to control the inverters by controlling their switching pulse [85]. These methods have drawbacks of their own, so there is a need for intelligent computing technology which opens the way for AI. While executing their targets, intelligence algorithms do their part in controlling the inverter. Further, in Refs. [86,87], a combined overview of AI methods with the conventional techniques is proposed, and various PQ issues like monitoring, compensation, harmonics reduction, conditioning, storage, and regulation are analysed; a statistical evaluation is made with pros and cons. Particularly, harmonics elimination in the power system is reviewed elaborately with various analyses like origin and location of harmonics, spectrum examination, harmonics clustering, and its elimination. Various techniques like fuzzy, ANN, decision tree, and SVM contribute to most of the papers in the review from the harmonic section alone. Most of the AI methods surpass the traditional methods but there is a need for an upgrade in the area of learning and training of the AI algorithm, application of hybrid methods, and optimum prediction. In Ref. [88], an Adaptive Fuzzy Neural Network (AFNN) algorithm is proposed to produce an efficient operation of a smart grid for various load and supply conditions. In addition, the PQ issue occurring in the MG is also addressed, and the improvement is done through hybrid filters.

In Ref. [89], a control approach based on an ANN is presented, which draws its utmost utilization from the grid by interfacing the MLI for a solar-based DG system. The ANN-based control algorithm is trained first, according to the switching angle of the inverter. The trained ANN with an active filter is merged with the distributed generation system to form a multitasking network. In Refs. [90,91], a FL control is proposed for a Z-source and a Quasi-Z-source-cascaded MLI. In general, the FL is used in the inverter circuit to decrease the total harmonic distortion value to a minimum extent and to boost the output dc voltage for the solar PV application. Further, in Refs. [92,93], an ANN technique is suggested for a single-stage and an H-bridge inverter. While controlling the ANN, models are trained and use the library for writing the program. An ANN library is very fast and uses backpropagation models. In Ref. [94], adaptive network-based fuzzy or adaptive neuro-fuzzy with an inference system is discussed and interfaced with the MLI for the grid-connected applications. This model combines the NN with FL rules and forms an adaptive circuit network. In Ref. [95], a PSO-based algorithm is proposed to optimize the fuzzy rules in a PV inverter connected to an electrical grid. The MPPT with fuzzy rules enhances the bus voltage of the inverter and reduces the Total Harmonic Distortion (THD) level compared to an MPPT with a fixed step. In Table 20.3, a detailed outlook of AI techniques for harmonics reduction is discussed.

20.3.4 Meteorological Data

In a grid-tied PV system, the integration of renewable energy sources will lead to uncertainty in the economy and operation of the power system. This intensifies some factors like fluctuation, harmonics, instability, etc. [96]. The nature of renewable sources is not a fixed one, which increases the importance of collecting, documenting, and processing the data from time to time. So, the meteorological or climatological data occupy a dominant position in the GTPVS. The data like an estimation of solar radiation, prediction of temperature, sun shining duration, and weather forecasting are acquired as varying parameters of GTPVS.

The traditional methods like ARMA and Markov Chain suffer from problems like missing data, being incapable of predicting the data in a particular location, incompetent to model, and forecasting data for longer period. Hence, AI-based techniques are applied for meteorological data collection and estimation [97]. In Ref. [98], one of the techniques of ML called regression is proposed with various types of NN, Support Vector Regression (SVR), and GPs. The study is experimented by training the regression with one year data of world solar radiation measured hourly. The variables like cloud index, solar radiation of clear sky, and reflectivity are observed. Further, in Ref. [99], the power forecasting of the PV system is examined by a technique called Extreme Learning Machine (ELM), where the model uses Incremental Conductance (IC) MPPT. The network is trained by

TABLE 20.3
AI Techniques for Harmonics Reduction

Algorithm & Ref. No.	Year	Author	Inverter Type	Objective
AFNN [88]	2021	Sowmiya Das et al.	Grid-connected inverter	AFNN is fast and time-convergent, so used to control the smart grid
ANN [89]	2020	Rajkumar et al	15-level cascaded H-bridge MLI	Interfacing inverters are used to append the power from the renewable source
FLC [90]	2016	Iniyaval et al.	Quasi-Z-source-cascaded MLI	Using FLC, voltage boosting is unified in a single inverter (single-stage)
ANN [94]	2015	Guellal et al.	H-bridge inverter	Control the fundamental wave by eliminating the unneeded harmonics
ANFIS [94]	2015	Karuppusamy et al.	Multilevel inverter	Absence of optimal PWM switching generator
ANN [93]	2014	Demirtas et al.	Single-stage inverter	Able to learn and make decisions on switching concerning ANN
FLC [91]	2013	Balachandran et al.	Z-source-cascaded MLI	Provide high accuracy. More fuzzy rules are used
PSO [95]	2012	Letting et al.	Grid-connected PV inverter	The design of the controller is simple. The simulation speed is fast

TABLE 20.4
AI Techniques for Meteorological Data

Algorithm & Ref. No.	Year	Author	Subject/Area Related to PV
Machine Learning Regression [98]	2019	Cornejo Buenoa et al.	Estimation of solar radiation
Modified PSO [99]	2018	Manoja Kumar et al.	Forecasting of power
GA [100]	2010	Zagroub et al.	Solar energy system
ANFIS [101]	2009	Moghaddamnia et al.	Solar radiation prediction
ANFIS [102]	2008	Chaabene et al.	Solar radiation prediction
ANN [103]	2010	Mellit et al.	Solar radiation prediction
ANN [104]	2009	Almonacid et al.	Si-crystalline PV module characteristics
ANN [105]	2009	Benghanem et al.	Si-crystalline PV module characteristics
FLC [106]	2008	Atlas et al.	Photovoltaic solar energy system
FLC [107]	2008	Paulescu et al.	Photovoltaic solar energy system

ELM, and its weights are upgraded by PSO techniques. In Ref. [100], a GA is suggested to know the parameters of a PV system such as cells, modules, and arrays. While formulating, GA formulates the optimization as a non-convex problem. Through this, the parameters (electrical) values of the PV system are obtained and they are more efficient. In Refs. [101,102], the ANFIS technique is proposed and five variables are utilized for the estimation of per day solar radiation. It includes maximum and minimum temperature, radiation, and forecasting. From Refs. [103–105], a Multilayer Perception (MLP) network is proposed to fore caste the solar radiation of a day. The mean value of daily irradiance and temperature is used as the source parameters of the PV system. In Refs. [106,107], a FL MPP algorithm is proposed, and the search for the maximum power points (MPPs), for various meteorological conditions are persist in the PV system. For controller design, Sugeno FIS is utilized. In Table 20.4, a detailed outlook of AI techniques for meteorological data is discussed.

20.3.5 MPPT during Partial Shading

Usually, the maximum output power is drawn from the PV system by placing an MPPT controller on the inverter side of the conversion process. IC and Perturb and Observe (P&O) are the two commonly used methods in the MPPT algorithm. Due to the drawbacks of these traditional methods, AI-based techniques are recommended, and the optimization and regulation are done with a variation. In Table 20.5, a detailed outlook of AI techniques for MPPT is discussed. In Ref. [108], the author briefed on the Pigeon-Inspired Optimization (PIO) technique for the PV system. The Pigeon-inspired technique improves the searching ability and obtains maximum power while tracking the solar system. This improves the efficiency and success outlay of the tracking system. In Ref. [109], a combined MPPT structure of Angle Incremental Conductance (AIC) and a type-2 TSK FLC is proposed, and the algorithm of AIC produces the error function value for the MPPT and defines the variables with a finite value. In Ref. [110], the author reviewed the ANN-based MPPT in the PV system mentioning its merits gathered from the articles over 6 years. He discussed the ANN-based MPPT and a hybrid structure of ANN with FLC. These algorithms produce an output of 98% with a high convergence speed. The oscillations due to harmonics are lesser in the MPP. Further, a P&O-PSO is suggested to excel the generated power of the PV system in all partial shading conditions of irradiance. In Ref. [111], the author explained an optimization technique Harris Hawk optimization (HHO) which tracks the maximum power for all climatic conditions. This method tracks the MPP with a quick convergence rate, which improves the oscillation reduction with low computation.

In Ref. [112], a hybrid combination of algorithms such as ANFIS with ABC is explained. Improved fuzzy rules are implemented to generate the signal and power the inverter. The adaptive (ANFIS) membership function (MF) with the bee algorithm is novel, in the field of hybrid MPPT controlling systems. Similarly, in Refs. [113], memetic reinforcement learning (MRL) and power-normalized Kernel Least Mean Fourth-based neural network (PNKLMF) techniques are discussed. In MRL, a memetic structure is included in the learning part, to magnify its searching capability. In PNKLMF-NN, a Hill Climbing (HC) algorithm is proposed and the time of load component extraction NN controls well and improves the PQ. This method meets the achievement of attaining the needed active power along with the chance for feeding the excess power into the electric grid.

TABLE 20.5
AI Techniques for MPPT

Algorithm and Ref. No.	Year	Author	Converter Type	Objective
PIO [108]	2022	Zhuoli Zhao et al.	Boost	PIO MPPT deals with partial shading conditions
FLC [109]	2020	Kececioglu et al.	Boost	AIC +IT2-TSK-FLC is proposed for MPPT
ANN-FLC [110]	2020	Cesar Villegas et al.	Buck	An efficient algorithm for an effective MPPT controller
P&O-PSO			Buck-Boost	Maximize the generated PV power in the same PCS
HHO [111]	2020	Majad Mansoor et al.	Boost	Track the maximum power in all climatic conditions
ANFIS-ABC [112]	2019	Sanjeevi Kumar et al.	Sepic	Minimize the root mean square error
MRL [113]	2019	Xiaoshun Zhang et al.	Boost	For the PV system under partial shading conditions
PNKLMF			Three-phase Voltage Source Converter (VSC)	To cope with the power need

TABLE 20.6
AI Techniques for Optimum PV Sizing

Algorithm & Ref. No.	Year	Author	Parameters Predicted
PSO-GA [115]	2021	Imene Khenissi et al.	Proper sizing of grid-connected PV
PSO [116]	2021	Kefale et al.	PV installation is done with proper size and location
GA [118]	2019	Kashyap et al.	Optimal placement of DGs
BAT [119]	2015	Othman et al.	Sizing of PV arrays
ANFIS+GA [120]	2014	Sanusi et al.	Solar cell count and battery size
PSO [117]	2010	Kornelakis et al.	Optimum installation of PV modules

20.3.6 OPTIMAL PV SIZING

At present, there is a necessity for finding the optimum values of PV systems, which assure the quality and quantity of the power supply. Estimating the component values of panels, batteries, and inverters through conventional methods is complicated, as it needs huge data and also and produces inaccurate results. In Ref. [114], the author reviewed the various techniques of AI in PV sizing with their merits and also briefed that these methods provide an optimum solution from the isolated location where the possibility of getting the weather report is difficult. Nowadays, AI techniques conquer the position by which it buried the difficulties in getting the data for sizing the PV systems even in isolated areas. In Table 20.6, a detailed outlook of AI techniques for optimum PV sizing is discussed. In Refs. [115,116,117], a PSO-based method is suggested for PV installation, which associates the array and module details like location, size tilt angle, etc. In Ref. [118], a GA is proposed to note the exact and efficient location of DGs in the electrical distribution system. While implementing, the system power loss is considered to be minimum, which improves the system voltage. In Ref. [119], Bat algorithm is proposed for GTPVS sizing optimized bat algorithm is proposed. It searches the PV module and the inverter from its database, and sizing of the array is also done considering its limits. In Ref. [120], NN is suggested to predict the parameters of the PV according to the location and geographical situations.

20.4 COMPARATIVE EVALUATION OF AI

In the previous section, various AI techniques with their objective, merits, and demerits are discussed for the grid-tied PV system in various aspects. Compared with the conventional methods, AI reduces the harmonics and oscillations to a low value and gave a reliable output for sudden parameter disturbances. Moreover, a comparative evaluation is done with the AI algorithms in terms of speed, complexity, tuning, real-time monitoring, and implementation. Table 20.7 briefs these features in detail.

20.4.1 SPEED

While compared with the conventional methods, the speed of several AI techniques is more. Some algorithms need more levels of iterations, which makes the speed a little low.

20.4.2 SYSTEM COMPLEX

Generally, the system complexity is dependent upon the algorithm complexity level used for the system. Moreover, AI algorithms are complex in nature, particularly GA and hybrid combinations of AI algorithms as mentioned in Table 20.7.

TABLE 20.7
AI Techniques Comparative Evaluation

Algorithm	Speed	Complex	Tuning	Monitoring	Implementation
GA	Fast	Complex	×	Easy	Yes
FL	Fast	Average	√	Easy	Yes
NN	Average	Average	√	Average	Difficult
PSO	Fast	Average	×	Average	Average
ANFIS	Fast	Average	×	Average	Yes
ACO	Average	Average	×	Average	Average
ANN	Average	Average	√	Average	Difficult
Hybrid	Fast	Complex	×	Difficult	Difficult

20.4.3 TUNING

The periodic tuning of the AI algorithm is based on the set of values close to the global optimum maximum value. Tuning is possible when there is a possibility of an MF of fuzzy rules and training data.

20.4.4 MONITORING AND IMPLEMENTATION

The real-time monitoring and hardware implementation details of possible AI techniques are listed in Table 20.7. The high-end processors are difficult to operate, as this algorithm needs more space, especially in ANN and hybrid combinations.

20.5 CONCLUSION

From this overview, it is concluded that enormous number of papers are published on the topic "Application of AI techniques in grid-tied PV systems" till date. In solar PV systems, the role of AI particularly ANN, PSO, GA, and optimization techniques are more likely used by the researchers for solving the issues which were discussed briefly. With the performance output, it is difficult to generalize the particular models for particular applications. In the future, advanced research has to be conducted, to improve the drawbacks like inconsistency, privacy and security of data, algorithm transparency, and its implementation.

REFERENCES

1. Sawin Janet, L.; Freyr Sverrisson, W.R. Renewables 2015 Global Status Report. France, 2015.
2. Masson, G.; Alice, D.; Kaizuka, I.; Jäger-Waldau, A. IEA-PVPS TCP; 2021; PVPS Report: Snapshot of Global PV Markets 2021 Task 1 Strategic PV Analysis and Outreach PVPS, 2021.
3. Masson, G.; Kaizuka, I.; Jäger-Waldau, A.; Donoso, J. IEA PVPS; 2022; Task 1 Strategic PV Analysis and Outreach Report IEA- PVPS T1-42:2022, 2022. ISBN 978-3-907281-31-4.
4. Murdock, H.E.; Gibb, D.; André, T.; Sawin, J.L.; Brown, A.; Appavou, F.; Ellis, G.; Epp, B.; Guerra, F.; Joubert, F. Renewables 2020-Global Status Report; REN21 Secretariat: Paris, France, 2020. *00-72020*.
5. Kouro, S.; Leon, J.I.; Vinnikov, D.; Franquelo, L.G. "Grid-connected photovoltaic systems: An overview of recent research and emerging PV converter technology," *IEEE Industrial Electronics Magazine*, vol. 9, no. 1, pp. 47–61, 2015.
6. Bortolini, M.; Gamberi, M.; Graziani, A. "Technical and economic design of photovoltaic and battery energy storage system," *Energy Conversion Management*, vol. 86, pp. 81–92, 2014.
7. Hoppmann, J.; Volland, J.; Schmidt, T.S.; Hoffmann, V.H. "The economic viability of battery storage for residential solar photovoltaic systems: A review and a simulation model," *Renewable and Sustainable Energy Reviews*, vol. 39, pp. 1101–1118, 2014.

8. Zeng, Z.; Yang, H.; Zhao, R.; Cheng, C. "Topologies and control strategies of multifunctional grid-connected inverters for power quality enhancement: A comprehensive review," *Renewable and Sustainable Energy Reviews*, vol. 24, pp. 223–270, 2013.

9. Boumaaraf, H.; Talha, A.; Bouhali, O. "A three-phase NPC grid-connected inverter for photovoltaic applications using neural network MPPT," *Renewable and Sustainable Energy Reviews*, vol. 49, pp. 1171–1179, 2015.

10. Chakraborty, A. "Advancements in power electronics and drives in interface with growing renewable energy resources," *Renewable Sustainable Energy Reviews*, vol. 15, pp. 1816–1827, 2011.

11. Magal, A.; Engelmeier, T.; Mathew, G.; Gambhir, A.; Dixit, S.; Kulkarni, A. "Grid integration of in India distributed solar photovoltaics (PV) in India distributed solar," 2014.

12. Zhao, S.; Blaabjerg, F.; Wang, H. "An overview of artificial intelligence applications for power electronics," *IEEE Transactions on Power Electronics*, vol. 36, pp. 4633–4658, 2021.

13. Wu, T.; Wang, Z.; Ozpineci, B.; Chinthavali, M.; Campbell, S. "Automated heatsink optimization for air-cooled power semiconductor modules," *IEEE Transactions on Power Electronics*, vol. 34, no. 6, pp. 5027–5031, 2019.

14. Raza, M.Q.; Khosravi, A. "A review on artificial intelligence-based load demand forecasting techniques for smart grid and building," *Renewable and Sustainable Energy Reviews*, vol. 50, pp. 1352–1372, 2015.

15. Zhan, X.; Wang, W.; Chung, H. "A neural-network-based color control method for multi-color LED systems," *IEEE Transactions on Power Electronics*, vol. 34, no. 8, pp. 7900–7913, 2019.

16. Sahoo, S.; Dragicevic, T.; Blaabjerg, F. "Cyber security in control of grid-tied power electronic converters–challenges and vulnerabilities," *IEEE Journal of Emerging and Selected Topics in Power Electronics*, vol. 9, no: 5, pp. 5326–5340, 2021.

17. Cyril, V.; Gilles, N.; Soteris, K.; Marie-Laure, N.; Christophe, P.; Fabrice, M.; Alexis, F.. "Machine learning methods for solar radiation forecasting: A review," *Renewable Energy*, vol. 105, pp. 569–82, 2017.

18. Elsayed, M.; Erol-Kantarci, M.; Kantarci, B.; Wu, L.; Li, J. "Low-latency communications for community resilience microgrids: A reinforcement learning approach." *IEEE Transactions on Smart Grid*, vol. 11, no. 2, pp. 1091–1099, 2019.

19. Kumar, A.; Gupta, N.; Gupta, V. "A comprehensive review on grid-tied solar photovoltaic system," *Journal of Green Engineering*, vol. 7, pp. 213–254, 2017.

20. Saez-de-Ibarra, A.; Milo, A.; Gaztanaga, H.; Debusschere, V.; Bacha, S. "Co-optimization of storage system sizing and control strategy for intelligent photovoltaic power plants market integration," *IEEE Transactions on Sustainable Energy*, vol. 7, pp. 1749–1761, 2016.

21. Molzahn, D.K.; Dörfler, F.; Sandberg, H.; Low, S.H.; Chakrabarti, S.; Baldick, R.; Lavaei, J. "A survey of distributed optimization and control algorithms for electric power systems. *IEEE Transactions on Smart Grid*, vol. 8, pp. 2941–2962, 2017.

22. Liu, C.-H.; Gu, J.-C.; Yang, M.T. "A simplified LSTM neural networks for one day-ahead solar power forecasting," *IEEE Access*, vol. 9, pp. 17174–17195, 2021.

23. Wang, Z.; Chen, Y.; Liu, F.; Xia, Y.; Zhang, X. "Power system security under false data injection attacks with exploitation and exploration based on reinforcement learning," *IEEE Access*, vol. 6, pp. 48785–48796, 2018.

24. Haque, A.; Bharath, K.V.S.; Khan, M.A.; Khan, I.; Jaffery, Z. A. "Fault diagnosis of photovoltaic modules," *Energy Science & Engineering*, vol. 7, pp. 622–644, 2019.

25. Semero, Y.K.; Zhang, J.; Zheng, D. North China electric power university PV power forecasting using an integrated GA-PSOANFIS approach and Gaussian process regression based feature selection strategy. *CSEE Journal of Power and Energy Systems* vol. 4, pp. 210–218, 2018.

26. Yao, C.; Chen, M.; Hong, Y.-Y. "Novel adaptive multi-clustering algorithm-based optimal ESS sizing in ship power system considering uncertainty," *IEEE Transactions on Power Systems*, vol. 33, pp. 307–316, 2018.

27. Sanchez-Garcia, R.; Fennelly, M.; Norris, S.; Wright, N.; Niblo, G.; Brodzki, J.; Bialek, J.W. "Hierarchical spectral clustering of power grids," *IEEE Transactions on Power Systems*, vol. 29, pp. 2229–2237, 2014.

28. Rodrigues, E.M.G.; Godina, R.; Marzband, M.; Pouresmaeil, E. "Simulation and comparison of mathematical models of PV cells with growing levels of complexity," *Energies*, vol. 11, p. 2902, 2018.

29. Kumar, R.; Singh, S.; Rodrigues, E.M.G. "Solar photovoltaic modeling and simulation: As a renewable energy solution," *Energy Reports*, vol. 4, pp. 701–712, 2018.

30. Petcut, F.M.; Leonida-Dragomir, T. "Solar cell parameter identification using genetic algorithms," *Control Engineering and Applied Informatics*, vol. 12, pp. 30–37, 2010.

31. Ebrahimi, S.M.; Salahshour, E.; Malekzadeh, M.; Gordillo, F. "Parameters identification of PV solar cells and modules using flexible particle swarm optimization algorithm," *Energy*, vol. 179, pp. 358–372, 2019.

32. Tong, N.T.; Pora, W. "A parameter extraction technique exploiting intrinsic properties of solar cells," *Applied Energy*, vol. 176, pp. 104–115, 2016.

33. Wang, R.; Zhan, Y.; Zhou, H. "Application of artificial bee colony in model parameter identification of solar cells," *Energies*, vol. 8, pp. 7563–7581, 2015.

34. Durrani, S.P.; Balluff, S.; Wurzer, L.; Krauter, S. "Photovoltaic yield prediction using an irradiance forecast model based on multiple neural networks," *Journal of Modern Power Systems and Clean Energy*, vol. 6, pp. 255–267, 2018.

35. Mayer, M.J.; Gróf, G. "Extensive comparison of physical models for photovoltaic power forecasting," *Applied Energy*, vol. 283, p. 116239, 2021.

36. Vatankhah, R.; Ghanatian, M. "Artificial neural networks and adaptive neuro-fuzzy inference systems for parameter identification of dynamic systems," *Journal of Intelligent Fuzzy System*, vol. 39, pp. 6145–6155, 2020.

37. Husein, M.; Chung, I.-Y. "Day-ahead solar irradiance forecasting for microgrids using a long short-term memory recurrent neural network: A deep learning approach," *Energies*, vol. 12, p. 1856, 2019.

38. Wang, F.; Yu, Y.; Zhang, Z.; Li, J.; Zhen, Z.; Li, K. "Wavelet decomposition and convolutional LSTM networks based improved deep learning model for solar irradiance forecasting," *Applied Science*, vol. 8, p. 1286, 2018.

39. Lubbe, F.; Maritz, J.; Harms, T. "Evaluating the potential of Gaussian process regression for solar radiation forecasting: A case study," *Energies*, vol. 13, p. 5509, 2020.

40. Tolba, H.; Dkhili, N.; Nou, J.; Eynard, J.; Thil, S.; Grieu, S. "Multi-horizon forecasting of global horizontal irradiance using online Gaussian process regression: A Kernel study," *Energies*, vol. 13, p. 4184, 2020.

41. De Paiva, G.M.; Pimentel, S.P.; Alvarenga, B.P.; Marra, E.; Mussetta, M.; Leva, S. "Multiple site intra-day solar irradiance forecasting by machine learning algorithms: MGGP and MLP neural networks," *Energies*, vol. 13, p. 3005, 2020.

42. Kwon, Y.; Kwasinski, A.; Kwasinski, A. "Solar irradiance forecast using Naïve Bayes classifier based on publicly available weather forecasting variables," *Energies*, vol. 12, p. 1529, 2019.

43. Chen, C.-R.; Kartini, U.T. "k-Nearest Neighbor neural network models for very short- term global solar irradiance forecasting based on meteorological data," *Energies*, vol. 10, p. 186, 2017.

44. Hosseinzadeh, M.; Salmasi, F.R.R. "Islanding fault detection in microgrids: A the survey," *Energies*, vol. 13, p. 3479, 2020.

45. Abyaz, A.; Panahi, H.; Zamani, R.; Alhelou, H.H.; Siano, P.; Shafie-Khah, M.; Parente, M. "An effective passive islanding detection algorithm for distributed generations," *Energies*, vol. 12, p. 3160, 2019.

46. Bakhshi-Jafarabadi, R.; Popov, M. "Hybrid islanding detection method of photovoltaic- based microgrid using reference current disturbance," *Energies*, vol. 14, p. 1390, 2021.

47. Merabet, A.L. "Control system for dual-mode operation of grid-tied photovoltaic and wind energy conversion systems with active and reactive power injection", Saint Mary's University, Halifax, NS, Canada, 2017.

48. Al-Shetwi, A.Q.; Sujod, M.Z.; Blaabjerg, F. "Low voltage ride-through capability control for single-stage inverter-based grid connected photovoltaic power plant," *Solar Energy*, vol. 159, pp. 665–681, 2018.

49. Al-Shetwi, A.Q.; Sujod, M.Z.; Blaabjerg, F. "Low voltage ride-through capability control for single-stage inverter-based grid connected photovoltaic power plant," *Solar Energy*, vol. 159, pp. 665–681, 2018.

50. Saad, N.H.; El-Sattar, A.A.; Mansour, A.E.-A.M. "Improved particle swarm optimization for photovoltaic system connected to the grid with low voltage ride through capability," *Renewable Energy*, vol. 85, pp. 181–194, 2016.

51. Kececioglu, O.F.; Gani, A.; Sekkeli, M. "Design and hardware implementation based on hybrid structure for MPPT of PV system using an interval Type-2 TSK Fuzzy logic controller," *Energies*, vol. 13, p. 1842, 2020.

52. Harrag, A.E.; Messalti, S. "Variable step size modified P&O MPPT algorithm using GA- based hybrid offline/online PID controller," *Renewable Sustainable Energy Reviews*, vol. 49, pp. 1247–1260, 2015.

53. Chen, Z.; Han, F.; Wu, L.; Yu, J.; Cheng, S.; Lin, P.; Chen, H. "Random forest based intelligent fault diagnosis for PV arrays using array voltage and string currents," *Energy Conversion and Management*, vol. 178, pp. 250–264, 2018.

54. Akram, M.N.; Lotfifard, S. "Modeling and health monitoring of DC side of photovoltaic array," *IEEE Transactions on Sustainable Energy*, vol. 6, pp. 1245–1253, 2015.

55. Huang, J.-M.; Wai, R.-J.; Yang, G.-J. "Design of hybrid artificial bee colony algorithm and semi-supervised extreme learning machine for PV fault diagnoses by considering dust impact," *IEEE Transactions on Power Electronics*, vol. 35, pp. 7086–7099, 2019.

56. Aziz, F.; Haq, A.U.; Ahmad, S.; Mahmoud, Y.; Jalal, M.; Ali, U. "A novel convolutional neural network-based approach for fault classification in photovoltaic arrays," *IEEE Access*, vol. 8, pp. 41889–41904, 2020.

57. Chowdhury, D.; Bhattacharya, M.; Khan, D.; Saha, S.; Dasgupta, A. "Wavelet decomposition based fault detection in cascaded H-bridge multilevel inverter using artificial neural network," In *Proceedings of the 2017 2nd IEEE International Conference on Recent Trends in Electronics, Information & Communication Technology (RTEICT)*, pp. 1931–1935, 2017.

58. Bharath, K.V.S.; Haque, A.; Khan, M.A. "Condition monitoring of photovoltaic systems using machine learning techniques," In *Proceedings of the 2nd IEEE International. Conference on Power Electronics, Intelligent Control and Energy Systems (ICPEICES)*, pp. 870–875, 2018.

59. Khan, M.A.; Haque, A.; Kurukuru, V.B. "Reliability analysis of a solar inverter during reactive power injection," In *Proceedings of the IEEE International Conference on Power Electronics, Drives and Energy Systems (PEDES)*, pp. 1–6, 2020.

60. Peyghami, S.; Dragicevic, T.; Blaabjerg, F. "Intelligent long-term performance analysis in power electronics systems," *Scientific Reports*, vol. 11, pp. 1–18, 2021.

61. Dragicevic, T.; Wheeler, P.; Blaabjerg, F. "Artificial intelligence aided automated design for reliability of power electronic systems," *IEEE Transactions on Power Electronics*, vol. 34, pp. 7161–7171, 2021.

62. Belhachat, F.; Larbes, C. "PV array reconfiguration techniques for maximum power optimization under partial shading conditions: A review." *Solar Energy*, vol. 230, pp. 558–582, 2021.

63. Rezazadeh, S.; Moradzadeh, A.; Pourhossein, K.; Akrami, M.; Mohammadi-Ivatloo, B.; Anvari-Moghaddam, A. "Photovoltaic array reconfiguration under partial shading conditions for maximum power extraction: A state-of-the-art review and new solution method," *Energy Conversion and Management*, vol. 258, p. 115468, 2022.

64. Karmakar, B.K.; Karmakar, G. "A current supported PV array reconfiguration technique to mitigate partial shading," *IEEE Transactions on Sustainable Energy*, vol. 12, no. 2, pp. 1449–1460, 2021.

65. Pachauri, R.K.; Mahela, O.P.; Sharma, A.; Bai, J.; Chauhan, Y.K.; Khan, B.; Alhelou, H.H. "Impact of partial shading on various PV array configurations and different modeling approaches: A comprehensive review," *IEEE Access*, vol. 8, pp. 181375–181403, 2020.

66. Durango-Flórez, M.; González-Montoya, D.; Trejos-Grisales, L.A.; Ramos-Paja, C.A. "PV array reconfiguration based on genetic algorithm for maximum power extraction and energy impact analysis," *Sustainability*, vol. 14, p. 3764, 2022.

67. Bouselham, L.; Rabhi, A.; Hajji, B.; Mellit, A. "Photovoltaic array reconfiguration method based on fuzzy logic and recursive least squares: An experimental validation," *Energy*, vol. 232, p. 121107, 2021.

68. Zhang, X.; Li, C.; Li, Z.; Yin, X.; Yang, B.; Gan, L.; Yu, T. "Optimal mileage-based PV array reconfiguration using swarm reinforcement Learning," *Energy Conversion and Management*, vol. 232, p. 113892, 2021.

69. Bouselham, L.; Rabhi, A.; Hajji, B.; Mellit, A.; Fouas, C.E. "An intelligent irradiance equalization approach based on Fuzzy logic for small reconfigurable PV architecture," *7th International Renewable and Sustainable Energy Conference (IRSEC)*, 2019.

70. Babu, T.S.; Ram, J.P.; Dragicevic, T.; Miyatake, M.; Blaabjerg, F.; Rajasekar, N. "Particle swarm optimization based solar PV array reconfiguration of the maximum power extraction under partial shading conditions," *IEEE Transactions on Sustainable Energy*, vol. 9, no. 1, pp. 74–85, 2018.

71. Solis-Cisneros, H.I.; Sevilla-Camacho, P.Y.; Robles-Ocampo, J.B.; Zuniga-Reyes, M.A.; Rodríguez-Resendíz, J.; Muniz-Soria, J.; Hernandez-Gutierrez, C.A. "A dynamic reconfiguration method based on neuro-fuzzy control algorithm for partially shaded PV arrays," *Sustainable Energy Technologies and Assessments*, vol. 52, p. 102147, 2022.

72. Cao, R.; Ding, Y.; Fang, X.; Liang, S.; Qi, F.; Yang, Q.; Yan, W. "Photovoltaic array reconfiguration under partial shading conditions based on ant colony optimization," *Chinese Control and Decision Conference*, 2020.

73. Khamis, A.; Shareef, H.; Bizkevelci, E.; Khatib, T. "A review of islanding detection techniques for renewable distributed generation systems," *Renewable and Sustainable Energy Reviews*, vol. 28, pp. 483–93, 2013.

74. Lin, C.; Cao, C.; Caon, Y.; Kuang, Y.; Zeng, L.; Fang, B. "A review of islanding detection methods for microgrid," *Renewable and Sustainable Energy Reviews*, vol. 35, pp. 211–220, 2014.

75. Hosseinzadeh, M.; Salmasi, F.R.R. "Islanding fault detection in microgrids: A survey," *Energies*, vol. 13, p. 3479, 2020.

76. Khan, M.A.; Kurukuru, V.S.B.; Haque, A.; Mekhilef, S. "Islanding classification mechanism for grid-connected photovoltaic systems," *IEEE Journal of Emerging and Selected Topics in Power Electronics*, vol. 9, pp. 1966–1975, 2021.

77. Bharath Kurukuru, V.S.; Haque, A.; Khan, M.A.; Sahoo, S.; Malik, A.; Blaabjerg, F. "A review on artificial intelligence applications for grid-connected solar photovoltaic systems," *Energies*, vol. 14, p. 4690, 2021.

78. Ananda Kumar, S.; Subathra, M.S.P.; Manoj Kumar, N.; Malvoni, M.; Sairamya, N.J.; Thomas George, S.; Suviseshamuthu, E.S.; Chopra, S.S. "A novel islanding detection technique for a resilient photovoltaic-based distributed power generation system using a Tunable-Q-Wavelet transform and an artificial neural network," *Energies*, vol. 13, p. 4238, 2020.

79. Admasie, S.; Ali Bukhari, S.B.; Gush, T.; Haider, R.; and Kim, C.H. "Intelligent islanding detection of multi-distributed generation using artificial neural network based on intrinsic mode function feature," *Journal of Modern Power Systems and Clean Energy*, vol. 8, no. 3, pp. 511–520, 2020.

80. Mlakic, D.; Baghaee, H.R.; Nikolovski, S. "A novel ANFIS-based islanding detection for inverter-interfaced microgrids," *IEEE Transactions on Smart Grid*, vol. 10, no. 4, pp. 4411–4424, 2018.

81. Merlin, V.; Santos, R.; Grilo, A.; Vieira, J.; Coury, D.; Oleskovicz, M. "A new artificial neural network-based method for islanding detection of distributed generators," *International Journal of Electrical Power & Energy Systems*, vol. 75, pp. 139–151, 2016.

82. Khamis, A.; Shareef, H.; Mohamed, A.; Bizkevelci, E. "Islanding detection in a distributed generation integrated power system using phase-space technique and probabilistic neural network," *Neurocomputing*, vol. 148, pp. 587–599, 2015.

83. Deshkar, S.N.; Dhale, S.B.; Mukherjee, J.S.; Babu, T.S.; Rajasekar, N. " Solar PV array reconfiguration under partial shading conditions for maximum power extraction using genetic algorithm," *Renewable and Sustainable Energy Reviews*, vol. 43, pp. 102–110, 2015.

84. Karakose, M.; Baygin, M.; Parlak, K.S. "A new real-time reconfiguration approach based on neural network in partial shading for PV arrays," *International Conference on Renewable Energy Research and Applications (ICRERA)*, 2014.

85. Martins, A.P.; Meireles, E.C.; Carvalho, A.S. "PWM-based control of a cascaded three phase multilevel inverter," In *Proceedings of the 14th European Conference Power Electron Applications*, pp. 1–10, 2011.

86. Eslami, A.; Negnevitsky, M.; Franklin, E.; Lyden, S. "Review of AI applications in harmonic analysis in power systems," *Renewable and Sustainable Energy Reviews*, vol. 154, p. 111897, 2022.

87. Kow, K.W.; Wong, Y.W.; Rajkumar, R.K. "A review on performance of artificial intelligence and conventional methods in mitigating PV grid-tied related power quality events," *Renewable and Sustainable Energy Reviews*, vol. 56, pp. 334–346, 2016.

88. Das, S.R.; Ray, P.K.; Sahoo, A.K.; Singh, K.K.; Dhiman, G.; Singh, A. "Artificial intelligence-based grid connected inverters for power quality improvement in smart grid applications," *Computers and Electrical Engineering*, vol. 93, p. 107208, 2021.

89. Rajkumar, R.; Ragupathy, U.S. "An ANN-based harmonic mitigation and power injection technique for solar-fed distributed generation system," *Soft Computing*, vol. 24, pp. 15763–15772, 2020.

90. Iniyaval, P.; Karthikeyan, S.R. "Fuzzy logic based quasi Z-source cascaded multilevel inverter with energy storage for photovoltaic power generation system," *International Conference on Emerging Trends in Engineering, Technology and Science*, 2016.

91. Balachandran, M.; Subramaniam, N.P. "Fuzzy logic controller for z-source cascaded multilevel inverter," *International Journal of Emerging Trends in Electrical and Electronics*, vol. 6, no. 1, pp. 30–34, 2013.

92. Guellal, A.; Larbes, C.; Bendib, D.; Hassaine, L.; Malek, A. "FPGA based on-line artificial neural network selective harmonic elimination PWM technique," *International Journal of Electrical Power & Energy Systems*, vol. 68, pp. 33–43, 2015.

93. Demirtas, M.; Cetinbas, I.; Serefoglu, S.; Kaplan, O. "ANN controlled single phase inverter for solar energy systems. In *16th International Power Electronics and Motion Control Conference and Exposition (PEMC)*, IEEE, pp. 768–72, 2014.

94. Karuppusamy, P.; Natarajan, A.; Vijeyakumar, K. "An adaptive Neuro-Fuzzy model to multilevel inverter for grid connected photovoltaic system," *Journal of Circuits, Systems, and Computers*, vol. 24, no. 5, p. 1550066, 2015.

95. Letting, L.; Munda, J.; Hamam, Y. "Optimization of a Fuzzy logic controller for PV grid inverter control using s-function based PSO," *Solar Energy*, vol. 86, no. 6, pp. 1689–1700, 2012.

96. Feng, C.; Liu, Y.; Zhang, J. "A taxonomical review on recent artificial intelligence applications to PV integration into power grids," *International Journal of Electrical Power and Energy Systems*, vol. 132, p. 107176, 2021.

97. Mellit, A.; Kalogirou, S.A. "Artificial intelligence techniques for photovoltaic applications: A review," *Progress in Energy and Combustion Science*, vol. 34, pp. 574–632, 2008.

98. Cornejo-Buenoa, L.; Casanova-Mateob, C.; Sanz-Justoc, J.; Salcedo-Sanza, S. "Machine learning regressors for solar radiation estimation from satellite data," *Solar Energy*, vol. 183, pp. 768–775, 2019.

99. Behera, M.K.; Majumder, I.; Nayak, N. "Solar photovoltaic power forecasting using optimized modified extreme learning machine technique," *Engineering Science and Technology, an International Journal*, vol. 21, pp. 428–438, 2018.

100. Zagrouba, M.; Sellami, A.; Bouaicha, M.; Ksouri, M. "Identification of PV solar and modules parameters using the genetic algorithms: Application to maximum power extraction," *Solar Energy*, vol. 84, no. 5, pp. 860–866, 2010.

101. Moghaddamnia, A.; Remesan, R.; Kashani, M.H.; Mohammadi, M.; Han, D.; Piri, J. "Comparison of LLR, MLP, Elman, NNARX and ANFIS models-with a case study in solar radiation estimation," *Journal of Atmospheric and Solar-Terrestrial Physics*, vol. 71, pp. 975–982, 2009.

102. Chaabene, M.; Ammar, M.B. "Neuro-Fuzzy dynamic model with Kalman filter to forecast irradiance and temperature for solar energy systems," *Renewable Energy*, vol. 33, pp. 1435–1443, 2008.

103. Mellit, A.; Pavan, A.M. "A 24-h forecast of solar irradiance using artificial neural network: Application for performance prediction of a grid-connected PV plant at Trieste, Italy," *Solar Energy*, vol. 84, no. 5, pp. 807–821, 2010.

104. Almonacid, F.; Rus, C.; Hontoria, L.; Fuentes, M.; Nofuentes, G. "Characterisation of Si-crystalline PV modules by artificial neural networks," *Renewable Energy*, vol. 34, pp. 941–949, 2009.

105. Benghanem, M.; Mellit, A.; Alamri, S.N. "ANN-based modelling and estimation of daily global solar radiation data: A case study," *Energy Conversion and Management*, vol. 50, pp. 1644–1655, 2009.

106. Altas, I.H.; Sharaf, A.M. "A novel maximum power Fuzzy logic controller for photovoltaic solar energy systems," *Renewable Energy*, vol. 33, pp. 388–399, 2008.

107. Paulescu, M.; Gravila, P.; Tulcan-Paulescu, E. "Fuzzy logic algorithms for atmospheric transmittances of use in solar energy estimation," *Energy Conversion and Management*, vol. 49, pp. 3691–3697, 2008.

108. Zhao, Z.; Zhang, M.; Zhang, Z.; Wang, Y.; Cheng, R.; Guo, J.; Yang, P.; Lai, C.S.; Li, P.; Lai, L.L. "Hierarchical pigeon-inspired optimization-based MPPT method for photovoltaic systems under complex partial shading," *IEEE Transactions on Industrial Electronics*, vol. 69, pp. 10129–10143, 2022.

109. Kececioglu, F.; Gani, A.; Sekkeli, M. "Design and hardware implement based on hybrid structure for MPPT of PV system using an interval type-2 TSK Fuzzy logic controller," *Energies*, vol. 13, no. 7, p.1842, 2020.

110. Villegas-Mier, C.G.; Rodriguez-Resendiz, J.; Álvarez-Alvarado, J.M.; Rodriguez-Resendiz, H.; Herrera-Navarro, A.M.; Rodríguez-Abreo, O. "Artificial neural networks in MPPT algorithms for optimization of photovoltaic power: A review," *Micromachines*, vol. 12, p. 1260, 2021.

111. Mansoor, M.; Mirza, A.F.; Ling, Q. "Harris hawk optimization-based MPPT control for PV systems under partial shading conditions," *Journal of Cleaner Production*, vol. 274, pp. 1–19, 2020.

112. Padmanabhan, S.K.; Priyadarshi, N.; Bhaskar, M.S.; Holm-Nielsen, J.B.; Ramachandaramurthy, V.K.; Hossain, E. "A hybrid ANFIS-ABC based MPPT controller for PV system with anti-islanding grid protection: Experimental realization," *IEEE Access*, vol. 7, pp. 103377–103389, 2019.

113. Zhang, X.; Li, S.; He, T.; Yang, B.; Yu, T.; Li, H.; Jiang, L.; Sun, L. "Memetic reinforcement learning based maximum power point tracking design for PV systems under partial shading condition," *Energy*, vol. 174, pp. 1079–1090, 2019.

114. Mellit, K.; Hontoria, S. "Artificial intelligence techniques for sizing photovoltaic systems: A review," *Renewable and Sustainable Energy Reviews*, vol. 13, pp. 406–419, 2009.

115. Khenissi, I.; Fakhfakh, M.A.; Sellami, R.; Neji, R. "A new approach for optimal sizing of a grid connected PV system using PSO and GA algorithms: Case of Tunisia," *Applied Artificial Intelligence*, vol. 35, no. 15, pp. 1930–1951, 2021.

116. Kefale, H.A.; Getie, E.M.; Eshetie, K.G. "Optimal design of grid-connected solar photovoltaic system using selective particle swarm optimization," *International Journal of Photoenergy*, vol. 2021, p. 9, 2021.

117. Kornelakis, A.; Marinakis, Y. "Contribution for optimal sizing of grid-connected PV systems using PSO," *Renewable Energy*, vol. 1, no. 12, pp. 1333–41, 2010.

118. Nath, V; Mandal, J.K. "Optimal placement of distributed generation using genetic algorithm approach," Lecture Notes in Electrical Engineering, *Proceeding of the Second International Conference on Microelectronics, Computing & Communication Systems (MCCS 2017)*, vol. 476, 2019.

119. Othman, Z.; Sulaiman, S.I.; Musirin, I, Mohamad, K. "Bat inspired algorithm for sizing optimization of grid-connected photovoltaic," In *Proceedings of the 2015 SAI Intelligent Systems Conference (IntelliSys)*, pp. 195–200, 2015.

120. Sanusi, Y.; Abisoye, S.; Awodugba, A. "Application of neural networks for predicting the optimal sizing parameters of standalone photovoltaic systems," *SOP Transactions on Applied Mathematics*, vol. 1, no. 1, pp. 1–5, 2014.

21 A Critical Review of IoT in Sustainable Energy Systems

S.S. Harish Raaghav, R. Lalitha Kala, and M. Thirumaran
Vellore Institute of Technology

Aayush Karthikeyan
University of Calgary

K.T.M.U. Hemapala
University of Moratuwa

O.V. Gnana Swathika
Vellore Institute of Technology

CONTENTS

ABSTRACT

In recent days, the need for an uninterrupted energy supply has become a challenge. Optimal and continuous demand for digital disruption magnified the need for technological advancements. Integrating the Internet of Things (IoT) with the energy sector solves energy optimization and enables sharing and distribution of renewable energy efficiently. Combining along with the cloud, a hybrid system of energy is formed. With the help of sensors and actuators, cloud infrastructure is created, where distribution and transmission of energy with the support of Internet are framed as sensors and actuators as a service. This work targets modernising renewable energy by encouraging sustainable change. IoT solutions are expected to revolutionise business markets by promoting disruption in the form of disruptive innovations that will promote sustainability, since cloud applications are accessed by both public and private sectors.

KEYWORDS

Internet of Things; Sustainable Energy; Smart Energy Systems; Data Analytics; IoT Applications

DOI: 10.1201/9781003331117-21

21.1   INTRODUCTION: BACKGROUND AND DRIVING FORCES

Smart cities are booming in every country, as well as the population is growing exponentially, and the demand for basic amenities and services is also increasing simultaneously. Smart cities generate and distribute huge numbers of data through IoT, but it is challenging to control these data. To overcome this problem, Deep Reinforcement Learning (DRL) is utilized, which is a successful time-variant control behavioural system. DRL is an effective schedule device and is cost- and time-efficient. A deep neural network classifies the energy devices currently available in a framework [1]. Smart cities are growing rapidly, renewable energy sources (RES) increase and improve waste management, and implementation of information and communication technology (ICT) infrastructures also increases. The main objective is to reduce the complexity of the smart power systems by using on-site and off-site resources. The Internet of Things (IoT)-based Smart Green Energy are proposed for smart cities. Smart cities detect energy consumption, secure communication, and predict demand [2]. IoT is finding a wide application in various fields of work in our day-to-day life. One such application is Energy Cooperation in the IoT of a smart city. These IoT gateways have interconnecting energy harvesting. This Energy Cooperation will help to balance the energy needs and transfers energy from rich gateways to energy scarce gateways. To find the optimal energy transfer between the gateways, convex optimization is used, which helps to find the solution for heterogeneous smart systems [3]. In the energy sector, day by day, the technology is developing, and existing resources get outdated, especially in digital disruption. Sustainable energy has developed faster in this decade. To generate strong, safe, and sustainable energy by implementing services like IoT and 5G wireless communication for faster communication is necessary. The technology Glowworm Swarm Optimization monitors the demand and supply of the energy and distributes the energy according to the load. Digital technologies are developing drastically in the energy sector, especially AI (artificial intelligence) technology. AI combines data of demands, supply, and renewable sources into the power grid to control. The three aspects of AI in power grids are implementing AI in solar and hydrogen power generation, AI demand and supply management control, and growing technologies in AI. The main motive of AI in power grid analysis is big data, which is the main role in the future energy market to increase efficiency and performance [4,5]. For emerging energy management systems, especially in 5G communications, a sustainable demand response (DR) model is required to treat fluctuations causing imbalance in green energy for IoT technology to manage average energy consumption, encouraging renewable energy using software-defined networking, network functions virtualization, and integer linear programming. DR is a request action model, acting in between power consumer and supplier. This will make the energy consumption itself a part of the management model, and prioritise them for consuming or rejecting, and making IOE [6].

Meeting the rising energy demand and limiting its environmental impact are the two intertwined issues faced in the 21st century. Governments in different countries have been engaged in developing regulations and related policies to encourage environmentally friendly renewable energy generation along with conservation strategies and technological innovations. It is important to develop sustainable energy policies and provide relevant and suitable policy recommendations for end-users. For the promotion of renewable energy, the development history of energy policy have been introduced in five countries, i.e., the United States, Germany, the United Kingdom, Denmark, and China. A literature survey of the articles aimed at promoting the development of sustainable energy policies and their modelling is carried out. It is observed that the energy-efficiency standard is one of the most popular strategies for building energy saving, which is dynamic and renewed based on the currently available technologies. Feed-in-tariff has been widely applied to encourage the application of renewable energy, which is demonstrated successfully in different countries. Building energy performance certification schemes should be enhanced in terms of reliable database systems and information transparency to pave the way for future net-zero-energy buildings and smart cities [7]. Sustainable energy management systems target energy efficiency, batteries, wave energy integration technology screening, and low carbon transitions. Anaerobic digestion, fermentation, and incineration convert the algae, seaweed, municipal solid waste, and food waste into biomass from bio-based renewable energy. Implementing AI, IoT, and Industry 4.0 for the chemical industries is discussed in reference [8]. The smart machine for sustainable energy systems is based on a comparative analysis of technology. The requirements of the smart machine to implement software, hardware, and design solutions are monitoring and diagnostics of the energy states, the state of household applications, and IoT devices. The main concept of smart machines is increasing the efficiency of households and industry with new types of information management services in the field [9]. Clean distributed energy gets electric power and energy systems (EPESs). When the IoT connects with existing EPES, EPES gets the features of cyber security, real-time monitoring, and control, and thereby transforms into an intelligent cyber-enabled EPES, which is secure, reliable, and sustainable, using IoT to improve, eliminate energy wastage, and save cost [10].

IoT plays a major role in smart monitoring systems and smart metering systems. Smart monitoring combines energy metres with smart equipment control and bidirectional communication. For control and monitoring of energy usage and also power quality issues, Demand-Side Management strategy is applied by using the Commercial Building Energy Management System using IoT-based Smart Compact Energy Meter (SCEM). An ARM microprocessor is used to render complex operations. These operations are controlled and monitored by the Blynk application. SCEM analyses the power quality at a low cost with accuracy [11]. Generally, universities have large buildings that have high energy demand. It incorporates renewable energy and monitors and controls the demand of the buildings by IoT-based wireless sensor network (WSN), which controls lighting systems, and heating, ventilation, and air conditioning (HVAC) in buildings. The goal is to transform buildings into Zero-Energy Buildings (ZEBs) at a low cost [12]. Photovoltaic (PV) array controls and analytics for solar farms and intelligent fault detection are elaborated about. For the software fault detection and solar array management to build the smart monitoring device (SMD). Later, including other sensors for monitoring current, irradiance, voltage, and temperature for each panel. Currently, solar energy IoT systems provide detection and remedy fault, mobile analytics, enable solar farm control, and optimise power under different shading conditions. The aim is to improve strength and efficiency using machine learning [13].

The solar market is abruptly increasing day by day. The reason behind this is sustainability and availability. Implementing the IoT framework in this sector can lead to upgrades and portability. PV modules of mono and poly-crystallines of combining solar cells, controlled solar water radiators which are ozone-friendly, solar syphons perfect for rural water conveyance, batteries, inverters to convert solar-powered alternating into direct current, regulators to control the intensity, and solar chargers are the components to integrate with IoT exploiting machine learning algorithms such as support vector machines (SVMs) to vitality yield [14]. The smart grid (SG) has big data, machine learning, and communication infrastructure to increase the management of power demand and supply. In emergencies, machine learning helps in the SG. The performance of smart grid is analysed using machine learning algorithms such as Artificial Neural Network (ANN), SVM, and Linear Regression (LR). These algorithms detect the power demand in the Home Area Network. ANN deals with large amounts of data and is superior to SVM and LR in predicting accuracy and time requirements [15].

21.2 DATA-DRIVEN SMART CITIES

To tackle the energy demand [16] and consume it wisely, an IoT-based data-driven methodology was implemented. CITyFiED (city in Spain) project (GA # 609129) of retrofitting buildings is the dataset characterised by static and dynamic training using machine learning. Developing a city strategy of digital architecture exploiting IoT, and big data analytics to ease mobility and governance is defined by the standard UNE 17804:2015. Using clusters to classify the buildings based on features in all parameters and regression to estimate the energy consumption.

A technique that can save energy and forecast its consumption can be developed by engineering, statistical, and artificial intelligence. Autoregressive integrated moving average (ARIMA) method is introduced in public building sectors with Generalized AutoRegressive Conditional Heteroskedasticity (GARCH), and wavelet theorems give the highest accuracy for energy consumption analysis, whereas in the case of smart offices, a multi-layer perceptron ANN and a SVM returned with a minimum of 4% errors. A cloud of stochastic wiener filter and k clustering of integrated reinforced learning approach results in saving of energy to the maximum of 18% [17]. Targeting the Internet of Medical Things in smart cities, a battery-less energy harvesting model from solar cells is developed for a low-power IoT node that supplies four different voltage ranges [18]. Charge pumps are integrated for buck-boost application of voltage harvested. Digital controllers and low dropout regulators are involved in load mechanisms suitable for monolithic integration. Bandgap reference generators employ harvested energy stored in supercapacitors producing constant voltage regardless of any parameters. To develop a quality of life, infrastructure in cities and urban areas need to be taken care of. Public lightning [19] plays a vital role in massive administration using neural networks and statistical assessments of the hybrid modelling approach. Lumiere project developed an ICT service, say, a data centre to collect the data from IoT nodes, store it, and process the statistical and dynamic parameters to provide uninterrupted service and analyse the consumption rate for better prediction and evaluation. Internet of Energy (IoE) joins hands with transferring computational sustainability for efficient IoT smart grid data management [20]. Sustainability movement along with data-driven methodologies is essential for resource allocation and optimization. Each forecasting horizon has a variety of issues. Using different equations and solving critical challenges with Betz limit, it is evident that the power coefficients efficiency has increased to 59%. Wireless sensors, radio frequency identifications (RFIDs), intelligent electronic devices (IEDSs) in sensor connectivity layer and network; along with Bluetooth, Zigbee in gateways, security control distribution process data in management service; DR, monitoring faults, and smart metres in the application layer sum up as the total setup. Integrating legacy and super old buildings with IoT sensors and protocols for retrofitting and sustainable development is required, since buildings are responsible for 30% carbon emission by consuming energy [21]. Installing sensor nodes chose to be wireless to avoid renovations. To develop a sustainable protocol, the initial step is to get cooperation from the occupants. Analysing the indoor parameters, say, occupants range and average renovation period already undertaken, is mandatory. To evaluate sensor quantity and variety required, a labelling protocol map is a key to forming a gateway network. Based on environmental constraints, required sensor module units are listed, say, power nodes and temperature. Before deploying, it is advised to take sensor tests for proximity, position, and delay, and evaluate the gateway to record its performance. Once everything is set, replication of nodes is carried out. It has a wide amount of positive influences on various concepts across the world. One such area of research it had imparted is EFW Nexus, i.e., Energy-Food-Water [22]. Nexus deals with sustainable energy management across various natural parameters such as water, soil, etc., and integrates its management and governance. EFW Nexus mainly focuses on the goal of reducing climate change. Implementing IoT along with EFW Nexus brings a lot of applications for various sectors. IoT applications on food are Food Transport, Food Packing, Intelligent Packaging and Delivery, etc., and on water are water quality and safety monitoring, wastewater management, etc. The proposed model of IoT architecture EFW Nexus has six layers, and they are Focus Layer, Cognizance Layer, Transmission Layer, Application Layer, Infrastructure Layer, and Competence Business Layer. All of these layers contribute individually towards service improvement. Thus, this proposed model of approach brings in effective sustainable management of resources [23]. A Specific Energy Consumption (SEC) factor is being generated for a ready-to-drink-juice production line up. SEC factors play an important role in bringing out the efficiency of the output by measuring the amount of energy that is being used on the input side. The energy consumption index which is generated contributes towards the efficiency and cost of energy that is being utilised in the production line up, and also this energy consumption index does fetch the necessary details about the energy consumption for a particular production line up. Calculation of this data was made by indexing the electricity consumption of the respective plant according to the power consumption of the plant, which is used by the production line up, and the weight of the product is obtained using IoT. So, based on this effective data, which is gained in the previous step, we come out with the energy index value of the respective production line up and hence the specific energy consumption factor. We have relied on fossil fuels for a longer period and considered them a primary source of energy for many years. But now due to the increased environmental impacts and lesser availability of fossil fuels, more of the focus is getting shifted to green and sustainable energy sources. One such sustainable energy source which is elucidated in this proposed model of work is solar power. Using solar energy, an active 2-axis solar tracker IoT system is proposed. The advantages of the proposed system are that it is affordable, easy to construct, require less maintenance, intelligent, and 30% more powerful than the traditional mode of systems. Components that are used in this proposed work are Sensors, Grid-Tie Inverter, ESP8266, Cloud Server, NODE-RED software, and solar panel. This proposed model of work will help to track efficiently and sustainably. Thus, the concept of green energy and sustainable development can be achieved [24]. Generating wind energy in the form of distributed networks offers an off-grid methodology. A grid connected with turbines and auxiliaries overcomes power fluctuations with the aid of a small sub-turbine connected along a battery bank (of non-renewable). To attain the optimum power and to manage energy, a circuit breaker is attached to the grid with a generator. The circuit breaker is connected with an ac transmission line of three-phase, and with respective ac bus. This ac bus is again coupled with a step-down transformer. An induction-type generator is used to convert variable speed of wind into a bus voltage. Also, a fault detector is set up. A Simulink model is developed under the category of the vertical and horizontal axis of wind turbines. Integrating IoT here makes the application more sustainable and discussed in future works [25].

To achieve sustainable energy management, wireless-network adaptive hybrid devices and IoT frameworks were introduced. With big data analytics, a hidden pattern inside the energy consumption data is found. Schick lear python library algorithms are used to control and manage the power flow. Controlling and managing heat ventilation and air conditioning is feasible, since they are the culprits consuming tons of energy in every bound. A microcontroller with power sensors forms a back-end connecting sensors to the master hub which processes and stores data using a system on chip microcomputers with the help of machine learning. A subunit is simply a user handy of having control switches and Wi-Fi modules for transmission of data. All three back ends, master hub, and subunit are interconnected. This machine learning IoT model yields 35% increased performance for accumulating energy management [26].

21.3   COMMUNICATION AND AI

Digital technologies are developing drastically in the energy sector, especially AI technology. AI combines data of demands, supply, and renewable sources into the power grid to control. The three aspects of AI in power grids are implementing AI in solar and hydrogen power generation, AI demand and supply management control, and growing technologies in AI. The main motive of AI in power grid analysis is big data, which is the main role in the future energy market to increase efficiency and performance [27]. The global energy level increases rapidly every year, and according to the Energy Information Administration, in 2050 the demand will increase up to 50%. For this, governments start to move towards and invest in sustainable renewable energy systems supported by cyber-physical technologies. This leads to popping up prosumers who are capable of producing and consuming energy. By adapting to renewable resources can provide more sustainable and efficient operations [28]. Sustainable energy plays a major role in lifetime environmental existence. It combines nature and technology. The physical layer comprising power and cyber layer is for control, computation, and communication to be achieved for the goal of energy sustenance, with the help of network on-chips, edge computing, and the IoT. These layers are deployed in smart cities, villages and building applications [29–31]. This article aims to manage energy consumption in a particular city with a cloud environment. For communication, clouds play with IoT-Urban Data Management. Big data analysis is complex, so it implements different kinds of sensors like weather, smart, traffic congestion, and vehicle and water sensors to collect a huge amount of stored data. Tackle by using Hadoop Distributed File System, it converts huge amounts of data into tiny blocks, and decision making is done by a machine learning algorithm. IoT communication worldwide is using Long-Range Wide Area Network (LoRaWAN), for long life for the device and long-range, but the disadvantage of the LoRa is frequent battery replacements, and that it increases maintenance costs and disposal of huge numbers of batteries which damages the environment. Controlling End-Device configurations by using EH-CRAM is a centralised Kalman filter-based optimization algorithm based upon incoming solar energy and traffic and maintaining the balanced energy supply and demand. EH-CRAM maximises the energy efficiency and reliability for sustainable energy harvesting for LoRa End-Devices. To reduce carbon footprint, an energy-efficacious WSN of IoT communication systems was made [32]. High spatial data output from various sensors was collected and encoded using Bernoulli's trial process, coupled with silent signal communication. Once done, the data lines are modulated by performing hybrid frequency-shift keying (FSK) or amplitude-shift keying (ASK), which is perfect for low-cost and low-data rate nodes based on WSN. Since the 5G network is evolving to connect hyper-fast IoT, this mass energy-saving communication introduced in communication nodes paves ways for improved complex sensing-driven IoT applications. Connecting cyber technology with IoT, a high number of devices contend for spectrum. To handle this, a wireless model of spatiotemporal using stochastic and queuing geometrical theories was introduced [33]. These theories talk about the success rate from buffers, rate of generation of energy from batteries, and probability of transmission of that energy. Investigation of queuing parameters was done, and network performance was analysed. A mathematical model of a self-sustainable IoT network for radio frequency (RF) downlink cellular supply that is recycled is characterised. Using two dimensional (2D) discrete-time Markov chain (DTMC), the time of evolution starting from generating is monitored. To treat the energy consumption of IoT devices and their data centres, a decision tree-based classification and regression on software-defined networking were introduced [34]. A request–response model based on the Gini index and an optimal tree is developed to provide vast access, thereby ensuring storage minimization as well as maximisation of energy storage. The main formulation is modelling a decision tree and pruned to the optimum. With access to the Gini index formula for different nodes, all training sets are stored in the root node/class carrying different members or attributes. Based on the Gini index, splitting criteria were formulated and loss function was calculated. With that cutting off, the tree with the smallest Gini index value is considered to be the optimum result. The same is repeated recursively until the expectation meets. In SDN, IPV4 protocols are brought to each pocket entity. The WSN, a part of LNN, is low power and lossy due to routing radios and protocols. The low-power and lossy networks contribute to poor performance, high traffic, and slow data and delivery rates. With the help of object capacity, RPL cut-off sink systems, and transmission from child to parent node were considered for routing [35]. With battery discharge index sinking with load, data traffic gets minimised. Destination-oriented directed acyclic graph is responsible for every child node to pick its parent node. With the help of COOJA software simulators, a set of microcontrollers formed an IoT network. This paper also discusses about implementation of ETX-aid in average power consumption and controlled traffic behaviour. IoT is being focused on and used in many applications in our daily life. One such application where IoT is being implemented is smart wearable solutions. The Body Sensor Network (BSN) is implemented in smart wearables solutions, over the prolonged service cycle which is being produced by the BSN turns out as a negative point to it since it has limited battery capacity and energy supply for sensors. So improving energy efficiency and harvesting energy can be considered the ways to improve and bring sustainability. So for this, Lyapunov Optimization technique is used to solve the MMOP framework and an online energy optimization algorithm is used to achieve near-optimal system utility. Thus, the proposed system turns out to be a sustainable system. The IoT is pictured as an important enabler of the good town, that 5G little cell goes to hide. The proposed model of work deals with the design of a dense cellular M2M communication system employing a Ginibre determinant purpose method and studying the corresponding communication metrics, together with grid energy consumption and grid energy potency (EE). Millimetre wave-based hybrid energy harvest home (EH) mechanism is projected to keep up self-sustainable communication. Two EH modes are considered: one is appropriate for the long-run operation, whereas the other mode is well-liked in managing a brief energy shortage. Initially minimise the typical grid energy consumption by resolving a large-scale applied mathematics drawback. Once the battery in the machine device is charged with sufficient energy at the initial time, a closed-form formula describing the typical grid energy consumption of each of the modes may be derived beneath moderate output constraints [36,37]. Owing to the exponential proliferation of the Web of Things (IoT), it's anticipated that the quantity of little IoT devices can grow efficiently over the ensuing few years. These billions of little IoT detectors and devices can consume a large power for information transmission. During this fashion, RF energy gathering has been contemplated as an Associate in Nursing appealing resolution to the design of long and self-sustainable next-generation wireless systems like IoT networks. However, within the sensible setting, like IoT networks or systems, square measure is subjected to external interference factors which regularly lead to the loss of the system rate. In a model, the working is different from generic RF EH system. Wherever possible, solely a supply node information is relayed through intermediate EH relaying node. Transmitting the information of IoT relay node along with side supply node data mistreatment non-orthogonal multiple access protocols is utilized within the presence of Associate in Nursing meddling signal to their several destinations. Specifically, within the presence of a meddling signal, we tend to study the mix of two well-liked energy gather relaying architectures – time change relaying and power cacophonous relaying – with ulcer protocol for IoT relay systems. Considering the interference from the external entity, we've mathematically derived the outage likelihood, throughput, and sum-throughput for our planned system [38]. Rooftop PV systems are currently turning into the present with the reduction in prices of PV panels and connected technologies. Developing countries like the Asian nation have undertaken inexperienced energy transitions to make sure property and energy self-direction. The rise in shopper participation visible in the nation's goal necessitates technological advancements in condition observation of PV systems. Nowadays, state-of-the-art and sturdy observation systems are accessible and deployed for big star farms and have been tested to extend the operational potency of these systems, whereas there are not any commonplace observation systems for small-scale installations. This paper presents a survey on Operations and Maintenance of PV systems and an IoT-based Wireless Detector Network for PV panel observation [39].

21.4 SUSTAINABLE ENERGY MANAGEMENT

In today's rapidly developing economy, energy resources have been considered the most important players which contribute to the development. Hence, the efficient use and management of energy resources which are available are considered an important factor that leads to the sustainability of existing resources. In this proposed model of a work [40], authors have designed a lighting system based on fuzzy logic which is supported by IoT. This model detects and varies the light intensity according to the motion which is detected in the environment, and this system can be implemented in heavy light-consuming environments such as hospitals and schools. It aids to reach higher range of power efficiency. To analyse the various impacts which are imparted from the successful implementation of Wireless Smart Plugs [41] that are connected with hands-on Internet sources. In the implementation of this model of work, Smart Plugs are connected to the Central Server via Wi-Fi-Gateway. Once the data from the Plugs reach the Central Server, the data is analysed, processed as required, and will be imparted into the Web Application that is used to showcase the needful data to the end-user. By this implementation, the user is able to keep track of their energy consumption statistics and to switch on or off the switches remotely via the Internet sources. Thus, this model of work corresponds to the efficient and sustainable way of energy consumption by the end-user. As for the application [42], Energy Information Management System (EIMS) and its importance in the IoT system are discussed. EIMS keeps track of our energy consumption patterns and leaves out the effective planning and efficient management of resources in commercial and industrial sectors. For data acquisition, various technologies such as Wi-Fi, Bluetooth Low Energy, LORA, sigfo, and NB-IoT are used, and for data mining, methods such as classification, regression, clustering, and dependency modelling are used. Challenges that are faced in this respective management system are Heterogeneity of Data, Encryption, and Security. In terms of energy consumption, in our environment, we have various forms of energy with different sources as their origin. So, to reach sustainable energy management between these resources, we need to balance out the energy which is being consumed from it. Now, in this proposed model of a work [43], they have used water as the source of the energy. The proposed system creates water resources as a water energy system and various functions and properties it has are Converted Energy, Saved Energy, Generated Energy, Created Energy, and Stored Energy. The benefits from this model of work are urban flood control, water storage, and water supply assessment. Developing countries consider reducing the energy consumption values, and carbon footprint across the world can effectively bring back the situation and offer a better sustainable environment for future generations. In the proposed model of work [44], IoE is considered a viable option to achieve the above-mentioned goal of reduced power consumption. The IoE is considered a child of the existing IoT. The core values which an IoE offers towards the market standards are a reduced amount of energy consumption for a given application and the important characteristic which it possesses is used in predicting the behavioural pattern of energy consumption and suggestions. Increasing consumption of energy and diverse extraction obstacles that most nations worldwide presently face caused researchers and customers in several eras together with industrial and home domains to hunt for brand new and optimum solutions to extend productivity and reduce harmful effects of consumption. These obstacles include the completion of fossil fuels, tendency towards renewable energy, environmental changes, tendency towards network data systems, rising energy prices, and technological development. The most important goal to research and achieve is to decrease energy consumption and increase consumers' welfare in all areas. IoE technology together with the most recent methodologies within the world has created an intelligent atmosphere with advanced sensors [45]. These technologies give intelligent association between users, suppliers, and suppliers with energy consumption atmosphere and observance systems that consequently result in potency in energy consumption. Besides, users' behaviours and their consumption behaviour patterns will be expected to be changed by victimisation enablers of this technology that is predicated on sensors. Utilising every available technology aids in improving the effectiveness of sustainable energy. Enhancing sustainable energy schemes is fundamental to improving the rate of global energy efficiency coupled with decreasing the share of carbon systems in the global energy mix. With the integration of modern technologies like the IoT, machine learning, and AI, energy systems are generally smarter, more efficient, reliable, and also possess the potential to mitigate global energy problems. The IoT finds application in wind farms and the energy sector at large. Integrating IoT into wind farms has the potential to save about 50–230 million US dollars of their operating expenses, improve the current 743 GW global wind power capacity to an amount substantial to enable it to supply 20% of the world's electricity by 2030, and also allow wind energy technology to hold a good share in the 100 billion objects that are expected to be connected to the Internet by 2025 [46]. In addition to this, it reviews the challenges associated with the deployment of IoT in wind farms and the future of blockchain technology and green IoT in energy systems. The main goal of the IoT is to ensure better efficiency, simplify processes of the system, and improve quality of life. The sustainability of the growing population becomes an issue, and the fast development of IoT must be carefully evaluated and monitored by the environment. The four main areas of IoT technologies are IoT in sustainable energy and environment, Smart city, E-health and smart transportation, and low carbon emission. In this proposed model of work, they have been focusing on the higher energy demands which are newly arising due to the implementation of AI and IoT into a vast number of fields and have suggested a sustainable energy management scheme "REcache" to solve this problem effectively and efficiently in an eco-friendly manner. This proposed "REcache" management scheme consists of a dedicated design circuitry and energy-aware management policies and principles to coordinate with various energy parameters such as energy harvesting, power management, and workload Scheduling [47]. Energy management is a topic that is gaining a huge amount of concern in this rapidly developing environment. The reason behind this concern is the high amount of energy demands and needs which are arising on a daily basis. So, in this current scenario, the invention and development of smart metres plays a vital part and will be used to monitor energy consumption in a much better way. Smart energy metres create an intelligent track on the information such as usage of the appliances, data consumption, and energy flow for various appliances connected to a network. The proposed model of work implements a cost-effective three-phase smart energy metre, i.e., IoT-enabled, multi-protocol, and modular, which is capable of collecting, processing, and transmitting various information regarding electrical energy on both the consumer side and the distributer side [48].

21.5 EDGE COMPUTING

IoT plays a vital role in many applications all around the world. So, by this rapid growth of IoT itself, we are able to understand its positive side and the impacts which it is creating on our society; on the other hand, IoT has some limitations like high maintenance costs for batteries, delay constraints on IoT systems due to centralised mode of operation, etc. To bring in an effective solution for the above-mentioned vulnerabilities, "The Green and Sustainable Mobile Edge Computing (GS-MEC)" framework can be considered a good alternative. The above-mentioned concept is based on Energy Harvesting Technologies (EHTs), which utilise the Green Energy from the atmosphere and make these respective devices self-powered devices. This above-mentioned [49] framework will try to minimise the response time and packet losses and make the device more reliable; additionally, this "Dynamic Parallel Computing Offloading and Energy Management (DPCOEM)" algorithm along with the Lyapunov set will optimise the system outcomes. Implementing IoT in day-to-day life scenarios and working conditions gives us a higher number of applications and its use cases, but similarly, it does have a down-sight and has some challenges which are left out to be resolved. One such challenge is providing an uninterrupted power supply for the IoT nodes which are operating. A power supply is considered a main source for the IoT node to carry forward its operation and allow it to transmit the resources it gathered. Bringing in the concept of sustainable energy resources can be considered a successor for the above-mentioned issues. This proposed model of work [50] uses an ultra-low-power solar energy harvesting system; this respective solar cell is considered the input source for the IoT node. The low input voltage from the solar panel is being boosted by a DC-DC convertor. Further tuning of the values is made with the help of frequency tuning and capacitor value modulation techniques. So, by this method, we are able to power IoT smart nodes and bring sustainable development. IoT has evolved in many use cases and is showing successful development over them. But still, the concept of sustainable IoT is hard to achieve because of the latency and the energy consumption issues it possesses [51]. The value of energy that is being consumed increases when a user's request is running in the local mobile device rather than that a Cloud server and latency will emerge as a problem when a task is being executed in a cloud environment rather than the mobile devices. Hence, a baseline between these two above-mentioned parameters is required to build a successful sustainable IoT application; thus, the concept of edge computing arises. This proposed model of work uses an optimization technique to adapt these two parameters. Hence, it improves the performance and efficiency of the model and brings a sustainable IoT application. The nascent paradigm of edge computing advocates that machine and storage resources will be extended to the sting of the network so that the impact of knowledge transmission latency over the web will be effectively reduced for a time-constrained IoT applications. With the widespread readying of edge computing devices, the energy demand of those devices has exaggerated and commenced to become an obvious issue for the acceptable development of urban systems. For this, a unified energy management framework is being proposed for sanctioning a property edge computing paradigm with distributed renewable energy resources [52]. This framework supports cooperation between the energy provider system and therefore the edge ADP system so that renewable energy will be utilised, while providing improved quality of service for time-constrained IoT applications. An imaging system is additionally enforced by exploitation of microgrid (solar-wind hybrid energy system) and edge computing devices along. Traffic management system, issues by traffic jam delay in emergencies, unpredictable local transport, and consumption of fuel increased. A Hierarchical Peer Connected Fog Architecture (HPCFA) where hierarchy is organised by fog node and peer fog node interconnects the same level with each other [53]. The data from IoT devices track the vehicle's position on the roads and transmits it to the nearest fog node. From the fog node, the information transfers to HPCFA and then to the user. HPCFA works both fog and without fog nodes and directly with the cloud. The simulations and results are displayed in applications indicated in reference [54]. The development of agriculture crops inside a greenhouse with a quick distribution in the market, to increase the efficiency and intelligent management system energy. The Cyber-Physical Systems (CPS) increases the renewable sources as a primary utilisation with sensing technologies including IoT and cloud computing (CC) supports ICT that leads to new features to energy systems. CPS controls and manages sensors to improve efficient energy systems [55]. Cloud-oriented convolutional neural network edge image recognition and visualisation of wireless low-power computing are presented. Here, an x86 class computes which is an end-to-end system is set up. Combination of energy harvesting from wind or solar and supercapacitors to overcome power supply demand. PMU of master-slave configuration with single inductor fitted for optimization and sequencing power. Implementing stack modules in every layer of IoT and EHPMIC handles computer and platform rails to harvest energy and charge the battery using supercapacitors, which is initially developed for agricultural applications to sense, capture, transmit, store, secure, and compute data with the help of edge and gateway in the cloud. Integration of simultaneous wireless information and power transfer with fog radio access networks meets the solution for a self-sustainable network [56]. With this energy and computing, users harvest energy and compute tasks. Power allocation, bandwidth management, and offloading operations are optimised for better performance. Minimising the consumption of energy at the same maintains the quality of service. Uplink and downlink transmissions are known to increase energy gain. Solving NP-hard mixed integer programming problems by processing latency of different transmission levels and tasks, and calculating energy consumption and harvesting with the help of offloading strategy and joint optimization algorithm, decoupled problem solved. According to our current life situation and working scenarios, high energy demand is being created and requested from the user side daily both from an industrial framework and a household perspective. To bring in an effective alternative for this existing scenario, we can try to incorporate some renewable energy resource concepts into the system. Bringing these renewable energy concepts into the existing system can reduce the effective load on the grid side and will avoid many problems, which are faced by the grid, such as brownout. The respective renewable energy which is being used in this model of work is solar power. In the implementation side of the work, a solar PV panel is installed in a house, and a respective battery source stores the energy which is being generated from the panel. Once the system identifies a higher amount of load on the grid side, the relay circuitry switches its consumption from the grid to the existing battery source; likewise, if the renewable energy is experiencing any kind of downfall due to natural conditions, the relay unit switches the power from battery circuit to the grid systems. Hence, an effective way of energy consumption is being achieved by this respective model of work [57]. RES are promoted to beat environmental constraints because of the insufficiency of fossil energy sources. Several initiatives are deployed by group action in these sources (e.g., electrical phenomenon panel, wind, fuel cell, etc.) to extend energy potency (i.e., electricity production) in buildings while limiting CO_2 emissions. In parallel to the present development, microgrids have emerged as suburbanised systems to permit connecting completely different suburbanised sources within which buildings become active electrical producers in addition to the most ancient electric grid. However, a better DR management system is needed to efficiently equalise energy production, storage, and also electricity consumption. During this study, a demand/response management approach is introduced by group action into a holistic platform with completely different RES along with storage devices for the provision of electricity to buildings' instrumentality (e.g. ventilation, lighting, etc.) [58].

21.6 ENERGY HARVESTING – A FUTURE

Green source energy harvesting [59] is setting trends for sustainable energy applications. With the help of value engineering, replacing traditional power supplies, and updating IoT nodes and their energy status. Considering transmission protocols in the network during transmissions, every node renders the access and performs energy harvest to do communications. Once the framework is set up with the help of the Markov method of discrete, nodes that are ready for transmission and harvesting are found. The system throughput is optimised according to the requirement. Transmission probability is trained by the multiagent reinforcement algorithm until the desired outcome is achieved. These are suitable for deep world applications. At the same time [60], Advancements in IoT are altering everyday life. Implementation of technology often through network traffic. Harvesting energy is considered to be the best solution for low-power energy transmitting IoT devices to turn sustainable. Achieving energy harvesting is either by renewable sources or wireless energy transfer where high-energy devices charge low by transmitting RF signals. Simultaneous wireless information and power transfer is a methodology adapted to encounter the attenuations of RF signals to overcome their short-range transmission. Two hop transmissions of data and energy were noted, while powering relay by renewable sources to transmit data and charge low-power nodes which are the ultimate goal. Using the Lyapunov optimization function, a power allocation algorithm was developed. The ultimate application in recent times [61], Saur Sikka, is a solar energy trading platform for harvesters to encash surplus energy. Based on the circuit switching framework, peer-to-peer networks developed an ability to sell and track generated energy using cloud and IoT. In a duplex power-sharing, p2p-connected model, a node of the network requests energy and a willing node responds to the requested network. Once a link is established, a multi-channel relay is activated. Until sharing and receiving over, a connection of the distribution network is on track by CC. A primary lithium polymer battery, TP4056 charging module, the microcontroller of configuration ESP32, and in-between inverters are employed.

To make this trend feasible [62] and to increase the energy output from IoT-based energy harvesting, power beacon pre-coded phase shifters were introduced for passive intelligent surfaces which are employed in controlling electromagnetic waves. A full-duplex multiple-user channel protocol is set up to calculate least squares for cascading of channels in it. Multiple antennas for multiple users are set to provide power transmission wirelessly. With governing equations and parameters, passive beamforming is achieved by arranging external elements to the multiple antennas. A multi-input single-output model is finally developed by close-bound equations with low-power constraints. In terms of connectivity, the harvested energy is converted from one form to another [63]. With IoT, data monitoring, say, data centres, is powered by two microcontroller units. The controller controls the overall circuit by multi-modal techniques, as the energy inputs from solar and radio frequencies are managed by integrated chips to supply loads whenever needed with the integration of dc-dc converter. Harvesting energy is achieved by charging capacitors with BQ25505 chips, and backup battery banks are set up. Calculating charging time, and protection threshold aids to achieve and alter expected sustainability. Expanding it to a certain extent, harvesting energy makes its wireless connectivity. To maintain road and rail lines, a piezoelectric sensor module is employed and connected to a microcontroller, for example, an Arduino to connect the overall framework with the web using a Wi-Fi network to capture, analyse, and visualise the real-time data. A strip clustering scheme is used to tackle the battery storage of nodes. A pedestrian-based EHT [64,65] is implemented with a PV grid, regulators, and various sensor modules with MQTT protocols. Once the situation triggers, it turns on the alarm, finds the spot where the emergency occurred, and replaces an alternative spot for pedestrians to reach their destination and post messages to the web platform. With the help of machine learning, the consumption of solar energy by IoT devices (both low-powered and industrial IoT) is estimated. To develop a use case for energy harvesting, for a fair trade-off between sensing and generation, it follows quasi-cyclic patterns. Based on the duration, the pattern is classified into short or long-time horizons. Once the IoT nodes sense their solar consumption, data are transmitted to the management unit where aggregation combines this data with weather forecasting. Consequently, this becomes the training set for prediction. A smart persistence model is employed in alternating diurnal variance and irregularity in weather. Sustainable energy harvesting and IoT devices can benefit from a longer operating lifespan, cheaper maintenance, and a cleaner environment. Taking into account variable energy availability and dynamic energy demands, it remains challenging to develop applications that provide sustainable operation despite this potential. The aim of the article [66] is to reduce the complexity of harvesting sustainable energy by implementing AsTAR. It is an energy-aware scheduler that adapts to match available environmental energy automatically and prioritises the task based on its energy consumption. Evaluations are AsTAR guarantees sustainability compared with conventional charging by maintaining an optimum level of charge that is defined by the user. AsTAR responds quickly to changes in the environment and platform. Its main benefits are minimising the performance overhead in terms of computation, memory, and energy. In our Current Industrial developing scenario, Industry 4.0 is the new kind of revolution that we all are getting into, and many of the existing industries are updating to those standards. As a part of the growth, we all feel happy about the new revolutions and their implementations, but, on the other hand, we are failing to acknowledge the negative impacts which it leaves on us as a part of the revolution. Sticking to these higher performance standards, a huge amount of energy is being utilised for processing and execution. Hence, the concept of "green IoT" has to be implemented, which represents efficiency and reduces the consumption of resources for a particular task. To achieve this, various optimization techniques are carried out between different layers such as perception and processing layers, transport and network layers, and application layers. Thus, this implementation leads to sustainable energy management [67].

This paper describes a low-power IoT sensing element node SoC that will be used for manufacturing automation and wearable aid applications. It utilized dynamic power programming technique appropriate for the surroundings that work with Associate in Nursing energy harvester. The proposed SoC design consists of low-power (60 nA) RTC, a normally off RF that has a quick start-up time (0.5 μs) and alternative low-power parts. A mixed-signal SoC has been a fictitious victimisation of the TSMC 65 nm record method. The sensing element node system equipped with dynamic power programming attains magnitude relation of zero.1%. Analysis results show that the planned technique will scale back concerning 51 of the ability consumption compared with the case while not dynamic power programming in sensing element node SoC. Also, the parameters of the system in energy harvest surroundings are often adjusted beneath the trade-off with power consumption and activity accuracy in keeping with the applying necessities [68].

This sensible world is an associate era, during which things (e.g., cars, buses, computers, and mobile phones) will serve individuals efficiently and cooperatively. IoT connects several devices. IoT devices sense, gather, process, and transmit information from their surroundings. This transmission and exchange of an outsized quantity of information between billions of devices cause large consumption of energy. The central aim of green IoT is to reduce the power consumption of IoT devices to make a secure and property setting for IoT. During this study, we tend to bestow a summary regarding IoT and the green IoT is made, and then classified the techniques of the greenest into three taxonomies that are mostly software-based IoT techniques, Hardware-Based green IoT techniques, and Policy-based mostly green IoT techniques. Finally, the comparison between various green models, systems, and algorithms is done to enhance and cut back the energy consumption of IoT devices [69].

21.7 CONCLUSION

Generally, universities have large buildings that have high energy demand. It incorporates renewable energy and monitors and controls the demand of the buildings by IoT-based WSN which controls lighting systems, HVAC in buildings. The goal is to transform buildings into ZEBs at a low cost.

Connecting cyber technology with IoT, a high number of devices contend for spectrum. To handle this, a wireless model of spatiotemporal using stochastic and queuing geometrical theories was introduced. These theories talk about the success rate from buffers, rate of generation of energy from batteries, and probability of transmission of that energy. Investigation of queuing parameters was done, and network performance was analysed. A mathematical model of a self-sustainable IoT network for RF downlink cellular supply that is recycled is characterised. Using 2D DTMC, the time of evolution starting from generating is monitored.

Earth has several renewable sources. IoT in the sustainable energy system has challenges in the security of energy. To reduce the carbon emissions through the current infrastructure of microgrids, renewable energy technologies, and power-to-gas hydrogen systems, IoT enhances the bidirectional communication service in the grid.

Power system engineering is slowly adapting to digital development with the revolution of IoT and cloud. To achieve autonomy and a self-accessing decentralised structure, the cloud IoT technique is applied. Based on the grid information, the type of cloud is chosen (say hybrid, community, public). Then, the service model is found (interface as a service, platform as a service, etc.). IoT as informatics integrated with cloud networks acts an interface. The addressing limit, quality of service, contextual and big data, and security are the challenges encountered while integrating. However, once developed, a sustainable, v2g-reliable, better power system is found.

Human kinetic energy is a sustainable energy source. Only a little power is arrived at from human kinetic energy using IoT in power management strategies to increase energy efficiency. Determined to maximise the Inertial harvester model by human's daily motion. The proposed algorithm and jointly considered optimal sink selection improve the energy efficiency.

REFERENCES

1. G. Muhammad and M.S. Hossain, 2021. Deep-reinforcement-learning-based sustainable energy distribution for wireless communication. *IEEE Wireless Communications*, *28*(6), pp. 42–48.
2. X. Zhang, G. Manogaran, and B. Muthu, 2021. IoT enabled integrated system for green energy into smart cities. *Sustainable Energy Technologies and Assessments*, *46*, p. 101208.
3. Á.F. Gambin, E. Gindullina, L. Badia, and M. Rossi, 2018. Energy cooperation for sustainable IoT services within smart cities. *2018 IEEE Wireless Communications and Networking Conference (WCNC)*, pp. 1–6, doi: 10.1109/WCNC.2018.8377450.
4. D. Hemanand, D.S. Jayalakshmi, U. Ghosh, A. Balasundaram, P. Vijayakumar, and P.K. Sharma, 2021. Enabling sustainable energy for smart environment using 5G wireless communication and internet of things. *IEEE Wireless Communications*, *28*(6), pp. 56–61.
5. T. Ahmad, D. Zhang, C. Huang, H. Zhang, N. Dai, Y. Song, and H. Chen, 2021. Artificial intelligence in sustainable energy industry: Status Quo, challenges and opportunities. *Journal of Cleaner Production*, *289*, p. 125834.
6. C. Tipantuña and X. Hesselbach, 2020. NFV/SDN enabled architecture for efficient adaptive management of renewable and non-renewable energy. *IEEE Open Journal of the Communications Society*, *1*, pp. 357–380, doi: 10.1109/OJCOMS.2020.2984982.
7. Y. Lu, Z.A. Khan, M.S. Alvarez-Alvarado, Y. Zhang, Z. Huang, and M. Imran, 2020. A critical review of sustainable energy policies for the promotion of renewable energy sources. *Sustainability*, *12*(12), p. 5078.
8. P.Y. Liew, P.S. Varbanov, A. Foley, and J.J. Klemeš, 2021. Smart energy management and recovery towards sustainable energy system optimisation with bio-based renewable energy. *Renewable and Sustainable Energy Reviews*, *135*, p. 110385.
9. A.I. Vlasov, V.A. Shakhnov, S.S. Filin, and A.I. Krivoshein, 2019. Sustainable energy systems in the digital economy: Concept of smart machines. *Entrepreneurship and Sustainability Issues*, *6*(4), p. 1975.

10. G. Bedi, G.K. Venayagamoorthy, R. Singh, R.R. Brooks, and K.C. Wang, 2018. Review of Internet of Things (IoT) in electric power and energy systems. *IEEE Internet of Things Journal*, 5(2), pp. 847–870.

11. T. Karthick, and K. Chandrasekaran, 2021. Design of IoT based smart compact energy meter for monitoring and controlling the usage of energy and power quality issues with demand side management for a commercial building. *Sustainable Energy, Grids and Networks*, 26, p. 100454.

12. P. Moura, J.I. Moreno, G. López López, and M. Alvarez-Campana, 2021. IoT platform for energy sustainability in university campuses. *Sensors*, 21(2), p. 357.

13. A.S. Spanias, 2017. Solar energy management as an Internet of Things (IoT) application. In *2017 8th International Conference on Information, Intelligence, Systems & Applications (IISA)* (pp. 1–4). IEEE.

14. R.R. Batcha and M.K. Geetha, 2020. A survey on IoT based on renewable energy for efficient energy conservation using machine learning approaches, *2020 3rd International Conference on Emerging Technologies in Computer Engineering: Machine Learning and Internet of Things (ICETCE)*, pp. 123–128, doi: 10.1109/ICETCE48199.2020.9091737.

15. J. Managre and N. Khatri, 2022. A review on IoT and ML enabled smart grid for futuristic and sustainable energy management. *In 2022 International Conference for Advancement in Technology (ICONAT)* (pp. 1–8). IEEE.

16. S. Mulero, J.L. Hernández, J. Vicente, P.S. de Viteri, and F. Larrinaga, 2020. Data-driven energy resource planning for Smart Cities. *2020 Global Internet of Things Summit (GIoTS)*, pp. 1–6, doi: 10.1109/GIOTS49054.2020.9119561.

17. A. Verma and Y. Kumar, 2020. Study on machine learning based energy efficiency in developed countries. *2020 Fourth International Conference on I-SMAC (IoT in Social, Mobile, Analytics and Cloud) (I-SMAC)*, pp. 895–899, doi: 10.1109/I-SMAC49090.2020.9243421.

18. S.K. Ram, S. Chourasia, B.B. Das, A.K. Swain, K. Mahapatra, and S. Mohanty, 2020. A solar based power module for battery-less IoT sensors towards sustainable smart cities. *2020 IEEE Computer Society Annual Symposium on VLSI (ISVLSI)*, pp. 458–463, doi: 10.1109/ISVLSI49217.2020.00-14.

19. M. Leccisi, F. Leccese, F. Moretti, L. Blaso, A. Brutti, and N. Gozo, 2020. An IoT application for Industry 4.0: A new and efficient public lighting management model. *2020 IEEE International Workshop on Metrology for Industry 4.0 & IoT*, pp. 669–673, doi: 10.1109/MetroInd4.0IoT48571.2020.9138208.

20. N. Sakib, E. Hossain, and S.I. Ahamed, 2020. A qualitative study on the United States internet of energy: A step towards computational sustainability. *IEEE Access*, 8, pp. 69003–69037, doi: 10.1109/ACCESS.2020.2986317.

21. S. Dzulkifly, H. Aris, B.N. Jorgensen, and A.Q. Santos, 2020. Methodology for a large scale building internet of things retrofit. *2020 8th International Conference on Information Technology and Multimedia (ICIMU)*, pp. 62–67, doi: 10.1109/ICIMU49871.2020.9243304.

22. N.M. Kumar, A. Dash, and N.K. Singh, 2018. Internet of Things (IoT): An opportunity for energy-food-water nexus. *2018 International Conference on Power Energy, Environment and Intelligent Control (PEEIC)*, pp. 68–72, doi: 10.1109/PEEIC.2018.8665632.

23. W. Muangjai, P. Thanin, W. Jantee, M. Ngaodet, and N. Nantakusol, 2018. An apply IoT for collection and analysis of specific energy consumption in production line of ready-to-drink juice at the second royal factory Mae Chan. *2018 International Conference and Utility Exhibition on Green Energy for Sustainable Development (ICUE)*, pp. 1–4, doi: 10.23919/ICUE-GESD.2018.8635775.

24. A. Sritoklin et al., 2018. A low cost, open-source IoT based 2-axis active solar tracker for smart communities. *2018 International Conference and Utility Exhibition on Green Energy for Sustainable Development (ICUE)*, pp. 1–4, doi: 10.23919/ICUE-GESD.2018.8635705.

25. T. Adefarati, B. Mokoena, R.C. Bansal, and R. Naidoo, 2020. Power management of a grid-connected wind energy system. *2020 International Conference on Power Electronics & IoT Applications in Renewable Energy and its Control (PARC)*, pp. 467–471, doi: 10.1109/PARC49193.2020.236657.

26. H. Shaikh, A.M. Khan, M. Rauf, A. Nadeem, M.T. Jilani, and M.T. Khan, 2020. IoT based linear models analysis for demand-side management of energy in residential buildings. *2020 Global Conference on Wireless and Optical Technologies (GCWOT)*, pp. 1–6, doi: 10.1109/GCWOT49901.2020.9391627.

27. L. Liu, H. Sun, P. Gao, N. Zheng, and T. Li, 2019. REcache: Efficient sustainable energy management circuits and policies for computing systems. *2019 IEEE International Symposium on Circuits and Systems (ISCAS)*, pp. 1–5, doi: 10.1109/ISCAS.2019.8702515.

28. M. Aloqaily, S. Kanhere, Y. Xiao, I. Al Ridhawi, and W. Guibene, 2021. Guest editorial: Empowering sustainable energy infrastructures via AI-assisted wireless Communications. *IEEE Wireless Communications*, 28(6), pp. 10–12, doi: 10.1109/MWC.2021.9690482.

29. S.K. Mazumder, J.H. Enslin, and F. Blaabjerg, 2021. Guest editorial: Special issue on sustainable energy through power-electronic innovations in cyber-physical systems. *IEEE Journal of Emerging and Selected Topics in Power Electronics, 9*(5), pp. 5142–5145.

30. D. Manu, S.G. Shorabh, O.G. Swathika, S. Umashankar, and P. Tejaswi, 2022. Design and realization of smart energy management system for Standalone PV system. *In IOP Conference Series: Earth and Environmental Science* (vol. 1026, no. 1, p. 012027). IOP Publishing.

31. O.G. Swathika, K. Karthikeyan, U. Subramaniam, K.U. Hemapala, and S.M. Bhaskar, 2022. Energy efficient outdoor lighting system design: Case study of IT campus. *In IOP Conference Series: Earth and Environmental Science* (vol. 1026, no. 1, p. 012029). IOP Publishing.

32. S. Sujeeth and O.G. Swathika, 2018. IoT based automated protection and control of DC microgrids. *In 2018 2nd International Conference on Inventive Systems and Control (ICISC)* (pp. 1422–1426). IEEE.

33. A. Patel, O.V. Swathika, U. Subramaniam, T.S. Babu, A. Tripathi, S. Nag, A. Karthick, and M. Muhibbullah, 2022. A practical approach for predicting power in a small-scale off-grid photovoltaic system using machine learning algorithms. *International Journal of Photoenergy, 2022*, pp. 1–21.

34. G.S. Odiyur Vathanam, K. Kalyanasundaram, R.M. Elavarasan, S. Hussain Khahro, U. Subramaniam, R. Pugazhendhi, M. Ramesh, and R.M. Gopalakrishnan, 2021. A review on effective use of daylight harvesting using intelligent lighting control systems for sustainable office buildings in India. *Sustainability, 13*(9), p. 4973.

35. O.V. Swathika and K.T.M.U. Hemapala, 2019. IOT based energy management system for standalone PV systems. *Journal of Electrical Engineering & Technology, 14*(5), pp. 1811–1821.

36. O.V. Swathika and K.T.M.U. Hemapala, 2019. IOT-based adaptive protection of microgrid. In *International Conference on Artificial Intelligence, Smart Grid and Smart City Applications* (pp. 123–130). Springer, Cham.

37. G.N. Kumar, and O.G. Swathika, 2022. Chapter 19: AI Applications to renewable energy an analysis. In: O.V. Gnana Swathika, K. Karthikeyan and S.K. Padmanaban (Eds.), *Smart Buildings Digitalization: IoT and Energy Efficient Smart Buildings Architecture and Applications* (p. 283). CRC Press: Boca Raton, FL

38. O.G. Swathika, 2022. Chapter 5: IoT-based smart. In: O.V. Gnana Swathika, K. Karthikeyan and S.K. Padmanaban (Eds.), *Smart Buildings Digitalization: IoT and Energy Efficient Smart Buildings Architecture and Applications* (p. 57). CRC Press: Boca Raton, FL.

39. P. Lal, V. Ananthakrishnan, O.G. Swathika, N.K. Gutha, and V.B. Hency, 2022. Chapter 14: IoT-based smart health. In: O.V. Gnana Swathika, K. Karthikeyan and S.K. Padmanaban (Eds.), *Smart Buildings Digitalization: IoT and Energy Efficient Smart Buildings Architecture and Applications* (p. 149). CRC Press: Boca Raton, FL.

40. S. Chowdhury, K.D. Saha, C.M. Sarkar, and O.G. Swathika, 2022. IoT-based data collection platform for smart buildings. In: O.V. Gnana Swathika, K. Karthikeyan and S.K. Padmanaban (Eds.), *Smart Buildings Digitalization: IoT and Energy Efficient Smart Buildings Architecture and Applications* (pp. 71–79). CRC Press: Boca Raton, FL.

41. A. Srinivasan, K. Baskaran, and G. Yann, 2019. IoT based smart plug-load energy conservation and management system. *2019 IEEE 2nd International Conference on Power and Energy Applications (ICPEA),* pp. 155–158, doi: 10.1109/ICPEA.2019.8818534.

42. J. Walia, A. Walia, C. Lund, and A. Arefi, 2019. The characteristics of smart energy information management systems for built environments. *2019 IEEE 10th International Workshop on Applied Measurements for Power Systems (AMPS),* pp. 1–6, doi: 10.1109/AMPS.2019.8897760.

43. K.-L. Weng, F.-C. Chen, and Y.-H. Lee, 2019. Application of water energy system to balance energy and environment in the whole world. *2019 IEEE Eurasia Conference on IOT, Communication and Engineering (ECICE),* pp. 21–23, doi: 10.1109/ECICE47484.2019.8942767.

44. H.D. Mohammadian, 2019. IoE: A solution for energy management challenges. *2019 IEEE Global Engineering Education Conference (EDUCON),* pp. 1455–1461, doi: 10.1109/EDUCON.2019.8725281.

45. H. Doost, 2018. Internet of energy: A solution for improving the efficiency of reversible energy. *2018 IEEE Global Engineering Education Conference (EDUCON),* pp. 1890–1895, doi: 10.1109/EDUCON.2018.8363466.

46. M.L. Adekanbi, 2021. Optimization and digitization of wind farms using internet of things: A review. *International Journal of Energy Research, 45*(11), pp. 15832–15838.

47. L. Liu, H. Sun, P. Gao, N. Zheng, and T. Li, 2019. REcache: Efficient sustainable energy management circuits and policies for computing systems. *2019 IEEE International Symposium on Circuits and Systems (ISCAS),* pp. 1–5, doi: 10.1109/ISCAS.2019.8702515.

48. D.B. Avancini, S.G.B. Martins, R.A.L. Rabelo, P. Solic, and J.J. P.C. Rodrigues, 2018. A flexible IoT energy monitoring solution. *2018 3rd International Conference on Smart and Sustainable Technologies (SpliTech)*, pp. 1–6.

49. Y. Deng, Z. Chen, X. Yao, S. Hassan, and A.M.A. Ibrahim, 2019. Parallel offloading in green and sustainable mobile edge computing for delay-constrained IoT system. *IEEE Transactions on Vehicular Technology*, 68(12), pp. 12202–12214, doi: 10.1109/TVT.2019.2944926.

50. S.K. Ram, B.B. Das, A.K. Swain, and K.K. Mahapatra, 2019. Ultra-low power solar energy harvester for IoT edge node devices. *2019 IEEE International Symposium on Smart Electronic Systems (iSES) (Formerly iNiS)*, pp. 205–208, doi: 10.1109/iSES47678.2019.00053.

51. S.K. Mishra, D. Puthal, B. Sahoo, S. Sharma, Z. Xue, and A.Y. Zomaya, 2018. Energy-efficient deployment of edge dataenters for mobile clouds in sustainable IoT. *IEEE Access*, 6, pp. 56587–56597, doi: 10.1109/ACCESS.2018.2872722.

52. W. Li et al., 2018. On enabling sustainable edge computing with renewable energy resources. *IEEE Communications Magazine*, 56(5), pp. 94–101, doi: 10.1109/MCOM.2018.1700888.

53. B. Suri, S. Taneja, and S. Kumar, 2022. A sustainable energy efficient IoT-based solution for real-time traffic assistance using fog computing. In: R. Tiwari, M. Mittal, and L.M. Goyal (Eds.), *Energy Conservation Solutions for Fog-Edge Computing Paradigms* (pp. 65–85). Springer: Singapore.

54. F.T. Oliveira, S.A. Leitão, A.S. Nabais, R.M. Ascenso, and J.R. Galvão, 2016. Greenhouse with sustainable energy for IoT. In *Doctoral Conference on Computing, Electrical and Industrial Systems* (pp. 416–424). Springer, Cham.

55. D. Kurian et al., 2020. Self-powered IOT system for edge inference, 2020. *21st International Symposium on Quality Electronic Design (ISQED)*, pp. 302–305, doi: 10.1109/ISQED48828.2020.9137027.

56. J. Hu, T. Shui, W. Zhang, F. Wu, and K. Yang, 2020. Spatial-domain resource scheduling for wireless computing and energy provision in energy self-sustainable F-RAN. *2020 IEEE International Conference on Parallel & Distributed Processing with Applications, Big Data & Cloud Computing, Sustainable Computing & Communications, Social Computing & Networking (ISPA/BDCloud/SocialCom/SustainCom)*, pp. 917–924, doi: 10.1109/ISPA-BDCloud-SocialCom-SustainCom51426.2020.00141.

57. M. Penna, Shivashankar, S. Mohammed Waseem, Avishek, A.G. Akshay and R. Manu, 2019. IOT based agile system for effective usage of reserved sustainable energy with unmanned switching. *2019 4th International Conference on Recent Trends on Electronics, Information, Communication & Technology (RTEICT)*, pp. 1379–1383, doi: 10.1109/RTEICT46194.2019.9016805.

58. A. Elmouatamid et al., 2018. Deployment and experimental evaluation of micro-grid systems. *2018 6th International Renewable and Sustainable Energy Conference (IRSEC)*, pp. 1–6, doi: 10.1109/IRSEC.2018.8703025.

59. M. Han et al., 2020. Multi-agent reinforcement learning for green energy powered IoT networks with random access. *2020 IEEE 92nd Vehicular Technology Conference (VTC2020-Fall)*, pp. 1–6, doi: 10.1109/VTC2020-Fall49728.2020.9348737.

60. X. Chen, Y. Liu, L.X. Cai, Z. Chen, and D. Zhang, 2020. Resource allocation for wireless cooperative IoT network with energy harvesting. *IEEE Transactions on Wireless Communications*, 19(7), pp. 4879–4893, doi: 10.1109/TWC.2020.2988016.

61. S. Salvi, S. Kumar, and N.D. Jacob, 2020. Saur Sikka: An IoT based prototype of basic solar power trading platform for independent distributed nano-grids. *2020 Fourth International Conference on I-SMAC (IoT in Social, Mobile, Analytics and Cloud) (I-SMAC)*, pp. 144–149, doi: 10.1109/I-SMAC49090.2020.9243560.

62. D. Mishra and E.G. Larsson, 2020. Passive intelligent surface assisted MIMO powered sustainable IoT. *ICASSP 2020-2020 IEEE International Conference on Acoustics, Speech and Signal Processing (ICASSP)*, pp. 8961–8965, doi: 10.1109/ICASSP40776.2020.9053628.

63. J.M. Williams, F. Gao, Y. Qian, C. Song, R. Khanna, and H. Liu, 2020. Solar and RF energy harvesting design model for sustainable wireless sensor tags. *2020 IEEE Topical Conference on Wireless Sensors and Sensor Networks (WiSNeT)*, pp. 1–4, doi: 10.1109/WiSNeT46826.2020.9037497.

64. M. Merenda, R. Carotenuto, F.G. Della Corte, F. Giammaria Praticò, and R. Fedele, 2020. Self-powered wireless IoT nodes for emergency management. *2020 IEEE 20th Mediterranean Electrotechnical Conference (MELECON)*, pp. 187–192, doi: 10.1109/MELECON48756.2020.9140503.

65. F.A. Kraemer, D. Palma, A.E. Braten, and D. Ammar, 2020. Operationalizing solar energy predictions for sustainable, autonomous IoT device management. *IEEE Internet of Things Journal*, 7(12), pp. 11803–11814, doi: 10.1109/JIOT.2020.3002330.

66. F. Yang, A.S. Thangarajan, G.S. Ramachandran, W. Joosen, and D. Hughes, 2021. AsTAR: sustainable energy harvesting for the internet of things through adaptive task scheduling. *ACM Transactions on Sensor Networks (TOSN)*, *18*(1), pp. 1–34.

67. V. Tahiliani and M. Dizalwar, 2018. Green IoT systems: An energy efficient perspective. *2018 Eleventh International Conference on Contemporary Computing (IC3)*, pp. 1–6, doi: 10.1109/IC3.2018.8530550.

68. Y. Yano et al., 2019. An IoT sensor node SoC with dynamic power scheduling for sustainable operation in energy harvesting environment. *2019 IEEE Asian Solid-State Circuits Conference (A-SSCC)*, pp. 267–270, doi: 10.1109/A-SSCC47793.2019.9056902.

69. S. Rawashdeh, W. Eyadat, A. Magableh, W. Mardini, and M.B. Yasin, 2019. Sustainable smart world. *2019 10th International Conference on Information and Communication Systems (ICICS)*, pp. 217–223, doi: 10.1109/IACS.2019.8809174.

Index